9

Topics in Heterocyclic Chemistry

Series Editor: R. R. Gupta

Topics in Heterocyclic Chemistry
Series Editor: R. R. Gupta

Recently Published and Forthcoming Volumes

Bioactive Heterocycles V
Volume Editor: M. T. H. Khan
Volume 11, 2007

Bioactive Heterocycles IV
Volume Editor: M. T. H. Khan
Volume 10, 2007

Bioactive Heterocycles III
Volume Editor: M. T. H. Khan
Volume 9, 2007

Bioactive Heterocycles II
Volume Editor: S. Eguchi
Volume 8, 2007

Heterocycles from Carbohydrate Precursors
Volume Editor: E. S. H. El Ashry
Volume 7, 2007

Bioactive Heterocycles I
Volume Editor: S. Eguchi
Volume 6, 2006

Marine Natural Products
Volume Editor: H. Kiyota
Volume 5, 2006

QSAR and Molecular Modeling Studies in Heterocyclic Drugs II
Volume Editor: S. P. Gupta
Volume 4, 2006

QSAR and Molecular Modeling Studies in Heterocyclic Drugs I
Volume Editor: S. P. Gupta
Volume 3, 2006

Heterocyclic Antitumor Antibiotics
Volume Editor: M. Lee
Volume 2, 2006

Microwave-Assisted Synthesis of Heterocycles
Volume Editors: E. Van der Eycken, C. O. Kappe
Volume 1, 2006

Bioactive Heterocycles III

Volume Editor: Mahmud Tareq Hassan Khan

With contributions by

M. Alamgir · N. Bianchi · D. S. C. Black · F. Clerici · F. Dall'Acqua
O. Demirkiran · R. Gambari · M. L. Gelmi · O. Kayser · M. T. H. Khan
N. Kumar · I. Lampronti · S. Pellegrino · D. Pocar · A.-M. Rydén
D. Vedaldi · R. P. Verma · G. Viola · C. Zuccato

Springer

The series *Topics in Heterocyclic Chemistry* presents critical reviews on "Heterocyclic Compounds" within topic-related volumes dealing with all aspects such as synthesis, reaction mechanisms, structure complexity, properties, reactivity, stability, fundamental and theoretical studies, biology, biomedical studies, pharmacological aspects, applications in material sciences, etc. Metabolism will be also included which will provide information useful in designing pharmacologically active agents. Pathways involving destruction of heterocyclic rings will also be dealt with so that synthesis of specifically functionalized non-heterocyclic molecules can be designed.
The overall scope is to cover topics dealing with most of the areas of current trends in heterocyclic chemistry which will suit to a larger heterocyclic community.
As a rule contributions are specially commissioned. The editors and publishers will, however, always be pleased to receive suggestions and supplementary information. Papers are accepted for *Topics in Heterocyclic Chemistry* in English.
In references *Topics in Heterocyclic Chemistry* is abbreviated *Top Heterocycl Chem* and is cited as a journal.

Springer WWW home page: springer.com
Visit the THC content at springerlink.com

ISSN 1861-9282

DOI 10.1007/978-3-540-73402-4

ISBN 978-3-642-09244-2 e-ISBN 978-3-540-73402-4

Springer is a part of Springer Science+Business Media

springer.com

Softcover reprint of the hardcover 1st edition 2007

Cover design: WMX Design GmbH, Heidelberg

Printed on acid-free paper 02/3100 YL – 5 4 3 2 1 0

Series Editor

Prof. R. R. Gupta

10A, Vasundhara Colony
Lane No. 1, Tonk Road
Jaipur-302 018, India
rrg_vg@yahoo.co.in

Volume Editor

Mahmud Tareq Hassan Khan

PhD School of Molecular and Structural Biology,
and Department of Pharmacology
Institute of Medical Biology
Faculty of Medicine
University of Tronsø
Tronsø, Norway
mahmud.khan@fagmed.nit.no

Editorial Board

Topics in Heterocyclic Chemistry Also Available Electronically

For all customers who have a standing order to Topics in Heterocyclic Chemistry, we offer the electronic version via SpringerLink free of charge. Please contact your librarian who can receive a password or free access to the full articles by registering at:

springerlink.com

If you do not have a subscription, you can still view the tables of contents of the volumes and the abstract of each article by going to the SpringerLink Homepage, clicking on "Browse by Online Libraries", then "Chemical Sciences", and finally choose Topics in Heterocyclic Chemistry.

You will find information about the

- Editorial Board
- Aims and Scope
- Instructions for Authors
- Sample Contribution

at springer.com using the search function.

Dedicated to my mentor, Prof. Ingebrigt Sylte, who cultivated maturity not only in my research, but in my thoughts, faith and writing as well.

Preface

"Bioactive Heterocycles III" provides readers with a comprehensive overview of the most recent breakthroughs in the field of heterocycles. This volume contains 8 chapters written by experts in their respective fields from all over the world. The chapters summarize years of extensive research in each area, and provide insight in the new themes of natural product research. Many of the contributors illustrate their laboratory experiences. It's obvious that readers will gain exciting and essential information from the volume.

In the first chapter, Kayser et al. describe the chemistry, biosynthesis and biological activities of artemisinin, one of the most promising antimalarial molecules, and its related natural peroxides. They present new strategies of producing artemisinin that utilize fascinating technologies. as Additionally, the pharmacokinetic profile and the development of new drug delivery systems on *Plasmodium* infected erythrocytes are presented .

Khan describes some aspects of sugar-derived heterocycles and their precursors which are utilized as the inhibitors against glycogen phosphorylases (GPs), and are responsible for the release of mono-glucose from poly-glucose (glycogen) in the second chapter. The inhibitors of GP could help to stop or slow down glycogenolysis as well as glucose production. Ultimately, the whole process will result in the recovery from diabetes of NIDDM patients.

In his contribution, Verma studies quantitative structure-activity relationship (QSAR) and proposes several interesting QSAR models of heterocyclic molecules having cytotoxic activities against different cancer cell lines, which could in turn clarify the chemical-biological interactions of such compounds.

Alamgir et al., in the fourth chapter, review the recent progress of the synthesis, reactivity and biological activities of benzimidazoles. Additionally, they describe several new techniques and procedures for the synthesis of the same scaffold.

In the fifth chapter, Khan reviews the essential role of the enzyme Tyrosinase in human melanin production, covering various related clinical problems. Finally, he describes the role of some inhibitors of this enzyme, themselves of heterocyclic origin, including biochemical features of the inhibition.

In the next chapter, Demirkiran describes the xanthones from *Hypericum* species and their synthesis and biological activities such as monoamine oxidase inhibition, antioxidant, antifungal, cytotoxic and hepatoprotective activities.

Within their contribution, Clerici et al. explain the chemistry of biologically active isothiazoles. They also present a range of different SAR studies, from well known to newly characterized compounds designed to improve their biological activities. In the same chapter, they also describe the agrochemical applications of the same pharmacophore.

In the final chapter, Gambari et al. summarize the structure and biological effects of furocoumarins. The authors mainly focus on linear and angular psoralens. Borrowing from their laboratory experiences, they describe the interesting biological effects of such compounds on cell cycle, apoptosis and differentiation as well as their use for the treatment of β-thalassemia.

Tromsø, Norway 2007 Mahmud Tareq Hassan Khan

Contents

Contents of Volume 10

Bioactive Heterocycles IV

Volume Editor: Khan, M. T. H.
ISBN: 978-3-540-73403-1

Contents of Volume 11

Bioactive Heterocycles V

Volume Editor: Khan, M. T. H.
ISBN: 978-3-540-73405-5

Top Heterocycl Chem (2007) 9: 1–31
DOI 10.1007/7081_2007_085

Published online: 7 September 2007

Chemistry, Biosynthesis and Biological Activity of Artemisinin and Related Natural Peroxides

Anna-Margareta Rydén · Oliver Kayser (✉)

Pharmaceutical Biology, GUIDE, University of Groningen, Antonius Deusinglaan 1, 9713 AV Groningen, The Netherlands
o.kayser@rug.nl

Abstract Artemisinin is a heterocyclic natural product and belongs to the natural product class of sesquiterpenoids with an unusual 1,2,4 trioxane substructure. Artemisinin is one of the most potent antimalarial drugs available and it serves as a lead compound in the drug development process to identify new chemical derivatives with antimalarial optimized activity and improved bioavailability. In this review we report about the latest status of research on chemical and physical properties of the drug and its derivatives. We describe new strategies to produce artemisinin on a biotechnological level in heterologous hosts and in plant cell cultures. We also summarize recent reports on its pharmacokinetic profile and attempts to develop drug delivery systems to overcome bioavailability problems and to target the drug to *Plasmodium* infected erythrocytes as main target cells.

Keywords Biosynthesis · Biochemistry · Pharmacokinetics · Synthesis · Analytics

Abbreviations

AACT	acetoacetyl-coenzyme A thiolase
A. annua	*Artemisia annua*
AMDS	amorpha-4,11-diene synthase
A. thaliana	*Arabidopsis thaliana*
CDP-ME	4-(Cytidine 5′-diphospho)-2-C-methyl-D-erythritol
CDP-MEP	4-(Cytidine 5′-diphospho)-2-C-methyl-D-erythritol 2-phosphate
cMEPP	2-C-Methyl-D-erythritol 2,4-cyclodiphosphate
CMK	4-(Cytidine 5′-diphospho)-2-C-methyl-D-erythritol kinase
CMS	2-C-Methyl-D-erythritol 4-phosphate cytidyl transferase
CoA	coenzyme A
CYP71AV1	cytochrome P450 71AV1
DMAPP	dimethylallyl diphosphate
*dpp*1	*S. cerevisiae* phosphatase dephosphorylating FPP (gene)
DXP	1-deoxy-D-xylulose 5-phosphate pathway
DXR	1-deoxy-D-xyluose 5-phosphate reductoisomerase
DXS	1-deoxy-D-xylulose 5-phosphate synthase
E. coli	*Escherichia coli*
erg9	*S. cerevisiae* squalene synthase (gene)
fpf1	flowering promoting factor (gene)
FPP	farnesyl diphosphate
FPPS	FPP synthase
G3P	glyceraldehyde 3-phosphate
GPP	geranyldiphosphate
GPPS	geranyldiphosphate synthase
HDS	1-Hydroxy-2-methyl-2-(*E*)-butenyl 4-diphosphate synthase
HMBPP	1-Hydroxy-2-methyl-2-(*E*)-butenyl 4-diphosphate
HMG-CoA	3*S*-Hydroxy-3-methylglutaryl-CoA
HMGR	3-hydroxy-3-methylglutaryl CoA reductase
HMGS	3-hydroxy-3-methylglutaryl CoA synthase
IDS	isopentenyl diphosphate/dimethylallyl diphosphate synthase
IPP	isopentenyl diphosphate
IPPi	isopentenyl diphosphate isomerase
ipt	isopentenyl transferase gene from *Agrobacterium tumefaciens*
MCS	2-C-Methyl-D-erythritol 2,4-cyclodiphosphate synthase
MDD	mevalonate diphosphate decarboxylase
MEP	2-C-Methy-D-erythritol 4-phosphate
MK	mevalonate kinase
MPK	mevalonate-5-phosphate kinase
MPP	mevalonate diphosphate
MS	medium Murashige and Skoog medium
MVA	3*R*-Mevalonic acid
MVAP	mevalonic acid-5-phosphate
OPP	paired diphosphate anion
P. falciparum	*Plasmodium falciparum*
S. cerevisiae	*Saccharomyces cerevisiae*
sue	*S. cerevisiae* mutation rendering efficient aerobic uptake of ergosterol
upc2-1	upregulates global transcription activity (mutation)

1 Chemistry

For thousands of years Chinese herbalists treated fever with a decoction of the plant called "qinghao", *Artemisia annua*, "sweet wormwood" or "annual wormwood" belonging to the family of Asteraceae. In the 1960s a program of the People Republic of China re-examined traditional herbal remedies on a rational scientific basis including the local qinghao plant. Early efforts to isolate the active principle were disappointing. In 1971 Chinese scientists followed an uncommon extraction route using diethyl ether at low temperatures obtaining an extract with a compound that was highly active in vivo against *P. berghei* in infected mice. The active ingredient was febrifuge, structurally elucidated in 1972, called mostly in China "qinghaosu", or "arteannuin" and in the west "artemisinin". Artemisinin, a sesquiterpene lactone, bears a peroxide group unlike most other antimalarials. It was also named artemisinine, but following IUPAC nomenclature a final "e" would suggest that it was a nitrogen-containing compound that is misleading and not favoured today.

Artemisinin and its antimalarial derivatives belong to the chemical class of unusual 1,2,4-trioxanes. Artemisinin is poorly soluble in water and decomposes in other protic solvents, probably by opening of the lactone ring. It is soluble in most aprotic solvents and is unaffected by them at temperatures up to 150 °C and shows a remarkable thermal stability. This section will focus on biological and pharmaceutical aspects; synthetic routes to improve antimalarial activity and to synthesize artemisinin derivatives with differ-

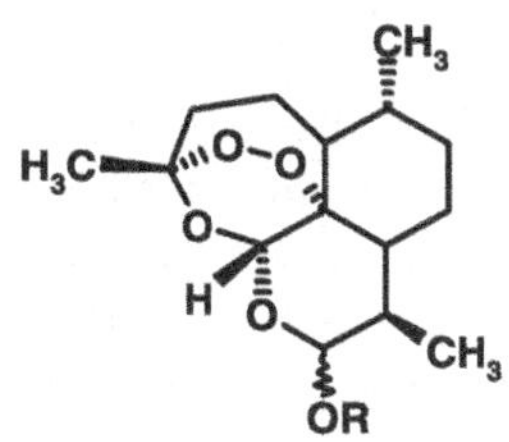

(1.1) dihydroartemisin; R = H (α + ß)

(1.2) artemether; R = CH_3 (ß)

(1.3) arteether; R = CH_2CH_3 (ß)

(1.4) artelinate; R = $CH_2C_6H_4COONa$ (ß)

(1.5) artesunate; R = $COCH_2CH_2COONa$ (α)

Fig. 1 Artemisinin and its derivatives

ent substitution patterns are reviewed elsewhere [1,2]. Most of the chemical modifications were conducted to modify the lactone function of artemisinin to a lactol. In general alkylation, or a mixture of dihydroartemisinin epimers in the presence of an acidic catalyst, it will give products with predominantly β-orientation, whereas acylation in alkaline medium preferentially yields α-orientation products (Fig. 1). Artemether (Fig. 1.2) as the active ingredient of Paluther® is prepared by treating a methanol solution of dihydroartemisinin with boron trifluoride etherate yielding both epimers. The main goal was to obtain derivatives that show a higher stability when dissolved in oils to enable parenteral use. The α-epimer is slightly more active ($EC_{50} = 1.02\ mg\,kg^{-1}$ b.w.) than the β-epimer ($EC_{50} = 1.42\ mg\,kg^{-1}$) and artemisinin itself ($EC_{50} = 6.2\ mg\,kg^{-1}$) [3]. Synthesis of derivatives with enhanced water solubility has been less successful. Sodium artesunate, Arsumax® (Fig. 1.5) has been introduced in clinics and is well tolerated and less toxic than artemisinin.

1.1 Trioxane and Peroxides in Nature

Besides artemisinin more than 150 natural peroxides are known in nature. The presence of the typical peroxide functions is not related to one natural product group and occurs as cyclic and acyclic peroxides in terpenoids, polyketides, phenolics and also alkaloids. The most stable are cyclic peroxides, even under harsh conditions and artemisinin is a nice example of this. Artemisinin can be boiled or treated with sodium borohydride without degradation of the peroxide function. In contrast, acyclic peroxides are rather unstable, form hydrogen peroxides and are easily broken by metals or bases.

Most natural peroxides have been isolated from plants and marine organisms, and terpenoids have attracted the most interest because of the structural diversity that they cover. In an excellent review by Jung et al. [4], an overview is given and it should be stressed that *Scapania undulata*, which is a bryophyte found in the northern parts of Europe, biosynthesizes amorphane like natural products with a cyclic peroxide (Fig. 2.1) structurally related to the well known artemisinin. There is less information about the biological activity of natural peroxides from plant origins, but some reports indicate its use against helminth infections, rheumatic diseases and antimicrobial activity. Natural cyclic peroxides from marine sources (Fig. 2) have been tested for a broad range of activities including antiviral (Aikupikoxide A), antimalarial, antimicrobial activity and cytotoxicity (Fig. 2.2). A second important natural product group are polyketides and it is interesting that all of the isolated polyketide-derived peroxides are from marine sources. Due to the high flexibility in the carbon chain and the presence of hydroxy substituents, a high chemical diversity can be documented ranging from simple and short

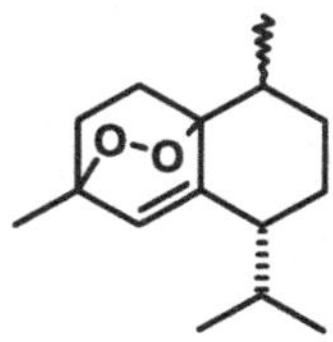

(2.1) Terpene type peroxide from *Scapania undulata*

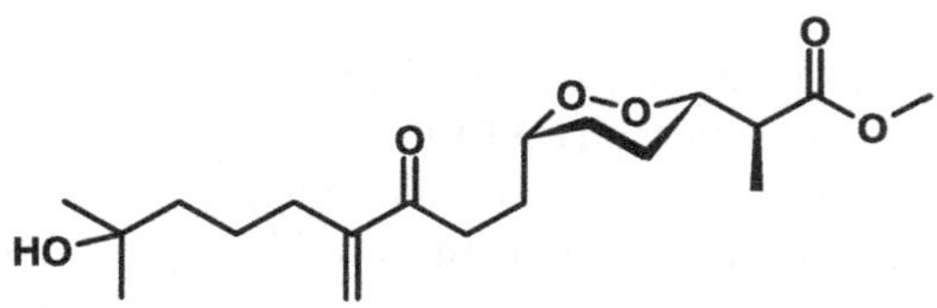

(2.2) Aikupikoxide-D from *Diacarnus erythraenus*

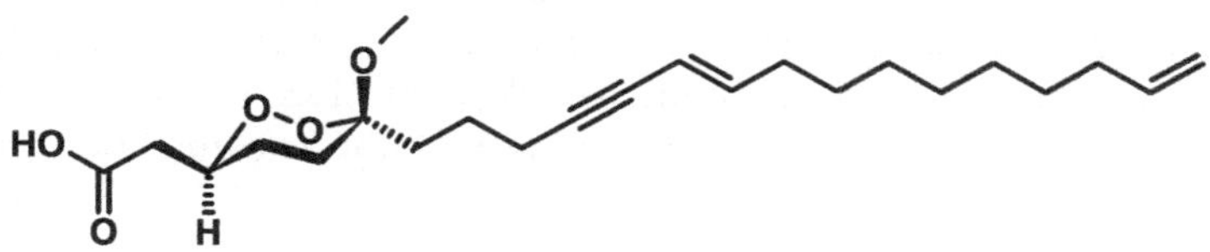

(2.3) Peroxyacarnoic acid A from *Acarnus bicladotyloata*

Fig. 2 Natural peroxides

peroxides like haterumdioins in Japanese sponge *Plaktoris lita* to more complex structures with long chain derivatives like peroxyacarnoic acids from the sponge *Acarnus bicladotylota* (Fig. 2.3). Most of the polyketide-derived peroxides show a high cytotoxic activity and moderate activity against microorganisms.

As expected due to chemical instability the number of acyclic peroxides is lower. Most of them occur as plant derived products, but also in soft corals like *Clavularia inflata*, hydroperoxides with potent cytotoxicity exist. Interestingly the bioactivity disappeared when the hydroperoxide function was deleted. It must be noted that most of natural hydroperoxides in plants are found in the group of saponins from *Panax ginseng* or *Ficus microcarpa*, which are used in ethnomedicine in South East Asia.

2 Biosynthesis

2.1 Biosynthesis in *Artemisia Annua*

2.1.1 Biochemistry

Two pathways are employed in plants for the production of isoprenoids, the 1-deoxy-D-xylulose 5-phosphate pathway (DXP) localized to the plastid and the mevalonate pathway present in the cytosol (Fig. 3) [5]. These pathways are normally used to produce different sets of isoprenoids, sesquiterpenoids, sterols and triterpenoids, among others being reserved for the mevalonate pathway, while the diterpenes and monoterpenes are produced by the DXP pathway. However, there is recent evidence that the pathways have some crosstalk on the isopentenyl diphosphate (IPP) level [5].

The first step taken in the biosynthetic pathway of artemisinin was the cyclization of the general mevalonate pathway originated sesquiterpenoid precursor farnesyl diphosphate (FPP) into (*1S*, *6R*, *7R*, *10R*)-amorpha-4,11-diene by amorpha-4,11-diene synthase (AMDS) (Fig. 4) [6–8]. The crystal structure of this sesquiterpene synthase is not known. From all plant

Fig. 3 Isoprenoid biosynthetic pathways in plant cells. The mevalonate pathway is represented in the cytosol; the MEP pathway in the plastid. Biosynthesis of artemisinin is depicted in detail. The *long dashed arrow* depicts transport. The dash punctured arrow depicts an unknown or putative enzymatic function. The *single arrow* depicts a single reaction step. *Multiple arrows* depict several reaction steps. Abbreviations of substrates: CDP-ME, 4-(Cytidine 5′-diphospho)-2-C-methyl-D-erythritol; CDP-MEP, 4-(Cytidine 5′-diphospho)-2-C-methyl-D-erythritol 2-phosphate; cMEPP, 2-C-Methyl-D-erythritol 2,4-cyclodiphosphate; DMAPP, Dimethylallyl diphosphate; DXP, 1-Deoxy-D-xylulose 5-phosphate; FPP, Farnesyl diphosphate; GPP, Geranyl diphosphate; HMBPP, 1-Hydroxy-2-methyl-2-(E)-butenyl 4-diphosphate; HMG-CoA, 3S-Hydroxy-3-methylglutaryl-CoA; IPP, Isopentenyl diphosphate; MEP, 2-C-Methy-D-erythritol 4-phosphate; MPP, Mevalonate diphosphate; MVA, 3R-Mevalonic acid; MVAP, Mevalonic acid-5-phosphate. Shortenings of enzymes: AACT, Acetoacetyl-coenzyme A (CoA) thiolase; AMDS, Amorpha-4,11-diene synthase; CMK, 4-(Cytidine 5′-diphospho)-2-C-methyl-D-erythritol kinase; CMS, 2-C-Methyl-D-erythritol 4-phosphate cytidyl transferase; CYP71AV1, Cytochrome P450 71AV1; DXR, 1-deoxy-D-xylulose 5-phosphate reductoisomerase; DXS, 1-deoxy-D-xyluose 5-phosphate synthase; FPPS, Farnesyl diphosphate synthase; GPPS, Geranyl diphosphate synthase; HDS, 1-Hydroxy-2-methyl-2-(E)-butenyl 4-diphosphate synthase; HMGR, 3-hydroxy-3-methylglutaryl CoA reductase; HMGS, 3-hydroxy-3-methylglutaryl CoA synthase; IPPi, Isopentenyl diphosphate isomerase; IDS, Isopentenyl diphosphate/Dimethylallyl diphosphate synthase; MCS, 2-C-Methyl-D-erythritol 2,4-cyclodiphosphate synthase; MDD, Mevalonate diphosphate decarboxylase; MK, mevalonate kinase; MPK, mevalonate-5-phosphate kinase ▶

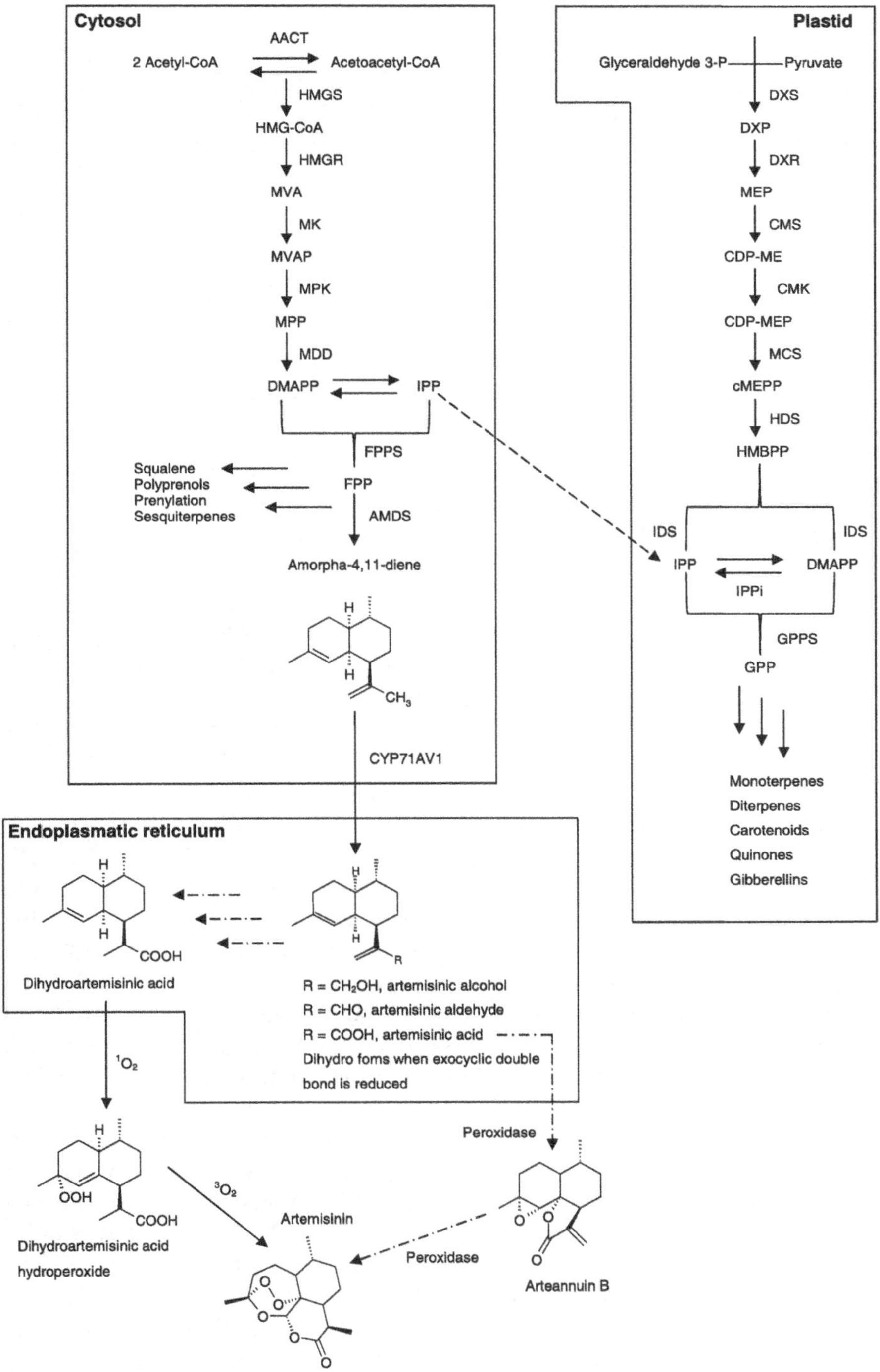
Cytosol
AACT
2 Acetyl-CoA
Acetoacetyl-CoA
HMGS
HMG-CoA
HMGR
MVA
MK
MVAP
MPK
MPP
MDD
DMAPP
IPP
FPPS
Squalene
Polyprenols
Prenylation
Sesquiterpenes
FPP
AMDS
Amorpha-4,11-diene
CYP71AV1
Plastid
Glyceraldehyde 3-P
Pyruvate
DXS
DXP
DXR
MEP
CMS
CDP-ME
CMK
CDP-MEP
MCS
cMEPP
HDS
HMBPP
IDS
IDS
IPP
DMAPP
IPPi
GPPS
GPP
Monoterpenes
Diterpenes
Carotenoids
Quinones
Gibberellins
Endoplasmatic reticulum
COOH
Dihydroartemisinic acid
R = CH2OH, artemisinic alcohol
R = CHO, artemisinic aldehyde
R = COOH, artemisinic acid
Dihydro foms when exocyclic double bond is reduced
1O2
Peroxidase
OOH
COOH
Dihydroartemisinic acid hydroperoxide
3O2
Artemisinin
Peroxidase
Arteannuin B

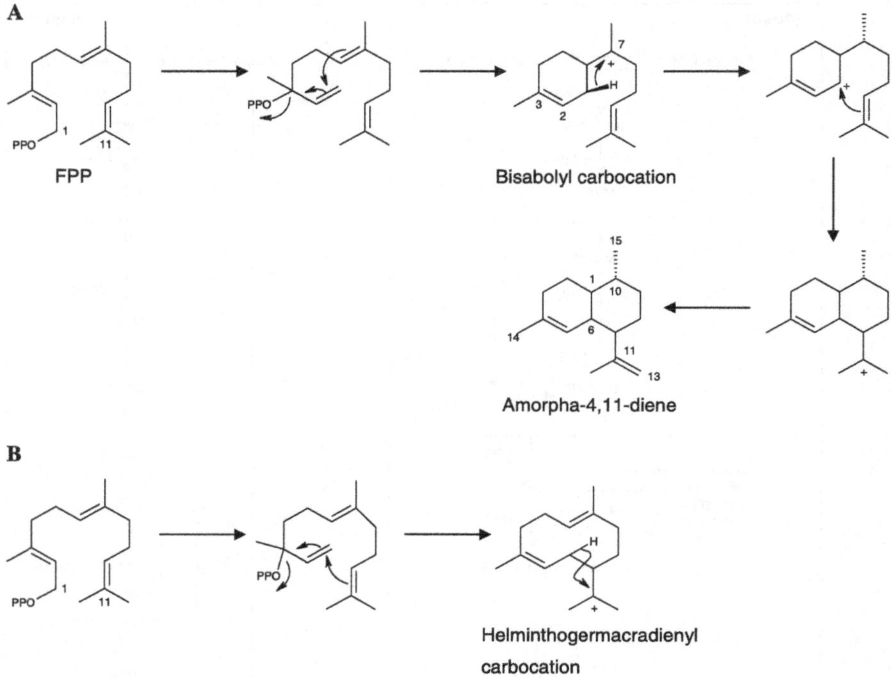

Fig. 4 **A** Cyclization of FPP to amorpha-4,11-diene by AMDS as described by Kim et al. and Picaud et al. [10, 11]. **B** Cyclization of FPP to helmonthogermabicradienyldiphosphate synthase carbocation

sesquiterpene synthases known, only the 5-*epi*-aristolochene synthase from tobacco has been elucidated [9]. In contrast, the mechanism behind the cyclization of FPP into amorpha-4,11-diene has been proven by Picaud et al. and Kim et al. through the use of deuterium labeled FPP (Fig. 4) [10, 11]. Differing from the bicyclic sesquiterpene cyclases δ-cadinene synthase from cotton [12] and pentalene synthase [13], which produce a germacrene cation as the first cyclic intermediate, AMDS produces a bisabolyl cation. FPP is ionized and the paired diphosphate anion (OPP) is transferred to C3 giving (3*R*)-nerolidyl diphosphate. This intermediate allows rotation around the C2–C3 bond to generate a cisoid form. The cisoid form brings C1 in close proximity to C6 allowing a bond formation between these two carbon atoms thus resulting in the first ring closure and a bisabolyl cation. The formed cation is in equilibrium with its deprotonized uncharged form, which is interesting because it implies a solvent proton acceptor and stands in contrast to studies discussing properties of the active site of an investigated trichodiene synthase [14]. Rynkiewicz and Cane came to the conclusion that the active site is completely devoid of any solvent molecule that would quench the reaction prematurely [14]. In a second report from the group of Vedula et al. the authors draw the conclusion from their results that terpene cyclization reactions in general are governed by kinetic rather than thermodynamic rules in the step leading to formation of the carbocation [15]. In the bisabolyl cation, an intermediate in the reaction towards amorpha-4,11-diene, a 1,3 hydride shift to C7 occurs, leaving a cation with a positive charge at C1 (FPP numbering). Through a nucleophilic attack on C1 by the double bond C10–C11 the second ring closes to give an amorphane cation. Deprotonation on C12 or C13 (amorphadiene numbering) gives amorpha-4,11-diene.

The three-dimensional structures of three non-plant sesquiterpene synthases reveals a single domain composed entirely of α-helices and loops despite the low homology on amino acid sequence level [14, 16, 17]. The secondary elements of 5-*epi*-aristolochene synthase, a plant sesquiterpene synthase, conform to this pattern with the exception of two domains solely composed of α-helices and loops. It is reasonable, but still a matter of debate, to extrapolate these data to the case of amorpha-4,11-diene synthase, which will probably only display α-helices and loops once the crystal structure has been solved.

A further element shared by all sesquiterpene synthases is the need for a divalent metal ion as cofactor. The metal ion is essential for substrate binding but also for product specificity. The metal ions stabilize the negatively charged pyrophosphate group of farnesyl diphosphate as illustrated by the crystal structure of 5-*epi*-aristolochene synthase [9]. The highly conserved sequence (I, L, V)DDxxD(E) serves to bind the metal ions in all known terpene and prenyl synthases (Fig. 5) [18–22]. A further interesting property among terpene synthases is that the active sites are enriched in relatively inert amino acids, thus it is the shape and dynamic of the active site that determines catalytic specificity [23].

Picaud et al. purified recombinant AMDS and determined its pH optimum to 6.5 [24]. Several sesquiterpene synthases show maximum activity in this range; examples are tobacco 5-*epi*-aristolochene synthase [25, 26], germacrene A synthase from chickory [26] and nerolidol synthase from maize [27]. Terpenoid synthases are, however, not restricted to a pH optimum in this range. Intriguing examples are the two (+)-δ-cadinene synthase variants from cotton, which exhibit maximum activity at pH 8.7 and 7–7.5, respectively [28] and 8-*epi*-cedrol synthase from *A. annua* [29] with the pH optimum around 8.5–9.0. The authors further investigated the metal ion required as cofactor for AMDS as well as substrate specificity. The kinetics studies revealed $k_{cat}K_m^{-1}$ values of 2.1×10^{-3} μM^{-1} s^{-1} for conversion of FPP at the pH optimum 6.5 with Mg^{2+} or Co^{2+} ions as cofactors and a slightly lower value of 1.9×10^{-3} μM^{-1} s^{-1} with Mn^{2+} as a cofactor. These very low efficiencies are common to several sesquiterpene synthases but substantial differences have been reported. The synthase reached a $k_{cat}K_m^{-1}$ value of 9.7×10^{-3} μM^{-1} s^{-1} for conversion of FPP at pH 9.5 using Mg^{2+} as a metal ion cofactor. This increase in efficiency is interesting and shows the broad window in which the enzyme

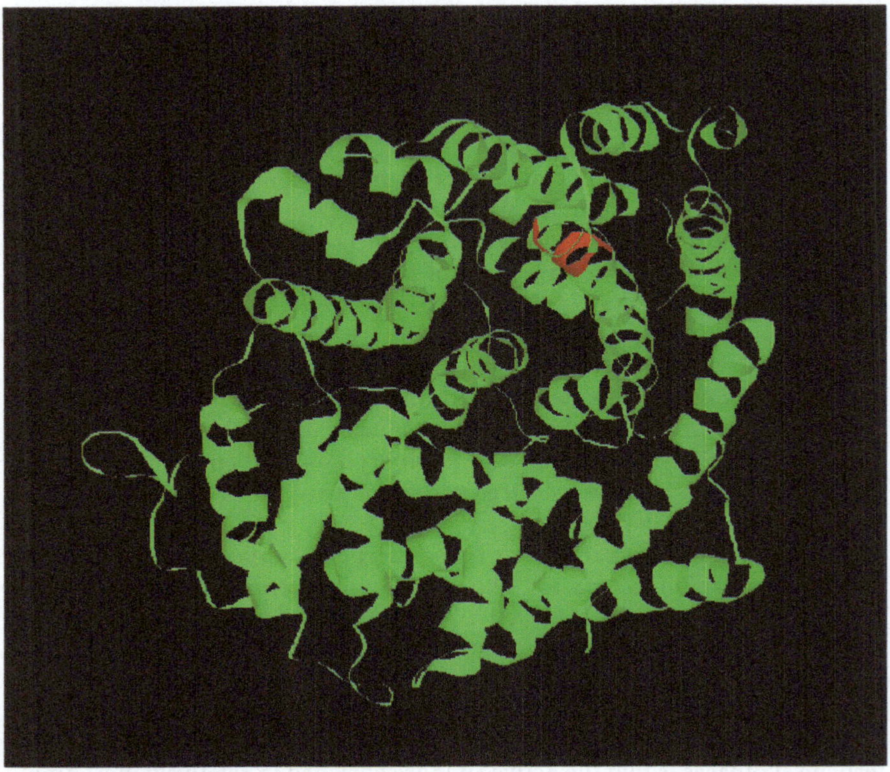

Fig. 5 Computerized 3D structure of amorpha-4,11-diene. Residues marked with red belong to the conserved metal ion binding amino acid sequence IDxxDD. The 3D model of the amorphadiene synthase (AMDS) courtesy of Wolfgang Brandt, Leibniz Institute of Plant Biochemistry Halle, Germany

can work, something that may prove to be industrially usable but that physiologically does not have a meaning in the plant. The increase in efficiency is not linear as the maximum activity of AMDS is around pH 6.5–7.0 with a minimum at pH 7.5. The established pH optimum of 6.5 is in line with the range established for AMDS isolated from *A. annua* leaves [30]. AMDS did not show any relevant activity in the presence of Ni^{2+}, Cu^{2+} or Zn^{2+}. In the presence of Mn^{2+} as cofactor, AMDS is capable of using geranyldiphosphate (GPP) as substrate although with very low efficiency (4.2×10^{-5} μM^{-1} s^{-1} at pH 6.5). Using Mn^{2+} as a cofactor also increased the product specificity of AMDS to $\sim$ 90% amorpha-4,11-diene with minor negative impact on efficiency. Under optimal conditions AMDS was proven to be faithful towards the production of amorpha-4,11-diene from FPP, converting $\sim$ 80% of the substrate into amorpha-4,11-diene, $\sim$ 5% amorpha-4,7 (11)-diene and $\sim$ 3.5% amorpha-4-en-7-ol together with 13 other sesquiterpenes in minute amounts.

Bertea et al. [31] postulated that the main route to artemisinin is the conversion of amorpha-4,11-diene to artemisinic alcohol, which is further oxidized to artemisinic aldehyde (Fig. 3). The C11–C13 double bond in artemisinic aldehyde was then proposed to be reduced giving dihydroartemisinic aldehyde, which would upon further oxidation give dihydroartemisinic acid. The authors supported their conclusion by demonstrating the existence of amorpha-4,11-diene, artemisinic alcohol, artemisinic aldehyde and artemisinic acid together with the reduced forms of the artemisinin intermediates in leaf- and glandular trichome microsomal pellets, by direct extraction from leaves and through enzyme assays. Interestingly, they could not show any significant conversion of artemisinic acid into dihydroartemisinic acid regardless of the presence of cofactors NADH and NADPH thus strengthening the hypothesis that reduction of the C11–C13 double bond occurs at the aldehyde level. In view of these results it is very likely that artemisinic acid is a dead end product that cannot be converted into artemisinin in contrast with some literature [32], unless reduced to dihydroartemisinic acid.

Recently, two research groups cloned the gene responsible for oxidizing amorpha-4,11-diene in three steps to artemisinic acid (Fig. 3) [33, 34]. This enzyme, a cytochrome P450 named CYP71AV1, was expressed in *Saccharomyces cerevisiae* (*S. cerevisiae*) and associated to the endoplasmatic reticulum. The isolation and application of this cytochrome P450 is described further below. Further research that will clarify whether additional cytochrome P450s or other oxidizing enzymes are present in the native biosynthetic pathway and where the reduction of the C11–C13 double bond occurs are still open fields of exploration.

Several terpenoids including artemisinin and some of its precursors and degradation products have been found in seeds of *A. annua* [35]. In its vegetative state, secretory glandular trichomes [36] are the site of production of artemisinin. Recently, Lommen et al. showed that the production of artemisinin is a combination of enzymatic and non-enzymatic steps [37]. The authors followed the production of artemisinin and its precursors on a level per leaf basis. The results showed that artemisinin is always present during the entire life cycle of a leaf, from appearance to senescence and that the quantity steadily increases as would be expected for an end product in a biosynthetic pathway. Interestingly, the immediate precursor to artemisinin, dihydroartemisinic acid [38] was more abundant than other precursors, indicating that the conversion of dihydroartemisinic acid into artemisinin is a limiting step. It was also shown that dihydroartemisinic acid is not converted to artemisinin directly. The authors argue, in line with other literature [38], that this might be due to a temporary accumulation of the putative intermediate dihydroartemisinic acid hydroperoxide (Fig. 3). The observation that artemisinin levels continued to increase at the same time as the numbers of glandular trichomes decreased further supports the idea that the final step of artemisinin formation is non-enzymatic. Wallaart et al. were able to show that conversion of dihydroartemisinic acid to artemisinin is possible when using mineral oil as reaction solvent instead of glandular oil (Fig. 3) [39]. By adding dihydroartemisinic acid and chlorophyll *a* to mineral oil and exposing the mixture to air and light, a conversion of 12% after 120 hours was achieved. In absence of mineral oil a conversion of 26.8% was achieved. Wallart et al. were later able to show that the hypothesized intermediate between dihydroartemisinic acid and artemisinin, dihydroartemisinic acid hydroperoxide, could be isolated from *A. annua* and upon exposure to air for 24 hours at room temperature yielded artemisinin and dihydro-*epi*-deoxyarteannuin B (Fig. 3) [40].

2.1.2 Genetic Versus Environmental Regulation of Artemisinin Production

The genetic regulation of the biosynthesis of artemisinin is poorly understood on the single pathway level. The situation is further complicated because there are several FPP synthase (FPPS) and 3-hydroxy-3-methylglutaryl CoA reductase (HMGR) isoforms making optimization options more versatile and complex. The active drug component in *A. annua* was isolated in the 1970s but it was only during the last eight years that key enzymes in the committed biosynthetic pathway of artemisinin have been cloned and characterized (Fig. 3) [6–8, 33, 34]. However, the genetic variation contributing to the level of artemisinin production has been investigated to some extent. The genetic variation is reflected in the existence of at least two chemotypes of *A. annua*. Wallaart et al. showed that plant specimens from different geographical origins had a different chemical composition of the essential oil during the vegetative period [41]. The authors distinguished one chemotype having a high content of dihydroartemisinic acid and artemisinin accompanied by a low level of artemisinic acid and a second chemotype represented by low artemisinin and dihydroartemisinic acid content together with a high level of artemisinic acid. With the aim of increasing the artemisinin production the authors induced tetraploid specimens from normal high producing diploids using colchicine [42]. This led to higher artemisinin content in the essential oil but to a 25% decrease in artemisinin yield per m^2 leaf biomass.

Only a few studies have investigated the effect of singular genes on artemisinin production. Wang et al. overexpressed the flowering promoting factor (*fpf1*) from *Arabidopsis thaliana* in *A. annua* and observed 20 days earlier flowering compared with the control plants but could not detect any significant change in artemisinin production [43]. From this it can be concluded that the event of flowering has no effect on artemisinin biosynthesis, an idea supported by a later study performed by the authors in which the early flowering gene from *A. thaliana* was overexpressed in *A. annua* [44]. In contrast, when an isopentenyl transferase gene from *Agrobacterium tumefa-*

ciens (*ipt*) was overexpressed in *A. annua*, the content of cytokinins, chlorophyll and artemisinin increased two- to three-fold, 20–60% and 30–70%, respectively [45]. By overexpression of endogenous FPP in *A. annua*, Han et al. established a maximum 34.4% increase in artemisinin content corresponding to 0.9% of the dry weight [46]. Similarly, a two- to three-fold increase in artemisinin production was obtained using a FPP from *Gossypium arboreum* [47].

To assess the genetic versus environmental contributions to artemisinin production, quantitative genetics was applied by Dealbays et al. [48]. Variance manifested in a phenotype or a trait like a chemotype is the sum of the genetic and environmental variance. The genetic variance can in its turn be divided into additive genetic variance, dominance variance and epistatic variance. Additive variance is a representation of the number of different alleles of a trait, dominance variance the relation between dominant and recessive alleles and epistatic variance the relation between alleles at different loci. Broad-sense heritability of a trait is defined as the variation attributed to genetic variance divided by the total variance in traits. In their experiments Ferreira et al. estimated a broad-sense heritability of up to 0.98 [49]. Delabays et al. confirm the broad-sense heritability of artemisinin to be between 0.95 and 1 [48] and that the dominance variance of 0.31 was present in the experiment. This implies that there are great variations between the same alleles, which besides a genetic based existence of chemotypes, support a mass-breading selection program of *A. annua* to produce a high yield artemisinin crop. With the breeding program CPQBA-UNICAMP aiming at improvement of biomass yields, rates between leaves and stem, artemisinin content, and essential oil composition and yield in *A. annua*, genotypes producing 1.69 to 2.01 g m^{2-1} have been obtained [50].

2.1.3 Cell Culture

One biotechnological research focus is to utilize hairy root cultures as a model of study and for the production of artemisinin. Hairy roots are genetically and biochemically stable, capable of producing a wide range of secondary metabolites, grow rapidly in comparison with the whole plant and can reach high densities [51, 52]. It is an interesting approach but is currently hampered by the difficulties in scaling up the production to industrial proportions. Scaling-up of *A. annua* hairy root cultures has been shown to produce complex patterns of terpenoid gene expression pointing towards the difficulty of obtaining a homogeneously producing culture [53]. In their study Souret et al. compare the expression levels of four key terpenoid biosynthetic genes, HMGR, 1-deoxy-D-xylulose 5-phosphate synthase (DXS), 1-deoxy-D-xyluose 5-phosphate reductoisomerase (DXR) and FPPS (Fig. 3), in three different culture conditions: shake flask, mist bioreactor and bubble column bioreactor. In shake flask conditions all key genes were temporally expressed but only FPPS had a correlation with artemisinin production. This is not surprising because the terpenoid cyclase has often proven to be the rate limiting step in a terpenoid biosynthetic pathway. Expression of the genes in both bioreactor types were similar or greater than the levels in shake flask cultures. In the bioreactors, the transcriptional regulation of all the four key genes were affected by the position of the roots in the reactors, but there was no correlation with the relative oxygen levels, light or root packing densities in the sample zones. Medium composition and preparation has been proven to affect the production of artemisinin in hairy root cultures. Jian Wen and Ren Xiang showed that the ratio of differently fixed nitrogen in Murashige and Skoog medium (MS medium) had a great impact on artemisinin level [54]. The optimal initial growth condition of 20 mM nitrogen in the ratio 5 : 1 NO^{3-}/NH^{4+} (*w*/*w*) produced a 57% increase in artemisinin production compared to the control in standard MS medium. Weathers et al. determined optimal growth at 15 mM nitrate, 1.0 mM phosphate and 5% *w*/*v* sucrose with an eight-day old inoculum but the production of artemisinic acid was not detected using phosphate at higher concentrations than 0.5 mM [55]. This implies that it is very difficult, if even possible, to optimize hairy root growth and terpenoid production at the same time; there has to be a trade off between biomass and product formation. As artemisinin is a secondary metabolite it is reasonable to assume that this compound will only be produced in significant amounts when the primary needs of the tissue have been covered. An extended phase of biomass formation would mean a procrastinated production of secondary metabolites. Interestingly, Weathers et al. found that artemisinic acid was not detected when arteannuin B was produced (Fig. 3). They suggest that artemisinic acid is degraded by a peroxidase to arteannuin B, which can be converted into artemisinin [55, 56]. This together with another observation that an oligosaccharide elicitor from the mycelial wall of an endophytic *Colletotrichum* sp. B501 promoted artemisinin production in *A. annua* hairy roots together with greatly increased peroxidase activity and cell death makes it tempting to see a peroxidase in the biosynthetic pathway from (dihydro)artemisinic acid to artemisinin [57]. Dhingra and Narasu purified and characterized an enzyme capable of performing the peroxidation reaction converting arteannuin B to artemisinin (Fig. 3) [56]. Conversion was estimated to be 58% of the substrate on molar basis. Sangwan et al. were able to show conversion of artemisinic acid into arteannuin B and artemisinin using horse radish peroxidase and hydrogen peroxide on cell free extracts from unmature *A. annua* leaves [58]. On the other hand, it has been found that in chickory the lactone ring formation in (+)-costunolide is dependent on a cytochrome P450 hydroxylase using germacrene acid as the substrate [59].

Sugars are not only energy sources but also function as signals in plants [60, 61]. Westers et al. performed a study in which autoclaved versus filter sterilized media were used with the conclusions that filter sterilized media give higher biomass and more consistent growth results, as well as better replicable terpenoid production results, although the yields of these secondary metabolites decreased [62]. The authors explain the inconsistent results accompanying autoclaved media with variable hydrolysis of sucrose. By carefully choosing nutrient composition, light quality and type of bioreactor the artemisinin production level can reach up to approximately 500 mg L^{-1} [63, 64]. Ploidity is another factor to consider. De Jesus-Gonzales and Weathers produced tetraploid *A. annua* hairy roots by treating normal diploid parents with colchicine and thereby obtained a tetraploid hairy root producing six times more artemisinin than the diploid versions. Tetraploid plants have also been made using colchicine, which led to a 39% increase in artemisinin production averaged over the whole vegetation period compared to diploid wild type plants [42].

2.2 Heterologous Biosynthesis

There are currently two main research strategies for production of artemisinine that are being intensively pursued. One is to increase production in the plant by bioengineering or through breeding programs, the second strategy is to utilize microorganisms in artificial biosynthesis of artemisinin. The group focusing on plant improvement brings forward the advantages of low production costs and easy handling, disregarding infestation and pest problems and additional costs for containment to prevent ecological pollution. The group favouring heterologous production of artemisinin in microoranisms admits higher production costs at the moment compared with artemisinin isolated from the plant but points to the advantages of efficient space versus production ratio, complete production and quality control, and a continuous supply of artemisinin possible only with sources not dependent on uncontrollable factors such as weather. In the two following chapters we give examples of the progress of the heterologous production of isoprenoid and artemisinin precursors in microorganisms.

2.2.1 Heterologous Production in *Escherichia Coli*

Of the two existing isoprenoid biosynthetic pathways (Fig. 3), DXP is used by most prokaryotes for production of IPP and dimethylallyl diphosphate (DMAPP) [65, 66]. With the available knowledge of the genes involved in the DXP pathway, several groups have studied the impact of changed expression levels of these genes on the production of reporter terpenoids. Farmer and Liao reconstructed the isoprene biosynthetic pathway in *Escherichia coli* (*E. coli*) to produce lycopene, which was used as an indication of an increase or decrease in isoprenoid production levels [67]. By overexpressing or inactivating the enzymes involved in keeping the balance of pyruvate and glyceraldehyde 3-phosphate (G3P), the authors established that directing flux from pyruvate to G3P increased lycopene production making the available pool of G3P the limiting precursor to isoprenoid biosynthesis. Kajiwara showed that overexpression of IPP lead to increased production of their terpenoid reporter molecule beta carotene [68]. Kim and Keasling investigated the influence of DXS, DXR, plasmid copy number, promoter strength and strain on production of the reporter terpenoid lycopene and were able to show a synergistic positive effect upon overexpression of both genes [69]. These kinds of strategies have all led to a moderate increase in production of terpenoid reporter molecules. Martin et al. hypothesized that the limited increase in isoprene production may be attributed to unknown endogenous control mechanisms [70]. By introducing the heterologous mevalonate pathway from *S. cerevisiae* into *E. coli* these internal controls were bypassed and isoprenoid precursors reached a toxic level. The introduction of a codon optimized AMDS alleviated this toxicity and led to production of amorpha-4,11-diene at the level of 24 μg caryophyllene equivalent ml^{-1} [70, 71]. Genes, such as transcriptional regulators, that are not directly involved in the isoprenoid biosynthetic pathway have also been shown to have a similar impact on production levels of terpenoid reporter molecules [72]. This can be expected because the isoprenoid biosynthetic pathway is tightly intertwined with the energy metabolism of the cell. The design strategy of the construct used can have a great influence on precursor production, as shown by Pfleger et al. [73]. By tuning intergenic regions in the mevalonate operon constructed by Martin et al. a seven-fold increase in mevalonate production compared with the starting operon conditions was recorded [70]. Brodelius et al. went a step beyond manipulating isolated genes, singular or multiple, in the biosynthesis of isoprenoids. By fusing FPPS isolated from *A. annua* and *epi*-aristolochene synthase from tobacco, the extreme proximity and, therefore, very short diffusion path led to a 2.5 fold increase in *epi*-aristolochene compared to solitary *epi*-aristolochene synthase [74]. Heterologous production of cyclized terpenoids is efficient but the following modification to form oxygenated plant terpenoids in *E. coli* seems to be a great bottleneck. Carter et al. engineered GPP biosynthesis coupled with the monoterpene cyclase limonene synthase, cytochrome P450 limonene hydroxylase, cytochrome P450 reductase and carveol dehydrogenase in *E. coli* with the expectation of producing the oxygenated limonene skeleton (–)-carvone [75]. Production of the unoxygenated intermediate limonene reached 5 mg L^{-1} but no oxygenated product was detected. The authors argue that this limitation may be due to cofactor limitations and membrane structural limitations in *E. coli* compared to plants. Hence, several research groups have turned to yeast for the heterologous expression of complex biosynthetic pathways.

2.2.2
Heterologous Production in *Saccharomyces Cerevisiae*

Fungi use the mevalonate pathway to produce all their isoprenoids. Lessons learned on the manipulation of the genes involved in the yeast mevalonate pathway were useful to increase the production of isoprenoids in *E. coli* as discussed above. Jackson et al. used *epi*-cedrol synthase converting FPP as a reporter gene for ispoprenoid production [76]. By overexpressing a truncated version of HMGR in a *S. cerevisiae* mutant (*upc2-1*, upregulates global transcription activity) taking up sterol, in the production, an increase from 90 $\mu g\,L^{-1}$ to 370 $\mu g\,L^{-1}$ of *epi*-cedrol was obtained. Overexpression of a native FPPS gene did not, however, improve the levels of *epi*-cedrol. As in the attempt of heterologous production of the oxygenated terpenoid *epi*-cedrol [75] in *E. coli*, an attempt to reconstruct early steps of taxane diterpenoid (taxoid) metabolism in *S. cerevisiae* produced taxadiene but did not proceed with cytochrome P450 hydroxylation steps [77]. Structural limitations of the membrane or co-factor limitations such as NADPH do explain this result. The authors discussed that poor expression of the heterologous plant cytochrome P450 genes might be an explanation to this pathway restriction. Another angle mentioned by the authors is a possible inefficient coupling and interaction between the endogenous yeast NADPH-cytochrome P450 reductase and the plant cytochrome P450 hydroxylase. This severely limits the transfer of electrons to the cytochrome P450 hydroxylase and leads as a consequence to premature termination of the pathway. Ro et al. introduced several genetic modifications in *S. cerevisiae* and were able to produce the oxygenated terpenoid artemisinic acid [34] at 100 $mg\,L^{-1}$ titre. This was achieved by optimized oxygen availability, downregulation of squalene synthase (*erg9*), which thus reduced endogenous consumption of the FPP pool, introduction of the *upc2-1* mutation, overexpression of FPPS and a catalytic form of HMGR, inducible expression of AMDS, cytochrome P450 71AV1 and a cytochrome P450 reductase from *A. annua*. More than 50 $mg\,L^{-1}$ amorpha-4,11-diene was produced in yeast engineered for overexpression of truncated HMGR and AMDS in a *upc2-1* yeast mutant genetic background. An additional two-fold to three-fold increase in the amorphadiene level was obtained through knock out of squalene synthase, but a marginal increase was harvested with additional overexpression of FPPS. Teoh et al. showed that oxygenation of amorphadiene to artemisinic alcohol and artemisinic alcohol to artemisinic acid was possible at proof of principle levels using cytochrome P450 71AV1 and a cytochrome P450 reductase from *A. thaliana* [33, 78]. Takahashi et al. chose a similar strategy to create a yeast platform for the production and oxygenation of terpenes [79]. A yeast mutant in squalene synthase (*erg9*), which is capable of efficient aerobic uptake of ergosterol from the culture media, produced 90 $mg\,L^{-1}$ farnesol, which is the dephophorylated form of FPP unaccessible for cyclization through terpene synthases. This mutant, when engineered with various single terpene synthases, was capable of producing around $\sim 80-\sim 100\,\mathrm{mg\,L^{-1}}$ sesquiterpene varying with the terpene synthase introduced. After additional engineering with hydroxylases, up to $50\,\mathrm{mg\,L^{-1}}$ hydroxylated terpene and $50\,\mathrm{mg\,L^{-1}}$ unmodified terpene product were obtained. Knocking out a phosphatase (*dpp*1) known to dephosphorylate FPP [80] and additional upregulation of the catalytic activity of HMGR did not yield an increase in terpene production compared to the *erg9/sue* yeast mutant. The authors note that a larger part of the farnesol is phosphorylated in a *dpp*1 mutant with a FPP function as a negative feedback signal on the mevalonate pathway suppressing the flux of carbon through the isoprene pathway [79, 81]. Inserting a terpene cyclase diverts the pool of FPP and relieves the feedback inhibition, which leads to an increase in carbon flux through the pathway almost matching the *erg9/sue* yeast mutant. In the *erg9/sue* mutant, a low but continuous flow through the mevalonate pathway led to higher production of terpenoids. Takahashi et al. also illustrate the importance of the design strategy of the expression vectors for optimal terpene production. Physically separating the cytochrome P450 reductase and the cytochrome P450 hydroxylase led to a very low yield of oxygenated terpenoid. On the other hand, expression vectors where the reductase preceded the hydroxylase gene on the same plasmid yielded approximately 50% coupling of oxygenation to hydrocarbon. Physically linking the terpene synthase with the hydroxylase was unsuccessful using both N-terminal and C-terminal fusion. Lindahl et al. showed that there are great differences in the production of amorphadiene depending on genomic or episomal expression [82]. The authors compared the production of amorpha-4,11-diene using the terpenoid synthase cloned in the high-copy number glactose inducible yeast plasmid pYeDP60 with the terpenoid synthase using the same galactose inducible promoter integrated into the genome of *S. cerevisiae* CEN PK113-5D. It was found that the yeast with an integrated AMDS grew at the same rate as the wild type, while the yeast carrying AMDS episomally had a slightly lower growth rate; yet the episomal system produced $600\,\mu\mathrm{g\,L^{-1}}$ amorpha-4,11-diene compared to $100\,\mu\mathrm{g\,L^{-1}}$ for the integrated system. This is an expected result that shows that in the case of integrated AMDS the enzyme activity is the limiting factor, while in the episomal system substrate availability is the limiting factor.

2.3 Growth of *Artemisia Annua* in Fields and Controlled Environments

Studies have been made where the intrinsic capacity of *A. annua* to produce artemisinin under various environmental conditions was explored. Ram et al. [83] undertook a study in which *A. annua* was grown with varying plant densities during the winter–summer season of one year in a semiarid-subtropical climate with no interculture and no fertilization. At a population density of 2.22×10^5 plants ha^{-1} 7.4 kg of artemisinin were obtained and 91 kg of essential oil. By increasing the plant density two-, four- and eight-fold an increase of artemisinin by one-and-a-half-, two- and two-and-a-half-fold was observed at the same oil yield level. Interestingly, the suppression of weeds was positively correlated with the increase in artemisinin production. Weeds, however, do not seem to be a trigger for artemisinin production because treatment of *A. annua* with herbicides removed weeds but did not influence artemisinin yields [84]. Kumar et al. showed [85] that multiple harvesting of *A. annua* grown in the subtropical Indo-Gangatic plains unsurprisingly increased the total yield of artemisinin but also increased the production of artemisinin in leaves as averaged over the separate sampling events. This trend was more expressed the later in the year the seeds were sown, confirming a study performed by Ram et al. [86]. The effect of post-harvest treatment of *A. annua* on artemisinin content was investigated by Laughlin in a study using *A. annua* grown and harvested in temperate maritime environment in Tasmania [87]. The experiments included drying of the cut-off plants in situ, in the shadow, indoors in the dark or in a 35 °C oven (used as a comparison base). Drying in situ did not give any concentration difference in artemisinin content compared to oven treatment. The authors noted a trend for sun-, shade- and dark drying for 21 days to give higher artemisinin levels than oven drying although artemisinic acid levels were unaffected.

Under greenhouse controlled conditions, Ferreira investigated the impact of acidity and macronutrient deficiency on biomass and artemisinin yield [88]. Acidic soil and low levels of nitrogen, phosphor and potassium reduced, as expected, the leaf biomass to 6.18 g per plant. Providing lime to increase pH and addition of the macronutrients nitrogen, phosphor and potassium gave a biomass of 70.3 g per plant. Potassium deficiency was shown to have the least negative effect on biomass accumulation and the most positive effect on artemisinin production. Plants grown under potassium deficient conditions were compared with plants grown under full addition of lime and macronutrients. This comparison did not detect any significant change in artemisinin production between the two growth conditions. The author concludes that under mild potassium deficiency conditions, a similar production of artemisinin can be obtained per ha as when fertilizing the soil with potassium. Potassium fertilization can thus be omitted in acidic soil growth conditions, decreasing the production costs as stated by the author, but this would also decrease the environmental pressure.

3
Synthesis of Artemisinin, Derivatives and New Antiplasmodial Drugs

Ever since artemisinin was isolated as the active compound against malaria, organic chemists have been trying and succeeding to produce the drug in the reaction flask. This has been performed with variable success but the general conclusion is still that it is a great scientific achievement but economically not attractive. A recent synthetic route to artemisinin involves 10 reaction steps from (+)-isolimonene to (+)-artemisinin with a final yield of a few percent [89]. Yet this result is considered a success in terms of yield and stereochemistry precision. In contrast, conversion of artemisinic acid into artemisinin is simple and can be done with photooxygenation in organic solvent [90]. In their study Sy and Brown describe the role of the 12-carboxylic acid group in spontaneous autooxidation of dihydroartemisinic acid to artemisinin [91]. The mechanism is further developed in an accompanying paper by the authors [38]. Artemisinin, however, has very poor solubility in both oil and water and, therefore, despite its antiplasmodial activity it not suitable as a drug. The development of artemisinin derivatives and completely synthetic analogues is described in a review by Ploypradith [92]. In the first attempts to improve the solubility of artemisinin the ketone was replaced with other bigger polar groups forming ester derivates of artemisinin. Depending on the attached groups, the first generation derivates showed solubility in either oil or water. The derivates sodium artesunate and artelinic acid are still in use due to their efficiency in clearing severe malaria infections. However, these first generation derivates are labile in acid environments, have a short half-life and some derivates have been shown to have neurotoxic effects. The second generation of semi-synthetic analogues was produced from artemisinin or artemisinic acid with the goals of improvement in metabolic and chemical stability, bioavailability and half-life. Two main streams were developed in the second generation of semi-synthetics. One group retained the acetal C10-oxygen, a second strategy was to reduce the acetal to an aliphatic group with increased acid stability. Of these two groups there are monomers and dimers. The dimers are interesting not only because they have a high antiplasmodial activity, but also because of their antineoplastic features.

Artemisinin with its crucial endoperoxide bridge is not the only natural compound exhibiting antiplasmodial activity. An example of the biosynthesis of antiplasmodial endoperoxidic compounds is plakortin, a simple 1,2-dioxane derivative, which is produced by the marine sponge *Plakortis simplex* [93]. This compound shows activity against chloroquine-resistant strains of *Plasmodium falciparum* (*P. falciparum*) at submicromolar level.

Several synthetic simplifications have been made as the knowledge of the mode of action of artemisinin has developed. In a review by Ploypradith selected strategies are reported [92]. One line is to omit the lactone ring which is considered to be less important if at all for antiplasmodial activity. Molecules that completely abandon the structure of artemisinin and its precursor only retaining the peroxide bond as a crucial functional pharmacophore are numerous. These molecules are easy to make but unfortunately display significantly reduced activity against malaria compared with artemisinin. As discussed in the introduction, they have a short half-life and poor chemical stability. A further dimension added in the synthesis of synthetic antiplasmodial was the idea to add multiple endoperoxide bridges within a molecule ring rather than adding them up as dimers with a linker in between. These tetraoxacycloalkanes showed a several-fold increase in efficiency against malaria compared to artemisinin, yet had a lower toxicity in mouse models. Design and synthesis of selected tetraoxanes are described in an article by Amewu et al. [94].

4
Analytics

The detection and structural elucidation of terpenes has been hampered by the often very low amounts and complex mixtures formed in plants. The spectrum of extraction methods and analytical methods has increased the ease and speed with which these problems can be solved. The choice of the extraction protocol greatly influences the yield and composition of the isolated product, as well as cost and time factors [95]. Peres et al. compare soxhlet, ultrasound-assisted and pressurized liquid extraction of terpenes, fatty acids and vitamin E from *Piper gaudichaudianum* Kunth [96]. The authors conclude that the method pressurized liquid extraction decrease the total time of extraction, the solvent use and handling compared to the other two methods. Furthermore, it was determined that pressurized liquid extraction extracted terpenes more efficiently than the other two methods. Lapkin et al. compare extraction of artemisinin using hexane, supercritical carbon dioxide, hydrofluorocarbon HFC-134a, several ionic liquids and ethanol [97]. Hexane was found to be simple and at a first glance the most cost efficient but is characterized by lower rates and efficiency compared to all other methods, including safety and environmental impact issues. The new techniques based on supercritical carbon dioxide, hydrofluorocarbon HFC-134a and ionic liquids consistently showed faster extraction cycles with higher recovery in addition to enhanced safety and decreased negative impact on the environment compared to hexane and ethanol extraction. With some process optimization, the authors predict that ionic liquid and HFC-134a extraction can compete with hexane extraction also on economical terms. In their review article Christen and Veuthey compare the extraction techniques supercritical fluid extraction, pressurized solvent extraction and microwave-assisted extraction and the detection methods gas chromatography, tandem mass spectrometry, HPLC-UV, -EC and -MS, as well as ELISA and capillary electrophoresis [95]. The use of evaporative light scattering detector is mentioned as a tool for detection of non-volatile non-chromophoric compounds. Common to all these methods is the trend toward mild operating conditions in order to avoid degradation of the analytes, isolation of one compound in complex mixtures and time and price reduction compared to traditional extraction methods. ELISA is accurate and is usable for screening of large plant populations but is laboursome and expensive compared to standard GC and HPLC based methods [98]. It is likely that this method will win stronger support in assessing the drug susceptibility of *P. falciparum* [99]. A simple, fast and selective method of quantification of artemisinin and related compounds was developed by Van Nieuwerburgh et al. [100]. This method makes use of HPLC-ESI-TOF-MS/MS technology and has a recovery of > 97% for all measured analytes. Peng et al. compared the use of GC-FID and HLPC-ELSD for detection of artemisinin in leaves [101]. Both methods are valuable for routine measurements because they are cheap, easy to use and do not require derivatization of artemisinin for detection. Both methods had a high sensitivity at ng level and produced reproducible results of artemisinin from field plants with a correlation coefficient of $r^2 = 0.86$ between the two methods. Another interesting simple and rapid method circumventing the problems with thermolability, lack of chromophoric or fluorophoric groups, low concentration in vivo and interfering compounds in planta of artemisinin detection is the method developed by Chen et al. [102]. Artemisinin is converted on-line to the strongly absorbing compound Q292 through treatment with NaOH. The obtained product is analyzed with capillary electrophoresis in 12 minutes, allowing a sampling frequency of 8 h^{-1}. With this work, Chen et al. show that it is possible to determine the artemisinin content based on the unstable UV-absorbing compound Q292, thus omitting the traditional time-consuming step of acidic conversion of Q292 to the stable UV-absorbing compound Q260 before analysis. The HPLC-MS method in selective ion mode developed by Wang et al. is another interesting cheap, sensitive and fast method for the detection and quantification of artemisinin in crude plant extracts [103]. The obtained linearity of detection in this method is about 5–80 ng ml^{-1} for artemisinin with an analysis time of 11 min per sample.

An old method that has been revived is the use of thin layer chromatography plates for the detection of sesquiterpenoids [104, 105]. While this kind of detection is qualitative and preferably to be used as quick determination of yes/no cases, more comprehensive and qualitative methods are needed for research purposes.

Ma et al. made a fingerprint of the volatile oil composition of *A. annua* by using two-dimensional gas chromatography time-of-flight mass spectrometry. With this method, approximately 700 unique peaks were detected and 303 of these were tentatively identified [106]. As a comparison, only 61 peaks were detected using GC. This type of comprehensive metabolic fingerprinting will ease detection of genes that are directly or indirectly relevant for the biosynthesis of artemisinin in experiments utilizing gene upregulation or downregulation mechanisms.

There is some discussion about the synergistic effects on clearing of the parasite *P. falciparum* from infected patients using extracts from *A. annua*.

Bilia et al. describe the importance of flavonoids in interaction between artemisinin and hemin [107]. Hemin is thought to play a role in the activation of artemisinin. It is thus of value to develop a method that can analyze artemisinin and flavonoids simultaneously. Bilia et al. developed a method based on HPLC/diode-array-detector/MS delivering just that [108].

5 Medicinal Use

The mode of action of artemisinin is subject to intense research [109–116]. Currently, the hypothesis supporting radical ion formation from artemisinin on the peroxide bridge is favoured.

Traditionally, artemisinin is administered as a tea infusion. With the advent of combination therapies using artemisinin as an isolated compound it is necessary to compare the kinetic characteristics of each delivery method. Räth et al. studied the pharmacokinetics and bioavailability of artemisinin from tea and oral solid dosage forms [117]. Interestingly, artemisinin was absorbed faster from herbal tea preparations than from oral solid forms, supporting the importance of flavonoids as synergistic factors. Nevertheless, bioavailability was similar in both treatments. Because only about 90 mg artemisinin was contained in 9 g *A. annua* and uptake of artemisinin through the human gut is very poor, only about 240 ng ml^{-1} was detected in plasma, a tea infusion is not recommended by the authors as a replacement for modern formulations in malaria therapy. This confirms the study of pharmacokinetics of artemisinin performed by Duc et al. [118]. Duc et al. proposed to increase the dose of artemisinin until adequate plasma levels are reached to compensate for poor bioavailability and rapid elimination, as no adverse effects were detected. This might prove a risky strategy because artemisinin-induced toxic brainstem encephalopathy has been observed in a patient treated for breast cancer with artemisinin [119]. The adverse effects were reversible and no permanent damage was detected. Toxicity of antimalarials including artemisinin derivatives is described in a review article by Taylor et al. [120]. In a pilot study Mueller et al. studied the efficacy and safety of the use of *A. annua* as tea against uncomplicated malaria [36]. Treatments were efficient but still less efficient compared to the traditional quinine; an average of 74% were cleared after seven days of treatment compared to 91% treated with quinine. As noted by the authors, recrudescence rates were high in the groups treated with artemisinin and they therefore recommend combination therapies, which is in line with the recommendation from WHO. However, the choice of combination partner in the combination therapies is a delicate question, which is exemplified in the study of Sisowath et al. [121]. In a recent review article the mechanism behind antimalarial drug resistance is covered [122]. Interestingly, resistance can be reversed [123]. It is obvious that the clearance of the parasite through tea preparations will depend on the amount of artemisinin present in the plant. Only approximately 40% of the available artemisinin in the plant was recovered in tea infusions, as shown in another study by Müller et al. [124]. Here it was demonstrated that malaria infested patients who were given tea preparations for two to four days showed a recovery of 92% within four days, a remarkable improvement compared with the previously mentioned study [36].

An overview of older (up to 1999) artemisinin derivatives is given in the article by Dhingra et al. [125]. All these derivatives were developed with the aim of obtaining a more efficient remedy against malaria. However, more recently artemisinin and its derivatives have been attributed intriguing functions other than antiplasmodial activities. In a study on flaviviruses Romero et al. describe the antiviral property of artemisinin [126]. Zhou et al. observed the derivate 3-(12-β-artemisininoxy-phenoxyl) succinic acid (SM735) to be strongly immunosuppressive in vitro and in vivo [127]. Artemisinin derivatives have also been shown to have strong antineoplastic properties [128–131].

6 Pharmacokinetics

A characteristic of artemisinin and its related endoperoxide drugs is the rapid clearance of parasites in the blood in almost 48 hours. Titulaer obtained pharmacokinetic data for the oral, intramuscular and rectal administration of artemisinin to volunteers [132]. Rapid but incomplete absorption of artemisinin given orally occurs in humans with a mean absorption time of 0.78 h with an absolute bioavailability of 15% and a relative bioavailability of 82%. Peak plasma concentrations reached after one to two hours and the drug is eliminated after three hours. The mean residence time after intramuscular administration was three times that when given orally. Other routes of administration, for example rectal or transdermal, are of limited success, but for the treatment of convulsive malaria in children artemether in a rectal formulation is favoured. Artesunate acts as a prodrug that is converted to dihydroartemisinin. When given orally the first pass mechanism in the gut wall takes places metabolizing half of the administered dose. Oral artemether is rapidly absorbed reaching maximum blood levels (C_{max}) within two to three hours. Intramuscular artemether is rapidly absorbed reaching C_{max} within four to nine hours. It is metabolized in the liver to the demethylated derivative dihydroartemisinin. The elimination is rapid, with a half-life time ($T_{1/2}$) of four hours. In comparison, dihydroartemisinin has a $T_{1/2}$ of more than ten tours. The degree of binding to plasma proteins varies markedly according to the species considered. The binding of artemether to plasma protein was 58% in mice, 61% in monkeys and 77% in humans. Radioactive labelled artemether was found to be equally distributed in plasma as well as in red blood cells, indicating an equal distribution of free drug between cells and plasma.

From the toxicological point of view artemisinin seems to be a safe drug for use in humans. In animal tests neurotoxicity has been documented, but as yet this side effect has not been reported in humans [133]. A major disadvantage of the artemisinin drugs is the occurrence of recrudescence when given in short monotherapy. So far no resistance has been observed clinically, although it has been induced in rodent models in vivo. The mechanism of action is different from the other clinically used antimalarials. Artemisinin drugs act against the early trophozoite and ring stages, they are not active against gametocytes, and they affect blood-stage but not liver-stage parasites. The mode of action is explained by haem or Fe^{2+}, from parasite digested haemoglobin, catalysing the opening of the endoperoxide ring and forming free radicals. Malaria parasites are known to be sensitive to radicals because of their lack of enzymatic cleaving mechanisms. The mechanism of action and the metabolism of reactive artemisinin metabolites is shown in Fig. 6.

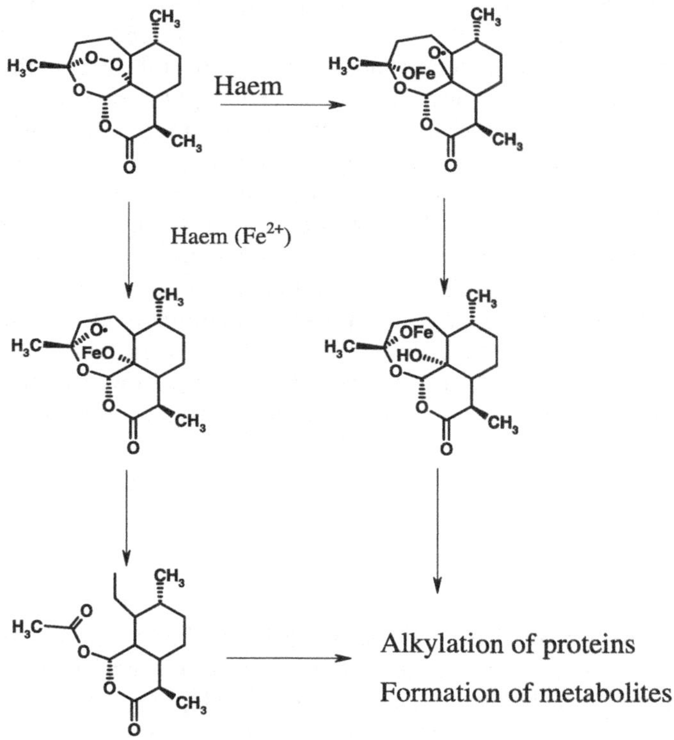

Fig. 6 Mechanism of the action of artemisinin drugs. Active metabolites and formation of reactive epoxide intermediates

7 Drug Delivery

Drug delivery of artemisinin and its derivatives is not as easy as known for intracellular microorganisms like *Leishmania* sp., *Mycobacterium tuberculosis* or *Listeria monogynes*. *P. falciparum* and related species are facultative intracellular parasites that mainly persist in erythrocyte as host cells. Drug targeting of infected erythrocytes is not well known and it does not seem to be a major area of interest for pharmaceutical technology to identify new strategies to deliver artemisinin or other antiplasmodial drugs to this target site. A literature search revealed no publication using liposomes, microemulsions, nanoemsulsions, microparticles or nanoparticles for targeting or drug delivery. Most formulation strategies have been focused on the improvement of the poor solubility of artemisinin ($< 5\ \mathrm{mg\,L^{-1}}\ H_2O$). One interesting approach has been published in detail, documenting the approach to increase solubility with cyclodextrines. Cyclodextrines are cyclic oligosaccharides consisting of six, seven or eight glucose molecules forming α-, β-, or γ-cyclodextrine, respectively. Cyclodextrines form pores with an inner diameter ranging from 0.5 to 0.8 nm where lipophilic drugs may be incorporated, thereby increasing their distribution in water. While the lipophilic compound is shielded inside, hydroxyl groups on the outer surfaces create an overall hydrophilic character for this inclusion complex. For experimental purposes, artemisinin has been formulated with different cyclodextrines to improve its solubility and oral absorption leading to increased bioavailability [134]. Solubility diagrams indicated that the complexation of artemisinin (85%, 40%, and 12%, α-, β-, or γ-cyclodextrine, respectively) and the three different types of cyclodextrines occurred at a molar ratio of 1 : 1, and showed a remarkable increase in artemisinin solubility [134]. In a bioavailability study by the same authors β-, or γ-cyclodextrines seem to be superior to commercial Artemisinin 250 and increased oral bioavailability with a mean of $782\ \mathrm{ng\,h\,ml^{-1}}$ to 1329 and $1131\ \mathrm{ng\,h\,ml^{-1}}$ (β-, or γ-cyclodextrine, respectively). However, the poor solubility was still a critical parameter for significantly improved oral bioavailability [135].

8 Conclusion

Artemisinin is a potent antimalarial drug belonging to the chemical class of sesquiterpenoid endoperoxide lactones. Its poor solubility in water and organic phases has led to a focus on the development of derivatives towards increased solubility, metabolic and chemical stability and bioavailability [92]. A common feature of the first generation of artemisinin derivatives was the replacement of the ketone with bigger polar groups to form ester derivatives (Fig. 1). Among these, sodium artesunate and artelinic acid are still in use. Unfortunately, other common features of the first generation artemisinin derivatives are instability in acid environment and a short half-life. Some derivatives also have a neurotoxic effect. The second generation of semi-synthetic artemisinin derivatives target improved metabolic and chemical stability, bioavailability and half-life. In parallel with the progressive understanding of the mode of action of artemisinin, synthetic simplified antimalarial compounds have been developed. Several promising candidates based on synthetic simplified molecules containing multiple peroxide bridges within one ring, which show higher activity against malaria and lower toxicity compared with artemisinin, have been reported [92].

Two genes have been isolated from the biosynthetic pathway of artemisinin: The first is the amorpha-4,11-diene synthase and the second enzyme in the pathway is cytochrome P450 71AV1, which catalyzes three consecutive oxygenation steps on the amorphane skeleton [33,34]. This opens up the way for molecular biotechnology strategies aiming towards artificial biology, making use of heterologous gene expression in optimized hosts and the improvement of artemisinin yield in transgene *A. annua* through genetic engineering. With the knowledge of nucleotide sequences, protein functions and characteristics, the evolution of the genes identified in the biosynthetic pathway is a possible and logical next step to follow for increased levels of the artemisinin precursors amorpha-4,11-diene and oxygenated forms thereof. Great improvement in the yield of amorpha-4,11-diene and other early precursors has been made with the aid of genetic engineering and optimization of culture conditions. There are currently two main research lines followed in parallel with a third line favouring artificial biology strategies, with the aim of increased artemisinin production compared to the wild type plant: The use of cell cultures is a field that combines culture optimization and genetic engineering and the second line employs traditional plant breeding through which the genetic dominance over environmental impact on artemisinin production can be exploited. All strategies show potential for substantial improvement and it is currently not settled which, if any, approach is better in terms of economy, environmental impact, yield, safety and production flow. The recent developments in detection and separation technologies of terpenoids should aid swift progress in screening mutants and complex networks in which the artemisinin biosynthesis pathway is embedded.

The traditional administration of artemisinin as a tea of the plant *A. annua* is a cheap, easily accessible source for malaria plagued countries but an unreliable cure due to the fact that the artemisinin level *in planta* is very low and varies considerably between plants and batches. Additionally, absorption through the human gut is rapid but inefficient and liver induction of cytochromes P450s will not allow repeated drug courses. The most efficient administration is intramuscular injection, as the drug then has a mean residence time three times longer than the orally administrated drug. Because intramuscular administration requires medical personnel, is painful and generally disliked by patients, other delivery strategies require urgent research, to increase the solubility of artemisinin.

References

1. Woerdenbag HJ, Pras N, van Uden W, Wallaart TE, Beekman AC, Lugt CB (1994) Pharm World Sci 16:169
2. Ziffer H, Highet RJ, Klayman DL (1997) Fortschr Chem Org Naturst 72:121
3. Chinese-Cooperative-Research-Group (1982) J Trad Chin Med 2:31
4. Jung M, Kim H, Lee K, Park M (2003) Mini Rev Med Chem 3:159
5. Liu Y, Wang H, Ye H-C, Li G-F (2005) J Integrat Plant Biol 47:769
6. Chang YJ, Song SH, Park SH, Kim SU (2000) Arch Biochem Biophys 383:178
7. Mercke P, Bengtsson M, Bouwmeester HJ, Posthumus MA, Brodelius PE (2000) Arch Biochem Biophys 381:173
8. Wallaart TE, Bouwmeester HJ, Hille J, Poppinga L, Maijers NC (2001) Planta 212:460
9. Starks CM, Back K, Chappell J, Noel JP (1997) Science 277:1815
10. Kim SH, Heo K, Chang YJ, Park SH, Rhee SK, Kim SU (2006) J Nat Prod 69:758
11. Picaud S, Mercke P, He X, Sterner O, Brodelius M, Cane DE, Brodelius PE (2006) Arch Biochem Biophys 448:150
12. Benedict CR, Lu JL, Pettigrew DW, Liu J, Stipanovic RD, Williams HJ (2001) Plant Physiol 125:1754
13. Cane DE, Watt RM (2003) Proc Natl Acad Sci USA 100:1547
14. Rynkiewicz MJ, Cane DE, Christianson DW (2001) Proc Natl Acad Sci USA 98:13543
15. Vedula LS, Rynkiewicz MJ, Pyun HJ, Coates RM, Cane DE, Christianson DW (2005) Biochemistry 44:6153
16. Lesburg CA, Zhai G, Cane DE, Christianson DW (1997) Science 277:1820
17. Caruthers JM, Kang I, Rynkiewicz MJ, Cane DE, Christianson DW (2000) J Biol Chem 275:25533
18. Colby SM, Alonso WR, Katahira EJ, McGarvey DJ, Croteau R (1993) J Biol Chem 268:23016
19. Hohn TM, Beremand PD (1989) Gene 79:131
20. Desjardins AE, Hohn TM, McCormick SP (1993) Microbiol Rev 57:595
21. Math SK, Hearst JE, Poulter CD (1992) Proc Natl Acad Sci USA 89:6761
22. Proctor RH, Hohn TM (1993) J Biol Chem 268:4543
23. Greenhagen BT, O'Maille PE, Noel JP, Chappell J (2006) Proc Natl Acad Sci USA 103:9826
24. Picaud S, Olofsson L, Brodelius M, Brodelius PE (2005) Arch Biochem Biophys 436:215
25. Vogeli U, Freeman JW, Chappell J (1990) Plant Physiol 93:182
26. Bouwmeester HJ, Kodde J, Verstappen FW, Altug IG, de Kraker JW, Wallaart TE (2002) Plant Physiol 129:134
27. Schnee C, Kollner TG, Gershenzon J, Degenhardt J (2002) Plant Physiol 130:2049
28. Chen XY, Wang M, Chen Y, Davisson VJ, Heinstein P (1996) J Nat Prod 59:944
29. Mercke P, Crock J, Croteau R, Brodelius PE (1999) Arch Biochem Biophys 369:213

30. Bouwmeester HJ, Wallaart TE, Janssen MH, van Loo B, Jansen BJ, Posthumus MA, Schmidt CO, De Kraker JW, Konig WA, Franssen MC (1999) Phytochemistry 52:843
31. Bertea CM, Freije JR, van der Woude H, Verstappen FW, Perk L, Marquez V, De Kraker JW, Posthumus MA, Jansen BJ, de Groot A, Franssen MC, Bouwmeester HJ (2005) Planta Med 71:40
32. Woerdenbag HJ, Lugt CB, Pras N (1990) Pharm Weekbl Sci 12:169
33. Teoh KH, Polichuk DR, Reed DW, Nowak G, Covello PS (2006) FEBS Lett 580:1411
34. Ro DK, Paradise EM, Ouellet M, Fisher KJ, Newman KL, Ndungu JM, Ho KA, Eachus RA, Ham TS, Kirby J, Chang MC, Withers ST, Shiba Y, Sarpong R, Keasling JD (2006) Nature 440:940
35. Brown GD, Liang GY, Sy LK (2003) Phytochemistry 64:303
36. Mueller MS, Runyambo N, Wagner I, Borrmann S, Dietz K, Heide L (2004) Trans R Soc Trop Med Hyg 98:318
37. Lommen WJ, Schenk E, Bouwmeester HJ, Verstappen FW (2006) Planta Med 72:336
38. Sy L-K, Brown GD (2002) Tetrahedron 58:897
39. Wallaart TE, van Uden W, Lubberink HG, Woerdenbag HJ, Pras N, Quax WJ (1999) J Nat Prod 62:430
40. Wallaart TE, Pras N, Quax WJ (1999) J Nat Prod 62:1160
41. Wallaart TE, Pras N, Beekman AC, Quax WJ (2000) Planta Med 66:57
42. Wallaart TE, Pras N, Quax W (1999) Planta Medica, p 723
43. Wang H, Ge L, Ye HC, Chong K, Liu BY, Li GF (2004) Planta Med 70:347
44. Wang H, Liu Y, Chong K, Liu BY, Ye HC, Li ZQ, Yan F, Li GF (2006) Plant Biol (Stuttg)
45. Sa G, Mi M, He-chun Y, Ben-ye L, Guo-feng L, Kang C (2001) Plant Sci 160:691
46. Han J-L, Liu B-Y, Ye H-C, Wang H, Li Z-Q, Li G-F (2006) J Integrat Plant Biol 48:482
47. Chen D-H, Ye H-C, Li G-F (2000) Plant Sci 155:179
48. Delabays N, Simonnet X, Gaudin M (2001) Curr Med Chem 8:1795
49. Ferreira JF, Simon JE, Janick J (1995) Planta Med 61:351
50. De Magalhaes PM, Pereira B, Sartoratto A (2004) Acta Horticulturae 629:421
51. Flores HE, Vivanco JM, Loyola-Vargas VM (1999) Trends Plant Sci 4:220
52. Shanks JV, Morgan J (1999) Curr Opin Biotechnol 10:151
53. Souret FF, Kim Y, Wyslouzil BE, Wobbe KK, Weathers PJ (2003) Biotechnol Bioeng 83:653
54. Jian Wen W, Ren Xiang T (2002) Biotechnol Lett 24:1153
55. Weathers PJ, Hemmavanh DD, Walcerz DB, Cheetham RD, Smith TC (1997) In Vitro Cell Dev Biol Plant 33:306
56. Dhingra V, Narasu ML (2001) Biochem Biophys Res Commun 281:558
57. Wang J, Xia Z, Tan R (2002) Acta Bot Sin 44:1233
58. Sangwan RS, Agarwal K, Luthra R, Thakur RS, Singh-Sangwan N (1993) Phytochemistry 34:1301
59. de Kraker JW, Franssen MC, Joerink M, de Groot A, Bouwmeester HJ (2002) Plant Physiol 129:257
60. Loreti E, Alpi A, Perata P (2000) Plant Physiol 123:939
61. Rolland F, Baena-Gonzalez E, Sheen J (2006) Annu Rev Plant Biol 57:675
62. Weathers PJ, DeJesus-Gonzalez L, Kim YJ, Souret FF, Towler MJ (2004) Plant Cell Rep 23:414
63. Liu C, Wang Y, Guo C, Ouyang F, Ye H, Li G (1998) Bioproc Engin 19:389
64. Liu CZ, Wang YC, Ouyang F, Ye HC, Li GF (1997) Biotechnol Lett 19:927
65. Boucher Y, Doolittle WF (2000) Mol Microbiol 37:703
66. Rohdich F, Hecht S, Gartner K, Adam P, Krieger C, Amslinger S, Arigoni D, Bacher A, Eisenreich W (2002) Proc Natl Acad Sci USA 99:1158

67. Farmer WR, Liao JC (2001) Biotechnol Prog 17:57
68. Kajiwara S, Fraser PD, Kondo K, Misawa N (1997) Biochem J 324(Pt2):421
69. Kim SW, Keasling JD (2001) Biotechnol Bioeng 72:408
70. Martin VJ, Pitera DJ, Withers ST, Newman JD, Keasling JD (2003) Nat Biotechnol 21:796
71. Newman JD, Marshall J, Chang M, Nowroozi F, Paradise E, Pitera D, Newman KL, Keasling JD (2006) Biotechnol Bioeng 95:684
72. Kang MJ, Lee YM, Yoon SH, Kim JH, Ock SW, Jung KH, Shin YC, Keasling JD, Kim SW (2005) Biotechnol Bioeng 91:636
73. Pfleger BF, Pitera DJ, Smolke CD, Keasling JD (2006) Nat Biotechnol 24:1027
74. Brodelius M, Lundgren A, Mercke P, Brodelius PE (2002) Eur J Biochem 269:3570
75. Carter OA, Peters RJ, Croteau R (2003) Phytochemistry 64:425
76. Jackson BE, Hart-Wells EA, Matsuda SP (2003) Org Lett 5:1629
77. Dejong JM, Liu Y, Bollon AP, Long RM, Jennewein S, Williams D, Croteau RB (2006) Biotechnol Bioeng 93:212
78. Urban P, Mignotte C, Kazmaier M, Delorme F, Pompon D (1997) J Biol Chem 272:19176
79. Takahashi S, Yeo Y, Greenhagen BT, McMullin T, Song L, Maurina-Brunker J, Rosson R, Noel JP, Chappell J (2006) Biotechnol Bioeng 97:170
80. Faulkner A, Chen X, Rush J, Horazdovsky B, Waechter CJ, Carman GM, Sternweis PC (1999) J Biol Chem 274:14831
81. Gardner RG, Hampton RY (1999) J Biol Chem 274:31671
82. Lindahl AL, Olsson ME, Mercke P, Tollbom O, Schelin J, Brodelius M, Brodelius PE (2006) Biotechnol Lett 28:571
83. Ram M, Gupta MM, Dwivedi S, Kumar S (1997) Planta Med 63:372
84. Bryson CT, Croom EM Jr (1991) Weed Technol 5:117
85. Kumar S, Gupta SK, Singh P, Bajpai P, Gupta MM, Singh D, Gupta AK, Ram G, Shasany AK, Sharma S (2004) Indust Crop Product 19:77
86. Ram M, Gupta MM, Naqvi AA, Kumar S (1997) J Essential Oil Res 9:193
87. Laughlin JC (2002) Acta Horticulturae 576:315
88. Ferreira JF (2007) J Agric Food Chem 55:1686
89. Yadav JS, Satheesh Babu R, Sabitha G (2003) Tetrahedron Lett 44:387
90. Roth RJ, Acton N (1989) J Nat Prod 52:1183
91. Sy L-K, Brown GD (2002) Tetrahedron 58:909
92. Ploypradith P (2004) Acta Trop 89:329
93. Fattorusso C, Campiani G, Catalanotti B, Persico M, Basilico N, Parapini S, Taramelli D, Campagnuolo C, Fattorusso E, Romano A, Taglialatela Scafati O (2006) J Med Chem 49:7088
94. Amewu R, Stachulski AV, Ward SA, Berry NG, Bray PG, Davies J, Labat G, Vivas L, O'Neill PM (2006) Org Biomol Chem 4:4431
95. Christen P, Veuthey JL (2001) Curr Med Chem 8:1827
96. Peres VF, Saffi J, Melecchi MIS, Abad FC, de Assis Jacques R, Martinez MM, Oliveira EC, Caramao EB (2006) J Chromatogr A 1105:115
97. Lapkin AA, Plucinski PK, Cutler M (2006) J Nat Prod 69:1653
98. Ferreira JF, Janick J (1996) Phytochemistry 41:97
99. Kaddouri H, Nakache S, Houze S, Mentre F, Le Bras J (2006) Antimicrob Agents Chemother 50:3343
100. Van Nieuwerburgh FC, Vande Casteele SR, Maes L, Goossens A, Inze D, Van Bocxlaer J, Deforce DL (2006) J Chromatogr A 1118:180
101. Peng CA, Ferreira JF, Wood AJ (2006) J Chromatogr A 1133:254

102. Chen HL, Wang KT, Pu QS, Chen XG, Hu ZD (2002) Electrophoresis 23:2865
103. Wang M, Park C, Wu Q, Simon JE (2005) J Agric Food Chem 53:7010
104. Klayman DL, Lin AJ, Acton N, Scovill JP, Hoch JM, Milhous WK, Theoharides AD, Dobek AS (1984) J Nat Prod 47:715
105. Bhandari P, Gupta AP, Singh B, Kaul VK (2005) J Sep Sci 28:2288
106. Ma C, Wang H, Lu X, Li H, Liu B, Xu GJ (2006) Chromatogr A 1102:11
107. Bilia AR, Lazari D, Messori L, Taglioli V, Temperini C, Vincieri FF (2002) Life Sci 70:769
108. Bilia AR, Melillo de Malgalhaes P, Bergonzi MC, Vincieri FF (2006) Phytomedicine 13:487
109. Drew MG, Metcalfe J, Dascombe MJ, Ismail FM (2006) J Med Chem 49:6065
110. O'Neill PM, Rawe SL, Borstnik K, Miller A, Ward SA, Bray PG, Davies J, Ho Oh C, Posner GH (2005) ChemBioChem 6:2048
111. Rafiee MA, Hadipour NL, Naderi-manesh H (2005) J Chem Inf Model 45:366
112. Messori L, Piccioli F, Temperini C, Bilia AR, Vincieri FF, Allegrozzi M, Turano P (2004) Inorg Chim Act 357:4602
113. Krishna S, Uhlemann A-C, Haynes RK (2004) Drug Resist Updates 7:233
114. Hoppe HC, van Schalkwyk DA, Wiehart UIM, Meredith SA, Egan J, Weber BW (2004) Antimicrob Agents Chemother 48:2370
115. Posner GH, O'Neill PM (2004) Acc Chem Res 37:397
116. Schmuck G, Roehrdanz E, Haynes RK, Kahl R (2002) Antimicrob Agents Chemother 46:821
117. Rath K, Taxis K, Walz G, Gleiter CH, Li SM, Heide L (2004) Am J Trop Med Hyg 70:128
118. Duc DD, de Vries PJ, Nguyen XK, Le Nguyen B, Kager PA, van Boxtel CJ (1994) Am J Trop Med Hyg 51:785
119. Panossian LA, Garga NI, Pelletier D (2005) Ann Neurol 58:812
120. Taylor WRJ, White NJ (2004) Drug Safety 27:25
121. Sisowath C, Stromberg J, Martensson A, Msellem M, Obondo C, Bjårkman A, Gil JP (2005) J Infect Diseas 191:1014
122. White NJ (2004) J Clin Invest 113:1084
123. Henry M, Alibert S, Orlandi-Pradines E, Bogreau H, Fusai T, Rogier C, Barbe J, Pradines B (2006) Curr Drug Target 7:935
124. Mueller MS, Karhagomba IB, Hirt HM, Wemakor E (2000) J Ethnopharmacol 73:487
125. Dhingra V, Vishweshwar Rao K, Lakshmi Narasu M (2000) Life Sci 66:279
126. Romero MR, Serrano MA, Vallejo M, Efferth T, Alvarez M, Marin JJ (2006) Planta Med 72:1169
127. Zhou W-l, Wu J-m, Wu Q-l, Wang J-x, Zhou Y, Zhou R, He P-l, Li X-y, Yang Y-f, Zhang Y, Li Y, Zuo J-p (2005) Acta Pharmacol Sin 26:1352
128. Efferth T, Olbrich A, Bauer R (2002) Biochem Pharmacol 64:617
129. Efferth T (2006) Curr Drug Target 7:407
130. Disbrow GL, Baege AC, Kierpiec KA, Yuan H, Centeno JA, Thibodeaux CA, Hartmann D, Schlegel R (2005) Cancer Res 65:10854
131. Liu Y, Wong VKW, Ko BCB, Wong MK, Che CM (2005) Org Lett 7:1561
132. Titulaer HA, Zuidema J, Kager PA, Wetsteyn JC, Lugt CB, Merkus FW (1990) J Pharm Pharmacol 42:810
133. Merali S, Meshnick SR (1991) Antimicrob Agents Chemother 35:1225
134. Wong JW, Yuen KH (2003) Drug Dev Ind Pharm 29:1035
135. Wong JW, Yuen KH (2001) Int J Pharm 227:177

Top Heterocycl Chem (2007) 9: 33–52
DOI 10.1007/7081_2007_067

Published online: 4 September 2007

Sugar-derived Heterocycles and Their Precursors as Inhibitors Against Glycogen Phosphorylases (GP)

Mahmud Tareq Hassan Khan[1,2]

[1]School of Molecular and Structural Biology, and Department of Pharmacology, Institute of Medical Biology, University of Tromsø, 9037 Tromsø, Norway
mahmud.khan@fagmed.uit.no

[2]*Present address:*
Pharmacology Research Lab., Faculty of Pharmaceutical Sciences, University of Science and Technology, Chittagong, Bangladesh

Abstract Non-insulin-dependent diabetes mellitus (NIDDM or Type II diabetes) is a multifactorial metabolic disorder in which hepatic glucose production is increased. Glycogenolysis and the main regulatory enzyme glycogen phosphorylase (GP) are responsible for the release of mono-glucose from poly-glucose (glycogen, as stored form in the liver). This protein possesses several binding pockets or cavities that regulate the catalytic functions of GP. So obviously, the inhibitors of GP will stop or slow down glycogenolysis as well as glucose production and ultimately the whole process will result in the recovery of diabetes in NIDDM patients. Glucose is one of most important regulators of GP, and glucose analog inhibitors (GAIs) have shown very promising activity for the inhibition of GP. There have been a large number of GAIs reported in last few decades that are promising for the control of NIDDM. This review briefly describes some aspects of GP and its relation with GAIs, mostly containing heterocyclic building blocks.

Keywords Glycogen phosphorylase · Glucose · Glucose analog inhibitors · NIDDM · Sugar-derived heterocycles

Abbreviations

AMP	Adenosine mono-phosphate
ATP	Adenosine tri-phosphate
CNS	Central nervous system
G1P	Glucose-1-phosphate
G6P	Glucose-6-phosphatase
GA	Glucose analogs
GAI	Glucose analog inhibitor
GK	Glucokinase
GlcNAc	*N*-Acetylglucosamine
GLP-1	Glucagon-like polypeptide 1
GP	Glycogen phosphorylase
GS	Glycogen synthase
NIDDM	Non-insulin-dependent diabetes mellitus
PDB	Protein databank
Pi	*o*-Phosphate
PK	Phosphorylase kinase
PLP	Pyridoxal 5′-phosphate
QSAR	Quantitative structure–activity relationship

1 Introduction

Type 2 diabetes (non-insulin-dependent diabetes mellitus, NIDDM) is a multifaceted metabolic disease with hyperglycemia as its recognizable feature. The liver is a key tissue in overall metabolic regulation and the hepatic glucose output is elevated in NIDDM patients. Experimental data recommends drugs that inhibit or lower the hepatic glucose production as efficient antihyperglycemic agents [1, 2]. Current treatments for NIDDM rely on diet, exercise, hypoglycemic drugs intended to reduce hyperglycemia, etc., and if these fail insulin itself, which ultimately suppresses glucose production [3, 4] in liver. The present medications have inadequate efficiency and acceptability and noteworthy mechanism-based side effects [4, 5]. As a consequence there is a sustained exploration for molecules that could improve treatment and give a better life to diabetic patients.

Glycogenolysis, which is the release of monomeric glucose from its polymeric storage form called glycogen, is a key contributor to hepatic glucose output. Glycogen phosphorylase (GP) is the key enzyme catalyzing this procedure [1, 2]. A molecular target intended at reducing unnecessary glucose assembly from liver engages the inhibition of GP of human carbohydrate metabolism, which is of particular significance for the mobilization of glycogen deposits [4].

The most modern technologies have been utilized to discover a way of exploiting the possibility of GLP-1 (glucagon-like polypeptide 1) for the treatment of NIDDM. This demonstrates how discoveries of novel binding pockets on GP and GK as the result of drug discovery programs have led to increased understanding of these key metabolic enzymes, and also to potential new therapies for NIDDM [1].

The recent advances, especially during last few years, in different heterocyclic glucose analog inhibitors (GAIs) and their precursors are briefly discussed in this chapter, including their biological activities against this very important enzyme GP, from different perspectives.

2
Glycogen Metabolism

Liver is the major source of blood glucose. Hepatic glucose is produced from two pathways [6, 7]:

Glycogenolysis – the breakdown of glycogen
Gluconeogenesis – de novo glucose synthesis

Glycogenolysis can explain why more than 70% of the hepatic glucose created by gluconeogenesis is cycled during the collection of glycogen preceding the start of efflux in the liver cells [6, 7]. Hepatic glucose productivity is synchronized by an intricate coordination of enzymes. The most important dogmatic enzyme of this coordination is glycogen phosphorylase (GP), and only the phosphorylated enzyme (known as GP*a*) has noteworthy biochemical activities in living systems [7]. Figure 1 show the pathway of gluconeogenesis in liver.

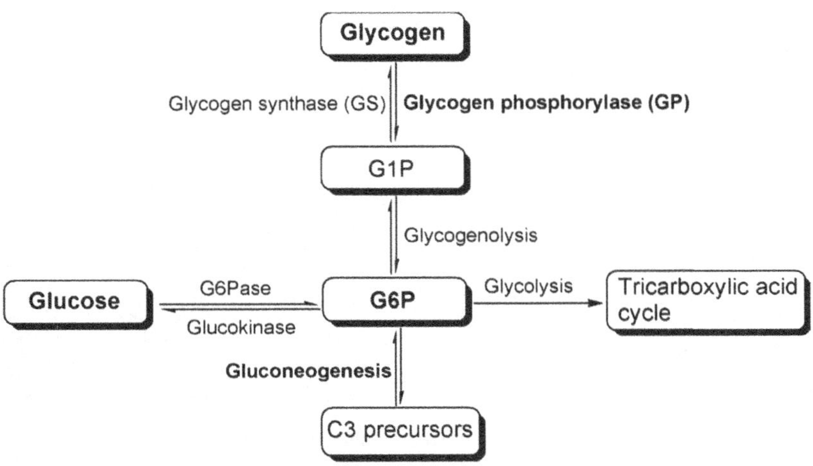

Fig. 1 Flowchart of gluconeogenesis in liver cells, which explains the role of GP and GS in the interconversion between glucose and glycogen (modified from Somsak et al. 2003) [7]

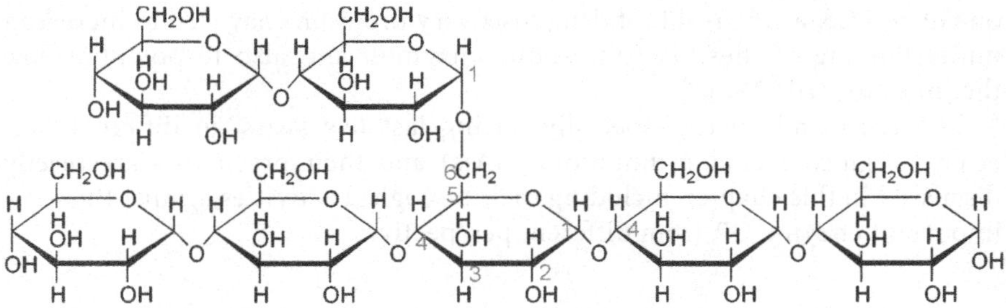

Fig. 2 Typical example of glycogen (*numbers in gray* show the numbering system)

Glycogens are polymers of glucose residues linked by α(1 → 4) glycosidic bonds, mainly, and α(1 → 6) glycosidic bonds at branch points. Figure 2 shows a typical example of glycogen, where the glycogen chains and branches are longer than shown. This is stored as glycogen predominantly in liver and muscle cells.

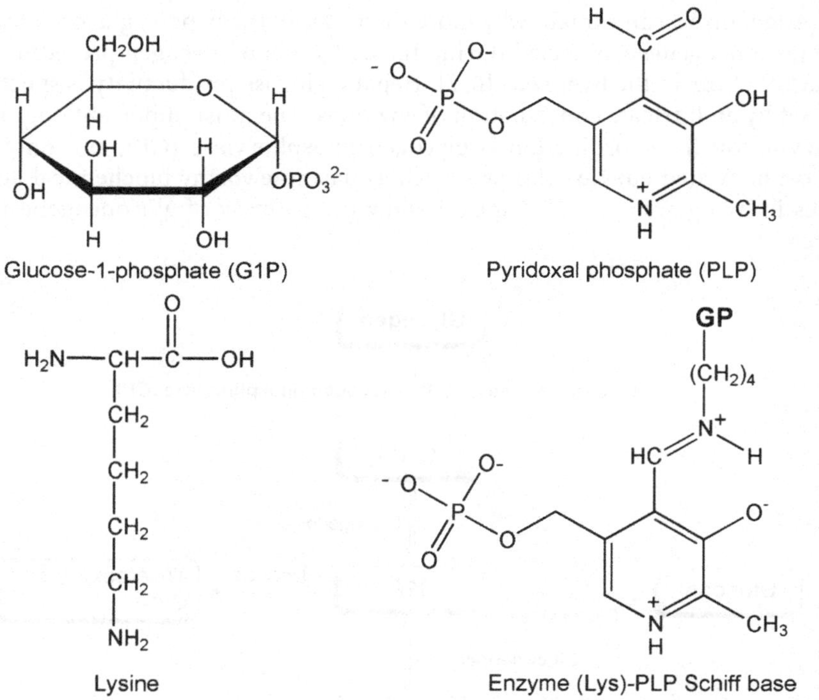

Fig. 3 Structures of the molecules involved in the glycogenolysis process. Here PLP is a derivative of vitamin B6 and serves as prosthetic group for GP; PLP is apprehended at the active site by a Schiff base linkage, formed by reaction of the aldehyde of PLP with the ε-amino group of a Lys residue

GP catalyzes phosphorolytic cleavage of the $\alpha(1 \rightarrow 4)$ glycosidic linkages of glycogen, releasing G1P as reaction product:

$$\text{Glycogen}(n\text{residues}) + P_i \rightarrow \text{glycogen}(n -- 1\text{residues}) + \text{G1P}$$

Figure 3 shows the structures of the molecules that are involved in the glycogenolysis process.

3
Glycogen Phosphorylase (GP) as a Novel Target and its Role

The carbohydrate reserve of most metabolically active cells in the animal kingdom is glycogen, a glucose polymer. The cellular demands to convert glycogen and *o*-phosphate (P_i) to G1P are met by GP, one of the most complex and finely regulated enzymes yet encountered [8]. In the liver, G1P is mostly converted by phosphoglucomutase and glucose-6-phosphatase (G6P), which is released for the benefit of other tissues, the CNS in particular relies on glucose as its major source of energy [4, 9].

GP is an allosteric enzyme that exists in two interchangeable forms [10]:
GP*a*High activity, high substrate affinity, and predominantly "R" state
GP*b*Low activity, low affinity toward the substrate, and predominantly "T" state [10].

The allosteric activators (like AMP) or inhibitors (like ATP, G6P, glucose or caffeine) can alter the equilibrium between a lower active "T" state to a more active "R" state or vice versa [4, 10]. The molecular structures of T and R states of GP have been illustrated through the X-ray diffraction studies, which have shown that the conformational transformations take place following the commencement of the muscle enzyme and its translation from the T to R state by phosphorylation or AMP [4, 11–15].

There are three mammalian GP isoenzymes, termed "muscle", "brain", and "liver" according to where they are expressed [7]. All are encoded by different genes, located on human chromosomes 11, 20, and 14, respectively [7, 16]. All the isoenzymes can be converted from the inactive (GP*b*) form to the active (GP*a*) form through the phosphorylation of Ser14 by phosphorylase kinase (PK) [7].
Several inhibitor binding sites have been identified in GP [4, 17–19]:

Ser14-phosphate recognition site
Allosteric site, which binds activator AMP and the inhibitor G6P,
Catalytic site, which binds substrates G1P and glycogen, and the inhibitor glucose and glucose analogs (GA)
Inhibitor site, which binds caffeine and related molecules
Glycogen storage site
Novel allosteric inhibitor site, which was discovered recently [4, 17–19]

In this chapter the main topic for discussion are the GAIs of GP, so of the above mentioned binding sites, the main concern is the *catalytic site*. This catalytic site is a deep cavity located at the center of the whole protein, 15 Å from the protein surface, and close to the essential cofactor pyridoxal 5′-phosphate (PLP) It has been probed with glucose and GAIs [4].

In 1976, Fletterick et al. [20] first reported a model of the polypeptide backbone of the dimer of GP*a*, which was built from a 3 Å resolution electron density map resulting from the X-ray diffraction analysis of native tetragonal crystals and two heavy atom isomorphous substitution derivatives [20]. The active site, of which there are two/dimer, is shared between the two subunits at their interface and comprises a pocket-like region within a "V"-shaped framework of two α-helices [20]. Within this region are found the binding sites for the substrates, G1P and arsenate, a competitive inhibitor UDP-glucose, and the allosteric effector AMP [20]. The site of metabolic control, Ser-14 phosphate, is hydrogen-bonded to a side chain on the outside of one of the α-helices forming the active site and is 15 Å from the AMP binding site [20].

Maltoheptaose, a glycogen analog and substrate for these enzymatically active crystals, binds in a second region of interest. Fletterick et al. suggested that this polysaccharide binding site may represent a storage site where phosphorylase is bound to the glycogen particle in the muscle cell. The polypeptide chain in a third region has the same topological structure as has been observed for the nucleotide binding domains in the dehydrogenases [20].

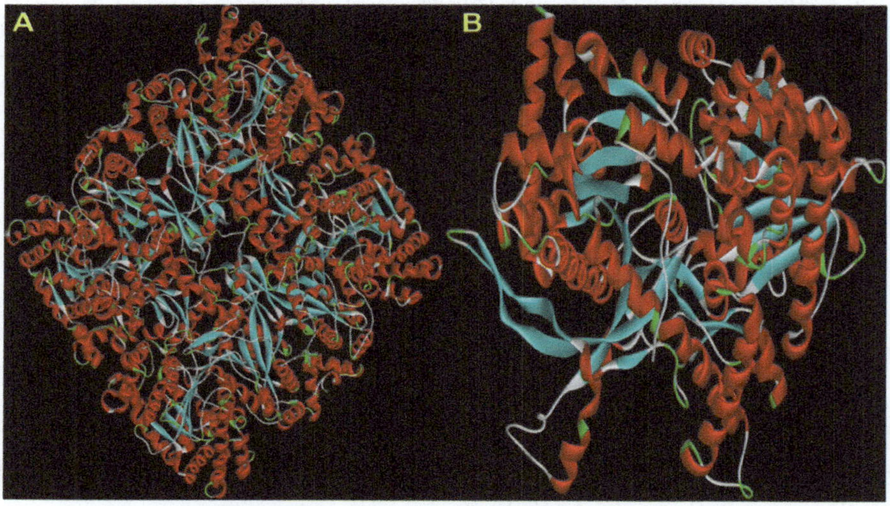

Fig. 4 Crystallographic structures of *A* GP*a* (PDB code 1gpa) [13], and *B* GP*b* (PDB code 1gpb) in 2.9 and 1.9 Å resolutions, respectively. Here, GP*a* contains four chains and GP*b* contains one [21]. The figure was created using the Accelrys DS Visualizer, version 1.6

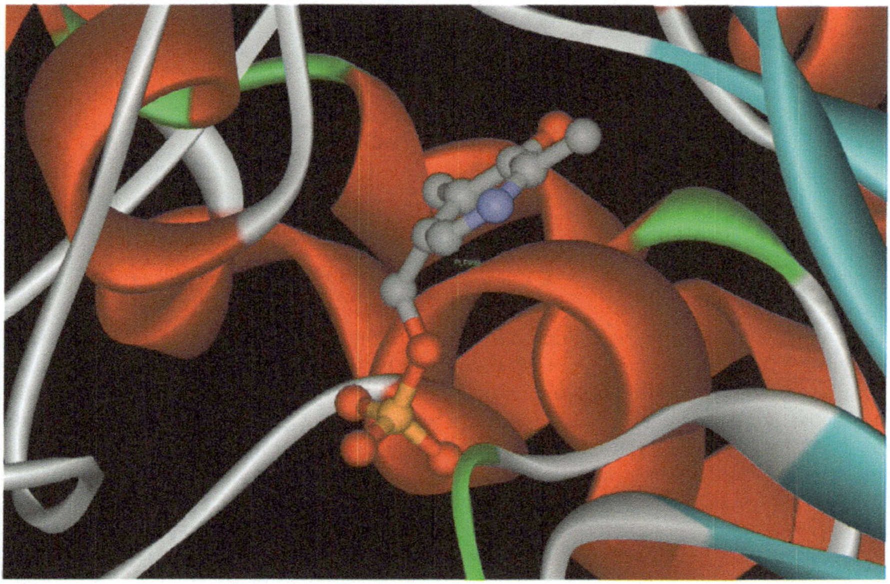

Fig. 5 A PLP molecule is bound in a catalytic domain of GP*a* (1pga) [13]. Here, the protein structure of GP*a* is shown in *solid ribbon model* and the GP*a*-bound PLP is shown in *ball and stick model*. The figure was created using the Accelrys DS Visualizer, version 1.6

Adenine or adenosine (but not the AMP) bind here in a position similar to the adenine ring of NAD in the dehydrogenases, while glucose binds 17 Å away in an interior crevice near the center of the monomer [20].

Figure 4 shows the 3D structures (X-ray crystallographic) of GP*a* (1gpa) [13] and *b* (1gpb) [21]. GP*a* contains four chains and is bound with PLP (shown in Fig. 5) at the catalytic domain [13].

4
The Role of GP Inhibitors

The isozymes of GP from muscle and liver are well characterized [22], but very less information is available about the brain-specific isozyme [23]. The crystal structures of human liver GP*a* and *b* are also known, which aids in the recognition of the binding modes of effector molecules [24, 25]. For convenience, most research has been accomplished with rabbit muscle GP; however, cloning and expression of human liver GP [26] has paved the way for investigations with the real target [7]. Figure 6 shows the crystal structure (3D) of human liver GP*a*, which contains several active sites, including the active site for the GP inhibitors. Figure 7 shows some early developed GP inhibitors [7].

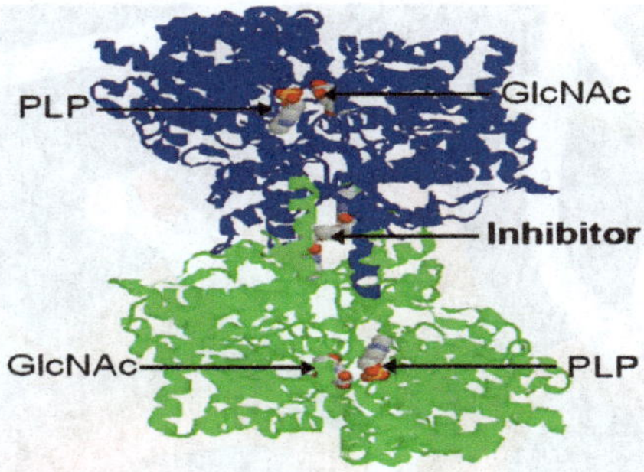

Fig. 6 Flat ribbon structure of human liver GP*a* (PDB code 1EM6). GP is a homodimeric enzyme, subject to allosteric control. It changes between "relaxed" (active) and "tense" (inhibited) conformations. *N*-Acetylglucosamine (GlcNAc, in space-filling CPK model), a glucose analog, is contiguous to PLP (in space-filling CPK model) at the active site in the 3D structure of the GP*a*. A class of molecules (inhibitors, in space-filling CPK model) developed for treating NIDDM inhibit the liver phosphorylase allosterically. These inhibitors bind at the interface of the dimer, stabilizing the inactive ("tense") conformation [20]

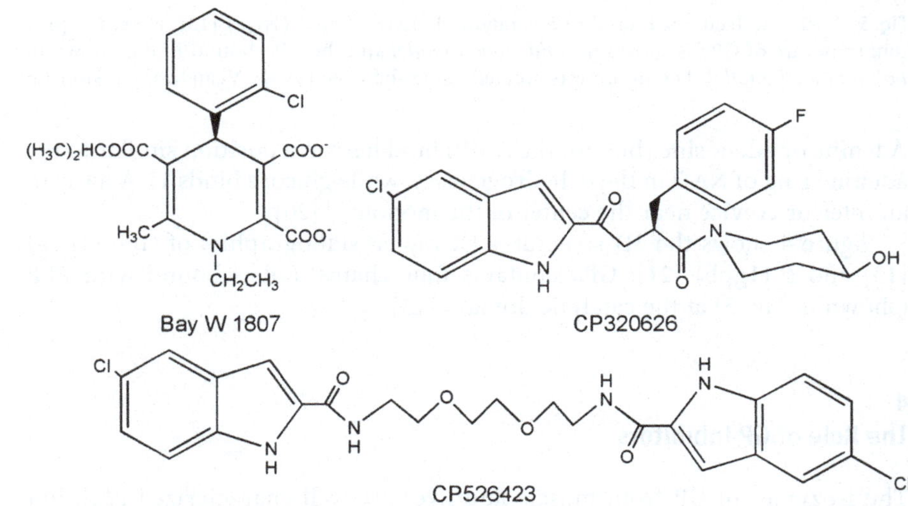

Fig. 7 Some potential GP inhibitors against GP*a*. Bay W 1807, CP320626, and CP526423 exhibited potent inhibition against GP*a* with IC_{50} values of 10.8 nM (against rabbit muscle GP*a*) [27], 205 nM (human liver GP*a*) [28], and 6 nM (human liver GP*a*) [24]

A large number of reviews have described the inhibition of GP, the role of GP inhibitors as oral antihyperglycemics (or oral hypoglycemics), as well as the GAIs [4, 7, 29–35]. Prof. Nikos G. Oikonomakos and his research team (National Hellenic Research Foundation, Athens, Greece) have reported huge numbers of discoveries of GP inhibitors over the last decades [27, 36–93].

4.1
Glucose

In certain physiological circumstances, glucose performs as a controller of GP by alleviating the less active T state of the enzyme through binding to the catalytic center [94]. It is an effective allosteric inhibitor for both the GP*a* and GP*b* with K_i values in the low millimolar range (2.0 and 1.7, respectively) [71, 76, 95]. Figure 8 shows the molecular structures of the α- and β-D-glucoses.

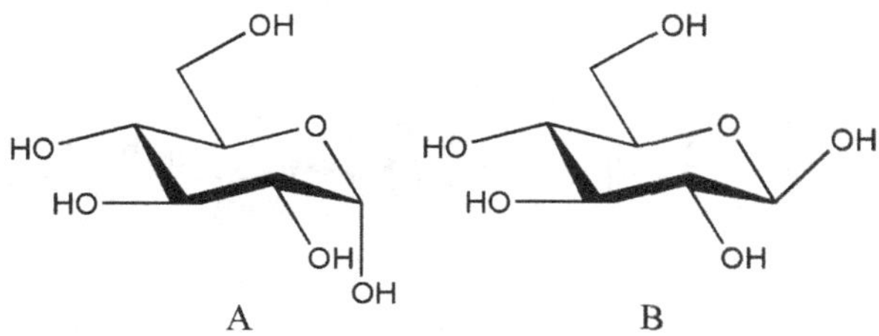

Fig. 8 Molecular structures of *A* α- and *B* β-D-glucose [71, 76, 95]

Glucose, on binding at the catalytic domain, upholds the less active T state by stabilization of the bunged pose of the 280s loop that obstructs the access for the substrate (e.g., glycogen) to the catalytic site [4].

5
Glucose Analog Inhibitors (GAIs) Against GPs

Figure 9 shows some of the early discovered GAIs against GPs, where experimental IC_{50} values calculated from A and B are 25.3 and 16.3 mM [71], respectively. For C, D, E, and F, the IC_{50} values were found to be 2.6, 0.032, 0.081, and 0.14 mM, respectively, [70]. For H, K, and L, the IC_{50} values were found to be 0.65, 0.44, and 0.37 mM, respectively, [72]. For compound I, the IC_{50} value was 0.014 mM [96], for compound J the IC_{50} value was 0.053 mM [66], and for compounds M and N the IC_{50} values were 0.0286 and 0.0031 mM, respectively [97].

5.1
D-Glucose Analogs

A number of β-D-glucose analogs were designed using the program GRID [98], which can predict energetically favorable substitutions and determine probable interaction sites between a functional group probe (e.g., hydroxyl, amino,

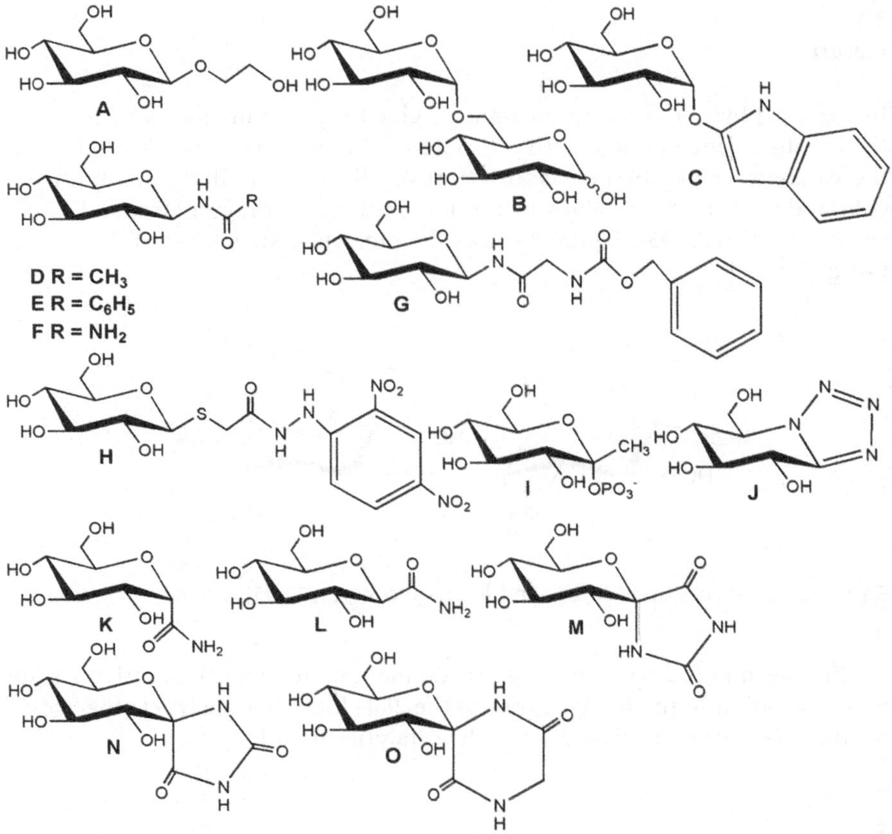

Fig. 9 Molecular structures of some of early discovered GAIs [7]

methyl, etc.) and the enzyme surface [4]. Figure 10 shows structures of some of the β-D-glucose analogs.

5.2 Glycosylidene Analogs

Several glycosilidene analogs with spiro-hydantoins and spioro-thiohydantoins exhibited potent inhibition against GPs. Figure 11 shows some of the examples.

For compounds A and B, the activities decreased due the replacement of S instead of O [97, 99]. Their K_i values were 3.1 and 5.1 μM, respectively. This means that O is more important for its binding with GP. But for C and D almost opposite results were observed, where their K_i values were > 11.5 and > 10 mM, respectively. The difference between A, B and C, D is the lack of the methoxy (= CH_2OH) group at the sugar ring, which resulted in almost com-

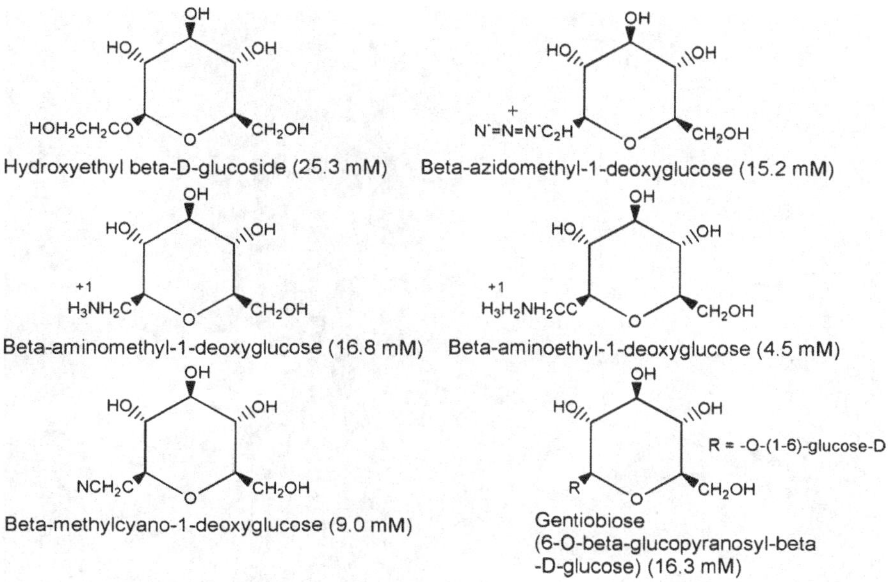

Fig. 10 Molecular structures of some of the β-D-glucose analogs (their IC_{50} values are shown in parenthesis) (modified from Oikonomakos et al. 2002) [4]

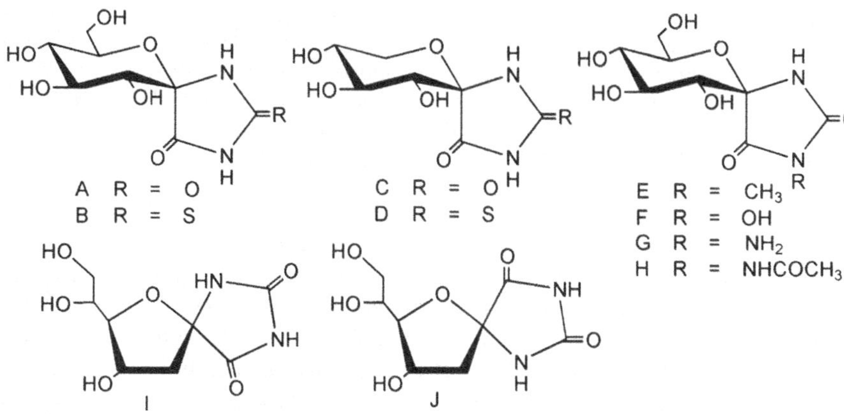

Fig. 11 Molecular structures of glycosylidene analogs experimentally proved as GP inhibitors [7]

plete loss of the inhibitory activity against GP. This shows that the = CH_2OH side chain at the sugar ring is much more important than that of the O or S at the R position [97, 99, 100].

Substitutions of N-9 in the spirohydantoin as shown in compounds E to H (see Fig. 11) fetched about no enhancement of the inhibition. Their K_i values

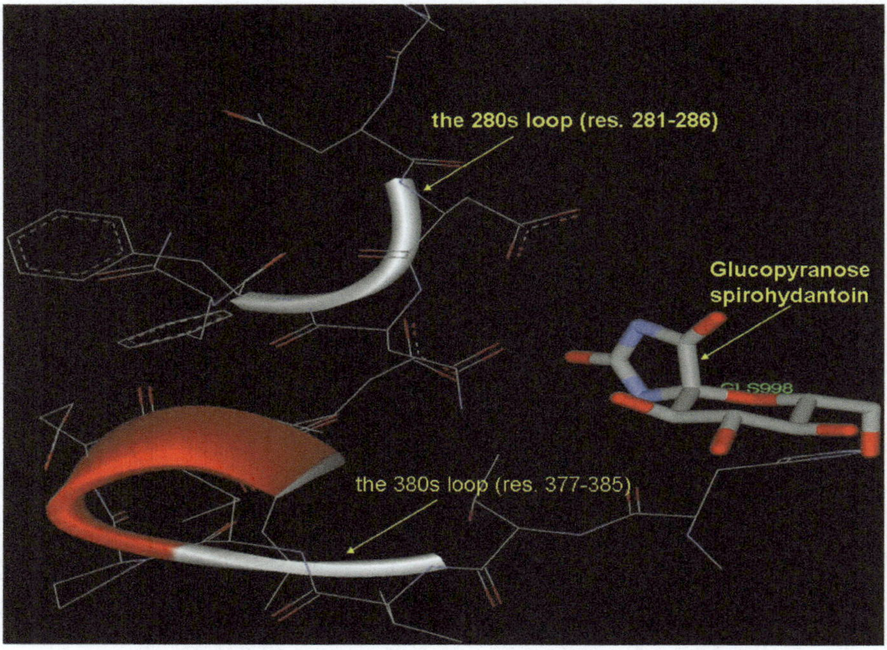

Fig. 12 Glucopyranose spirohydantoin (a pyranose analog of the potent herbicide, hydantocidin) has been identified as the most potent glucose analog inhibitor of GP*b* (PDB code 1A8I). Here, the molecule is bound in GP*b* at the small regions of the 280s and 380s loops [62]. The small molecule glucopyranose spirohydantoin is shown here as a *stick model*, the particular loops are shown in *solid ribbon*, and some other related parts of the protein GP*b* are shown in simple *line models*. The figure was created using the Accelrys DS Visualizer version 1.6

were 1.2, 0.039, 0.146 and 0.55 mM, respectively [101]. Alteration of the sugar ring to a furanoid structure as in I and J led to very weak binding, in agreement with previous interpretations with epimers and deoxy derivatives of D-glucose [7, 95, 102, 103]. The findings with C, D, I, and J, underline and confirm the high specificity of the GP active site towards a fully OH-substituted hexopyranoid sugar moiety of D-gluco configuration [7].

Figure 12 shows a crystal structure of a GP–glucopyranose spirohydantoin complex at 1.8 Å resolution (PDB code 1A8I) [62].

5.3
N-Acyl-*β*-D-glucopyranosylamides

Molecules having structural building blocks of the hydantoin moiety similar to A ($K_i = 0.032$ mM), in Fig. 13, can also bind with GPs [7]. Due to a novel synthetic approach [104], it has become possible to synthesize and

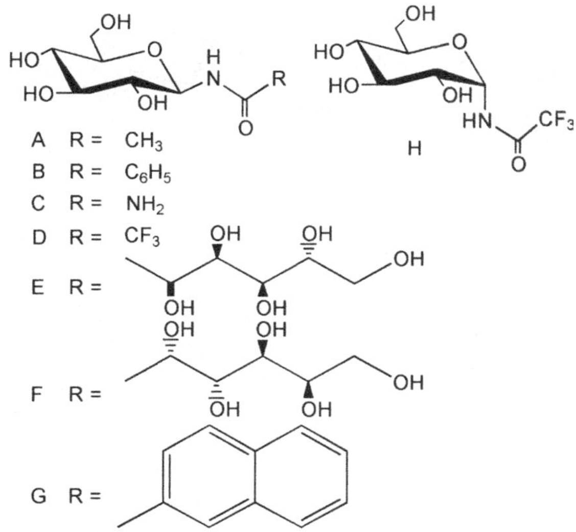

Fig. 13 Molecular structures of the *N*-acyl-β-D-glucopyranosylamide analogs [7]

explore anomeric pairs of *N*-trifluoroacetyl-D-glucopyranosylamine D (K_i = 0.71 mM) and H (*inactive*) (for structures see Fig. 13) [7]. Compared to A, the presence of the $-CF_3$ group in D significantly decreased the inhibitory activities against the enzyme GP, and the anomer H had totally lost the inhibition [7, 99]. These findings support the significance of structural building blocks of *N*-acyl-β-D-glucopyranosylamides. A hydrophobic amide side chain [104] in C (K_i = 0.341 mM) and F (K_i = 0.225 mM) exhibited weaker inhibitory profiles [7, 99]. Introduction of a naphthyl group, as in the case of G (K_i = 9.7 μM) resulted in a rather better inhibitor [7, 99]. These series of synthetic compounds and their in vitro experimental results recommend that analogous molecules with long aliphatic and/or hydrophobic side chains should be explored [7, 99].

5.4
N-Acyl-*N′*-*β*-D-glucopyranosyl Ureas

Analogs of *N*-acyl-*N′*-β-D-glucopyranosyl urea (compounds A–D in Fig. 14) [7, 53], can be regarded as "open" hydantoins [7]. Whereas the N-acetylated molecule A (K_i = 0.305 mM) [53] proved a weaker inhibitor than urea [7], increasing the hydrophobicity of the acyl side chain makes the binding stronger For example, the experimental K_i values of compounds B–D in Fig. 14were found to be 5.6, 13.0, and 0.4 μM, respectively [53]. The 2-naphthoyl urea (compound D, K_i = 0.4 μM) is to date the best glucose analog inhibitor against rabbit muscle GP*b* [7].

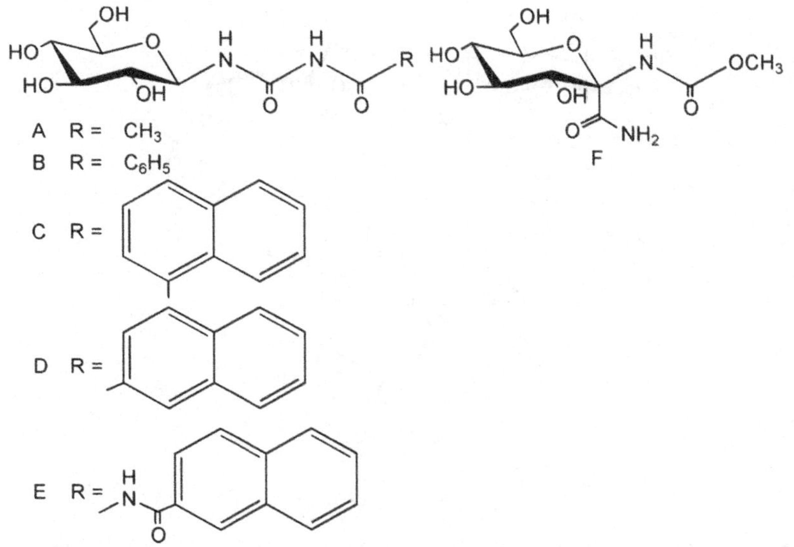

Fig. 14 Molecular structures of *N*-acyl-*N*′-β-D-glucopyranosyl ureas [7]

Placing the hydrophobic moiety farther from the sugar makes the binding stronger by more than one order of magnitude, although this distance must have an optimum, since compound E is a weak inhibitor. At 0.625 mM concentration, the observed inhibition was only ∼ 45% [7].

These discoveries suggest that properly situated large non-polar groups are able to fit into the β-channel next to to the active site of the enzyme GPs [7].

A different means of "opening" the hydantoin ring is presented by the molecule F ($K_i = 0.016$ mM) [105] (for molecular structure see Fig. 14). Though there has not yet been a synthetic counterpart of F in the *N*-acyl-glycosylamie series, a comparison with molecule A (from Fig. 13 (K_i = 0.032 mM) shows that the presence of the carboxamido group (– $CONH_2$) in the α position may be beneficial for binding with GP [7].

5.5 D-Gluco-heptulosonamide Analogs

Hydroxyamide (compound A in Fig. 15, $K_i = 3.1$ mM) can be considered a fusion of α-D-glucose and 2,6-anhydro-heptonamide, though the bifuctional anomeric center brought about a significant decrease in the inhibition in comparison to both these two molecules [99]. The carboxamidoglucoside B (see Fig. 15, ∼ 25% inhibition at 0.625 mM) of opposite anomeric configuration proved to have less inhibitory activity against rabbit muscle GP*b* [7].

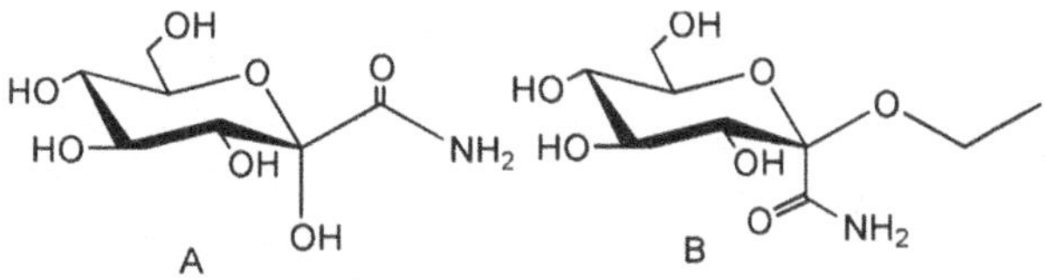

Fig. 15 Molecular structures of D-gluco-heptulosonamide analogs [7, 99]

6
Concluding Remarks, Future Challenges, and Recommendations

The inhibition of GP has established a potential approach in revising the exploitation of this enzyme, which is one of the most important key regulators of blood glucose levels. It suggests a new perception for combating NIDDM as an epidemically intensifying metabolic disorder. A motivating point about GP is that it has multiple binding pockets in its protein structure, which allows it to be targeted through diverse effectors [4, 7].

Theoretical (computational) calculations can also offer quantitative descriptors of physicochemical properties of the molecular structures, molecular interactions, and thermodynamics of interactions. Principally, extensive studies on the catalytic site of GP have been exploited in theoretical QSAR studies [4]. The techniques engaged correlate biochemical behaviors with the known crystallographic structures, and map regions around the inhibitor molecule and added water molecules to improve the in silico prediction [106–110].

Physicochemical explorations with glucose analogs have shown prospective blood glucose lowering in animal experimentations that are devoid of any noticeable signals of hypoglycemia. Since the inhibition of human liver GP also has metabolic significance, it is relatively certain that these compounds will lead to the development of novel inhibitors of GP and thus to a novel treatment for NIDDM or to an add-on to existing treatment [7]. Large numbers of molecules have been reported to possess moderate to potent inhibitory profiles against different types of GPs. Now it is time to prove their clinical efficacy and their specificity. Before going to clinical trials it is necessary to optimize the targeted molecule(s) for their safety profiles. So, there are questions of in vitro, and acute and chronic in vivo toxicity studies. As most of the potent inhibitors are synthetic in origin, the amounts of the particular molecules are not a big issue. There have been some reports published on the in vivo studies of some potential molecules on different mammalian models [22, 29, 111–120], but these data are not sufficient to lead to clinical trials. There is a need for long-term chronic toxicity studies on higher animals and also for genotoxic studies, which are also extremely important for long-term treatments.

References

1. McCormack JG (2006) Biochem Soc Trans 34:238
2. Henke BR, Sparks SM (2006) Mini Rev Med Chem 6:845
3. Zhang BB, Moller DE (2000) Curr Opin Chem Biol 4:461
4. Oikonomakos NG (2002) Curr Protein Pept Sci 3:561
5. Moller DE (2001) Nature 414:821
6. Andersen B, Rassov A, Westergaard N, Lundgren K (1999) Biochem J 342 Pt 3:545
7. Somsak L, Nagya V, Hadady Z, Docsa T, Gergely P (2003) Curr Pharm Des 9:1177
8. Fletterick RJ, Sprang SR (1982) Acc Chem Res 15:361
9. Newgard CB, Hwang PK, Fletterick RJ (1989) Crit Rev Biochem Mol Biol 24:69
10. Monod J, Wyman J, Changeux JP (1965) J Mol Biol 12:88
11. Sprang SR, Acharya KR, Goldsmith EJ, Stuart DI, Varvill K, Fletterick RJ, Madsen NB, Johnson LN (1988) Nature 336:215
12. Barford D, Johnson LN (1989) Nature 340:609
13. Barford D, Hu SH, Johnson LN (1991) J Mol Biol 218:233
14. Barford D (1991) Biochim Biophys Acta 1133:55
15. Sprang SR, Withers SG, Goldsmith EJ, Fletterick RJ, Madsen NB (1991) Science 254:1367
16. McCormack JG, Westergaard N, Kristiansen M, Brand CL, Lau J (2001) Curr Pharm Des 7:1451
17. Johnson LN (1992) FASEB J 6:2274
18. Johnson LN, Hajdu J, Acharya KR, Stuart DI, McLaughlin PJ, Oikonomakos NG, Barford D (1989) In: Herve G (ed) Allosteric enzymes. CRC, Boca Raton, FL, pp 81–127
19. Oikonomakos NG, Acharya KR, Johnson LN (1992) In: Harding JJ, Crabbe MJC (eds) Post-translational modification of proteins. CRC, Boca Raton, FL, pp 81–151
20. Fletterick RJ, Sygusch J, Semple H, Madsen NB (1976) J Biol Chem 251:6142
21. Berman HM, Westbrook J, Feng Z, Gilliland G, Bhat TN, Weissig H, Shindyalov IN, Bourne PE (2000) Nucleic Acids Res 28:235
22. Fosgerau K, Westergaard N, Quistorff B, Grunnet N, Kristiansen M, Lundgren K (2000) Arch Biochem Biophys 380:274
23. Waagepetersen HS, Westergaard N, Schousboe A (2000) Neurochem Int 36:435
24. Rath VL, Ammirati M, Danley DE, Ekstrom JL, Gibbs EM, Hynes TR, Mathiowetz AM, McPherson RK, Olson TV, Treadway JL, Hoover DJ (2000) Chem Biol 7:677
25. Rath VL, Ammirati M, LeMotte PK, Fennell KF, Mansour MN, Danley DE, Hynes TR, Schulte GK, Wasilko DJ, Pandit J (2000) Mol Cell 6:139
26. Coats WS, Browner MF, Fletterick RJ, Newgard CB (1991) J Biol Chem 266:16113
27. Oikonomakos NG, Tsitsanou KE, Zographos SE, Skamnaki VT, Goldmann S, Bischoff H (1999) Protein Sci 8:1930
28. Hoover DJ, Lefkowitz-Snow S, Burgess-Henry JL, Martin WH, Armento SJ, Stock IA, McPherson RK, Genereux PE, Gibbs EM, Treadway JL (1998) J Med Chem 41:2934
29. Kelland LR (2000) Expert Opin Investig Drugs 9:2903
30. Ragolia L, Begum N (1998) Mol Cell Biochem 182:49
31. Morand C, Remesy C, Demigne C (1992) Diabete Metab 18:87
32. Hers HG (1990) J Inherit Metab Dis 13:395
33. Cohen P (1989) Annu Rev Biochem 58:453
34. Exton JH (1987) Diabetes Metab Rev 3:163
35. Jenkins JA, Johnson LN, Stuart DI, Stura EA, Wilson KS, Zanotti G (1981) Philos Trans R Soc Lond B Biol Sci 293:23

36. Oikonomakos NG, Tiraidis C, Leonidas DD, Zographos SE, Kristiansen M, Jessen CU, Norskov-Lauritsen L, Agius L (2006) J Med Chem 49:5687
37. Whittamore PR, Addie MS, Bennett SN, Birch AM, Butters M, Godfrey L, Kenny PW, Morley AD, Murray PM, Oikonomakos NG, Otterbein LR, Pannifer AD, Parker JS, Readman K, Siedlecki PS, Schofield P, Stocker A, Taylor MJ, Townsend LA, Whalley DP, Whitehouse J (2006) Bioorg Med Chem Lett 16:5567
38. Hampson LJ, Arden C, Agius L, Ganotidis M, Kosmopoulou MN, Tiraidis C, Elemes Y, Sakarellos C, Leonidas DD, Oikonomakos NG (2006) Bioorg Med Chem 14:7835
39. Petsalakis EI, Chrysina ED, Tiraidis C, Hadjiloi T, Leonidas DD, Oikonomakos NG, Aich U, Varghese B, Loganathan D (2006) Bioorg Med Chem 14:5316
40. Lukacs CM, Oikonomakos NG, Crowther RL, Hong LN, Kammlott RU, Levin W, Li S, Liu CM, Lucas-McGady D, Pietranico S, Reik L (2006) Proteins 63:1123
41. Hadjiloi T, Tiraidis C, Chrysina ED, Leonidas DD, Oikonomakos NG, Tsipos P, Gimisis T (2006) Bioorg Med Chem 14:3872
42. Archontis G, Watson KA, Xie Q, Andreou G, Chrysina ED, Zographos SE, Oikonomakos NG, Karplus M (2005) Proteins 61:984
43. Watson KA, Chrysina ED, Tsitsanou KE, Zographos SE, Archontis G, Fleet GW, Oikonomakos NG (2005) Proteins 61:966
44. Anagnostou E, Kosmopoulou MN, Chrysina ED, Leonidas DD, Hadjiloi T, Tiraidis C, Zographos SE, Gyorgydeak Z, Somsak L, Docsa T, Gergely P, Kolisis FN, Oikonomakos NG (2006) Bioorg Med Chem 14:181
45. Klabunde T, Wendt KU, Kadereit D, Brachvogel V, Burger HJ, Herling AW, Oikonomakos NG, Kosmopoulou MN, Schmoll D, Sarubbi E, von Roedern E, Schonafinger K, Defossa E (2005) J Med Chem 48:6178
46. Oikonomakos NG, Kosmopoulou MN, Chrysina ED, Leonidas DD, Kostas ID, Wendt KU, Klabunde T, Defossa E (2005) Protein Sci 14:1760
47. Chrysina ED, Kosmopoulou MN, Tiraidis C, Kardakaris R, Bischler N, Leonidas DD, Hadady Z, Somsak L, Docsa T, Gergely P, Oikonomakos NG (2005) Protein Sci 14:873
48. Chrysina ED, Kosmopoulou MN, Kardakaris R, Bischler N, Leonidas DD, Kannan T, Loganathan D, Oikonomakos NG (2005) Bioorg Med Chem 13:765
49. Kosmopoulou MN, Leonidas DD, Chrysina ED, Bischler N, Eisenbrand G, Sakarellos CE, Pauptit R, Oikonomakos NG (2004) Eur J Biochem 271:2280
50. Pinotsis N, Leonidas DD, Chrysina ED, Oikonomakos NG, Mavridis IM (2003) Protein Sci 12:1914
51. Oikonomakos NG, Chrysina ED, Kosmopoulou MN, Leonidas DD (2003) Biochim Biophys Acta 1647:325
52. Oikonomakos NG (2002) Curr Protein Pept Sci 3:561
53. Oikonomakos NG, Kosmopoulou M, Zographos SE, Leonidas DD, Chrysina ED, Somsak L, Nagy V, Praly JP, Docsa T, Toth B, Gergely P (2002) Eur J Biochem 269:1684
54. Oikonomakos NG, Zographos SE, Skamnaki VT, Archontis G (2002) Bioorg Med Chem 10:1313
55. Oikonomakos NG, Skamnaki VT, Osz E, Szilagyi L, Somsak L, Docsa T, Toth B, Gergely P (2002) Bioorg Med Chem 10:261
56. Tsitsanou KE, Skamnaki VT, Oikonomakos NG (2000) Arch Biochem Biophys 384:245
57. Skamnaki VT, Oikonomakos NG (2000) J Protein Chem 19:499
58. Oikonomakos NG, Schnier JB, Zographos SE, Skamnaki VT, Tsitsanou KE, Johnson LN (2000) J Biol Chem 275:34566

59. Oikonomakos NG, Skamnaki VT, Tsitsanou KE, Gavalas NG, Johnson LN (2000) Structure 8:575
60. Skamnaki VT, Owen DJ, Noble ME, Lowe ED, Lowe G, Oikonomakos NG, Johnson LN (1999) Biochemistry 38:14718
61. Tsitsanou KE, Oikonomakos NG, Zographos SE, Skamnaki VT, Gregoriou M, Watson KA, Johnson LN, Fleet GW (1999) Protein Sci 8:741
62. Gregoriou M, Noble ME, Watson KA, Garman EF, Krulle TM, de la Fuente C, Fleet GW, Oikonomakos NG, Johnson LN (1998) Protein Sci 7:915
63. Zographos SE, Oikonomakos NG, Tsitsanou KE, Leonidas DD, Chrysina ED, Skamnaki VT, Bischoff H, Goldmann S, Watson KA, Johnson LN (1997) Structure 5:1413
64. Lowe ED, Noble ME, Skamnaki VT, Oikonomakos NG, Owen DJ, Johnson LN (1997) EMBO J 16:6646
65. Oikonomakos NG, Zographos SE, Tsitsanou KE, Johnson LN, Acharya KR (1996) Protein Sci 5:2416
66. Mitchell EP, Withers SG, Ermert P, Vasella AT, Garman EF, Oikonomakos NG, Johnson LN (1996) Biochemistry 35:7341
67. Oikonomakos NG, Zographos SE, Johnson LN, Papageorgiou AC, Acharya KR (1995) J Mol Biol 254:900
68. Oikonomakos NG, Kontou M, Zographos SE, Watson KA, Johnson LN, Bichard CJ, Fleet GW, Acharya KR (1995) Protein Sci 4:2469
69. Zographos SE, Oikonomakos NG, Dixon HB, Griffin WG, Johnson LN, Leonidas DD (1995) Biochem J 310(Pt 2):565
70. Watson KA, Mitchell EP, Johnson LN, Cruciani G, Son JC, Bichard CJ, Fleet GW, Oikonomakos NG, Kontou M, Zographos SE (1995) Acta Crystallogr D Biol Crystallogr 51:458
71. Oikonomakos NG, Kontou M, Zographos SE, Tsitoura HS, Johnson LN, Watson KA, Mitchell EP, Fleet GW, Son JC, Bichard CJ, et al. (1994) Eur J Drug Metab Pharmacokinet 19:185
72. Watson KA, Mitchell EP, Johnson LN, Son JC, Bichard CJ, Orchard MG, Fleet GW, Oikonomakos NG, Leonidas DD, Kontou M, et al. (1994) Biochemistry 33:5745
73. Johnson LN, Snape P, Martin JL, Acharya KR, Barford D, Oikonomakos NG (1993) J Mol Biol 232:253
74. Leonidas DD, Oikonomakos NG, Papageorgiou AC, Sotiroudis TG (1992) Protein Sci 1:1123
75. Leonidas DD, Oikonomakos NG, Papageorgiou AC, Acharya KR, Barford D, Johnson LN (1992) Protein Sci 1:1112
76. Martin JL, Veluraja K, Ross K, Johnson LN, Fleet GW, Ramsden NG, Bruce I, Orchard MG, Oikonomakos NG, Papageorgiou AC, et al. (1991) Biochemistry 30:10101
77. Papageorgiou AC, Oikonomakos NG, Leonidas DD, Bernet B, Beer D, Vasella A (1991) Biochem J 274(Pt 2):329
78. Papageorgiou AC, Oikonomakos NG, Leonidas DD (1989) Arch Biochem Biophys 272:376
79. Oikonomakos NG, Acharya KR, Melpidou AE, Stuart DI, Johnson LN (1989) Arch Biochem Biophys 270:62
80. Barford D, Schwabe JW, Oikonomakos NG, Acharya KR, Hajdu J, Papageorgiou AC, Martin JL, Knott JC, Vasella A, Johnson LN (1988) Biochemistry 27:6733
81. Oikonomakos NG, Acharya KR, Stuart DI, Melpidou AE, McLaughlin PJ, Johnson LN (1988) Eur J Biochem 173:569
82. Oikonomakos NG, Johnson LN, Acharya KR, Stuart DI, Barford D, Hajdu J, Varvill KM, Melpidou AE, Papageorgiou T, Graves DJ, et al. (1987) Biochemistry 26:8381

83. Johnson LN, Acharya KR, Stuart DI, Barford D, Oikonomakos NG, Hajdu J, Varvill KM (1987) Biochem Soc Trans 15:1001
84. Hajdu J, Acharya KR, Stuart DI, McLaughlin PJ, Barford D, Oikonomakos NG, Klein H, Johnson LN (1987) EMBO J 6:539
85. McLaughlin PJ, Stuart DI, Klein HW, Oikonomakos NG, Johnson LN (1984) Biochemistry 23:5862
86. Sotiroudis TG, Oikonomakos NG, Evangelopoulos AE (1984) Prog Clin Biol Res 144A:181
87. Melpidou AE, Oikonomakos NG (1983) FEBS Lett 154:105
88. Ktenas TB, Oikonomakos NG, Sotiroudis TG, Nikolaropoulos S, Evangelopoulos AE (1982) J Biochem (Tokyo) 92:2029
89. Sotiroudis TG, Oikonomakos NG, Evangelopoulos AE (1981) Biochim Biophys Acta 658:270
90. Ktenas TB, Oikonomakos NG, Sotiroudis TG, Nikolaropoulos SI, Evangelopoulos AE (1980) Biochem Biophys Res Commun 97:415
91. Oikonomakos NG, Sotiroudis TG, Evangelopoulos AE (1979) Biochem J 181:309
92. Sotiroudis TG, Oikonomakos NG, Evangelopoulos AE (1978) Eur J Biochem 88:573
93. Ktenas TB, Sotiroudis TG, Oikonomakos NG, Evangelopoulos AE (1978) FEBS Lett 88:313
94. Board M, Hadwen M, Johnson LN (1995) Eur J Biochem 228:753
95. Sprang SR, Goldsmith EJ, Fletterick RJ, Withers SG, Madsen NB (1982) Biochemistry 21:5364
96. Johnson LN, Acharya KR, Jordan MD, McLaughlin PJ (1990) J Mol Biol 211:645
97. Bichard CJF, Mitchell EP, Wormald MR, Watson KA, Johnson LN, Zographos SE, Koutra DD, Oikonomakos NG, Fleet GWJ (1995) Tetrahedron Lett 36:2145
98. Goodford PJ (1985) J Med Chem 28:849
99. Somsak L, Kovacs L, Toth M, Osz E, Szilagyi L, Gyorgydeak Z, Dinya Z, Docsa T, Toth B, Gergely P (2001) J Med Chem 44:2843
100. Somsak L, Nagya V, Hadady Z, Docsa T, Gergely P (2003) Curr Pharm Des 9:1177
101. de la Fuente C, Krülle TM, Watson KA, Gregoriou M, Johnson LN, Tsitsanou KE, Zographos SE, Oikonomakos NG, Fleet GWJ (1997) Synlett, p 485
102. Agasimundin YS, Mumper MW, Hosmane RS (1998) Bioorg Med Chem 6:911
103. Street IP, Armstrong CR, Withers SG (1986) Biochemistry 25:6021
104. Kovács L, Osz E, Domokos V, Holzer W, Györgydeák Z (2001) Tetrahedron 57:4609
105. Krülle TM, de la Fuente C, Watson KA, Gregoriou M, Johnson LN, Tsitsanou KE, Zographos SE, Oikonomakos NG, Fleet GWJ (1999) Synlett, p 211
106. Pastor M, Cruciani G, Watson KA (1997) J Med Chem 40:4089
107. Pastor M, Cruciani G, Clementi S (1997) J Med Chem 40:1455
108. So SS, Karplus M (1999) J Comput Aided Mol Des 13:243
109. Venkatarangan P, Hopfinger AJ (1999) J Med Chem 42:2169
110. So SS, Karplus M (2001) J Comput Aided Mol Des 15:613
111. Martin WH, Hoover DJ, Armento SJ, Stock IA, McPherson RK, Danley DE, Stevenson RW, Barrett EJ, Treadway JL (1998) Proc Natl Acad Sci USA 95:1776
112. Loxham SJ, Teague J, Poucher SM, De Schoolmeester J, Turnbull AV, Carey F (2007) J Pharmacol Toxicol Methods 55:71
113. Yu LJ, Chen Y, Treadway JL, McPherson RK, McCoid SC, Gibbs EM, Hoover DJ (2006) J Pharmacol Exp Ther 317:1230
114. Auerswald L, Gade G (2002) Insect Biochem Mol Biol 32:1793
115. Edgerton DS, Cardin S, Pan C, Neal D, Farmer B, Converse M, Cherrington AD (2002) Diabetes 51:3151

116. Ercan-Fang N, Taylor MR, Treadway JL, Levy CB, Genereux PE, Gibbs EM, Rath VL, Kwon Y, Gannon MC, Nuttall FQ (2005) Am J Physiol Endocrinol Metab 289:E366
117. Ercan-Fang NG, Nuttall FQ, Gannon MC (2001) Am J Physiol Endocrinol Metab 280:E248
118. Kaiser A, Nishi K, Gorin FA, Walsh DA, Bradbury EM, Schnier JB (2001) Arch Biochem Biophys 386:179
119. Schieferdecker HL, Schlaf G, Koleva M, Gotze O, Jungermann K (2000) J Immunol 164:5453
120. Tracey WR, Treadway JL, Magee WP, Sutt JC, McPherson RK, Levy CB, Wilder DE, Yu LJ, Chen Y, Shanker RM, Mutchler AK, Smith AH, Flynn DM, Knight DR (2004) Am J Physiol Heart Circ Physiol 286:H1177

Top Heterocycl Chem (2007) 9: 53–86
DOI 10.1007/7081_2007_059

Published online: 24 May 2007

Cytotoxicity of Heterocyclic Compounds against Various Cancer Cells: A Quantitative Structure–Activity Relationship Study

Rajeshwar P. Verma

Department of Chemistry, Pomona College, 645 North College Avenue,
Claremont, CA 91711, USA
rverma@pomona.edu

Abstract Cancer is an important area of interest in the life sciences because it has been a major killer disease throughout human history. Heterocyclic molecules are well known to play a critical role in health care and pharmaceutical drug design. A number of heterocyclic compounds are available commercially as anticancer drugs. Interest in the application of structure–activity relationships has steadily increased in recent decades. Thus, development of the quantitative structure–activity relationship (QSAR) model on the cytotoxicity data of heterocyclic compound series against various cancer cell lines should be of great importance. In this chapter, an attempt has been made to collect the cytotoxicity data on different series of heterocyclic compounds against various cancer cell lines. The data is discussed in terms of QSAR to understand the chemical–biological interactions.

Keywords Cancer · Heterocycles · Hydrophobicity · Molar refractivity · QSAR

Abbreviations

Clog *P*	Calculated hydrophobicity of the whole molecule (calculated logarithm of partition coefficient, *P*, in *n*-octanol/water)
Mlog *P*	Experimental hydrophobicity of the whole molecule
π	Calculated hydrophobicity of the substituent
CMR	Calculated molar refractivity of the whole molecule
MR	Calculated molar refractivity of the substituent
MgVol	Calculated molar volume of the whole molecule (McGowan volume)
MW	Molecular weight
NVE	Number of valence electrons
log $1/C$	Inverse logarithms of the biological activity
LOO	Leave-one-out
MRA	Multiple regression analysis
QSAR	Quantitative structure–activity relationship

1 Introduction

Cancer is becoming a very serious public health problem in the USA and other developed countries. The American Cancer Society estimates every year the number of new cancer cases and deaths expected in the USA and compiles the most recent cancer incidence data from the National Cancer Institute and mortality data from National Cancer for Health Statistics. A total number of 1 372 910 new cancer cases and 570 280 deaths were expected in the USA in 2005. Currently, about 25% of deaths in the USA are due only to cancer [1].

Cancer is not one disease, but a large group of diseases characterized by uncontrolled growth and spread of abnormal cells. Cells in different parts of the body may look and work differently but most of them go through a predictable life cycle – old cells die, and new cells arise to take their place. Normally, the division and growth of cells is orderly and controlled but if this process gets out of control for some reason, the cell will continue to divide and develop into a lump (abnormal swelling), which is called a tumor. These tumors are considered either benign or malignant. A benign tumor does not spread, or metastasize to the other parts of the body and is considered noncancerous. A malignant tumor, on the other hand, can spread throughout the body and is considered cancerous. When malignant cells break away from the primary tumor and spread into another part of the body, the resulting new tumor is called a metastasis tumor. Some cancers, like leukemia, do not form tumors. Instead, these cancer cells involve the blood and blood-forming organs, and circulate through other tissues, where they grow.

There are several major types of cancer:

- Carcinoma – cancer that begins in the skin or in tissues that line or cover internal organs, e.g., breast, colon, lung, mouth, prostate, throat etc.
- Sarcoma – cancer that begins in bones, cartilage, fat, blood vessels, muscles or other connective or supportive tissue
- Leukemia – cancer that starts in blood-forming tissues such as the bone marrow, and causes a large number of abnormal blood cells to be produced and enter the bloodstream
- Lymphoma – cancer that is found in the lymphatic system
- Germinoma – cancer derived from germ cells, normally found in the testicle and ovary
- Glioma – cancer related to the brain cells etc.

Most of the cancer forms are designated on the basis of their occurrence in the particular part of the body, e.g., breast cancer, brain cancer, bladder cancer, liver cancer, lung cancer, ovarian cancer, stomach cancer, vaginal cancer and so on. Different types of cancer can behave very differently. For example, lung cancer and breast cancer are very different diseases. They grow at different rates and respond to different treatments.

There are four major possibilities for the treatment of the cancer: (i) surgery, (ii) radiation, (iii) immunotherapy, and (iv) chemotherapy. Surgery cannot be applied when the disease is spread throughout the body, and radiation therapy damages not only the cancerous cells but also the normal cells. Thus, chemotherapy is of major importance, although immunotherapy (manipulation of the immune response) holds encouraging promise, but is still in an infant stage. Heterocyclic molecules are well known to play a critical role in health care and pharmaceutical drug design. A number of heterocyclic compounds are already available as anticancer drugs, e.g., anastrozole, cytarabine, cytoxan, etoposide, femara, fluorouracil, hexalen, ifosfamide, imatinib, methotrexate, paclitaxel, temodar, vincristine, xeloda (Table 1). However, clinical application of these drugs has encountered problems such as multidrug resistance, side effects, and secondary and/or collateral effects. Thus, there has been increasing interest in discovering and developing new heterocyclic molecules that are expected to be used either in place of, or in conjunction with, existing drugs.

QSAR analysis is an essential method for correlating the properties of a series of molecules with their biological activities and to predict the activities of new compounds. Thus, interest in the application of the QSAR paradigm has steadily increased in recent decades and we hope that it may be useful in the design and development of heterocyclic molecules as new anticancer drugs. In the present chapter, an attempt has been made to collect the cy-

Table 1 Heterocyclic compounds as anticancer drugs[a]

Name of drug	Structure of drug	Type of cancer
Anastrozole Arimidex (B)	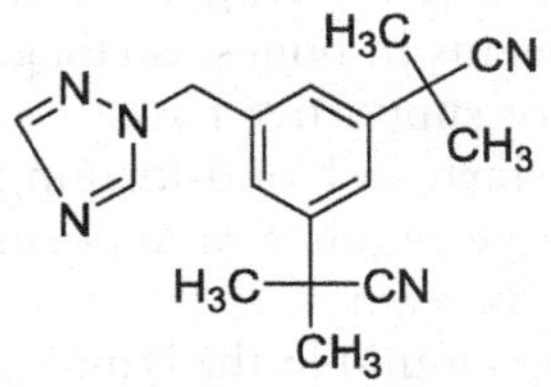	Breast cancers
Cytarabine Cytosar-U (B)	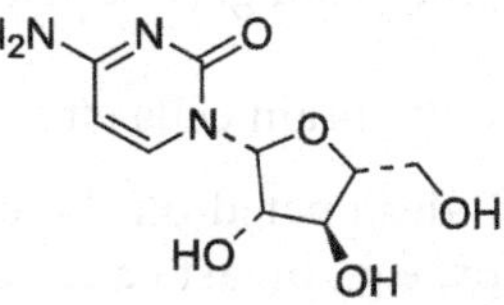	Variety of cancers
Cytoxan Cyclophosphamide (G)	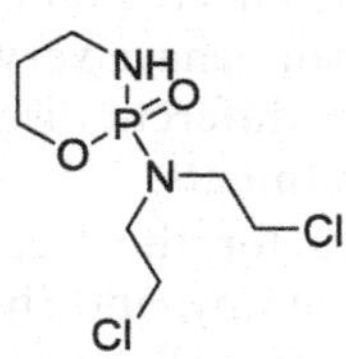	Variety of cancers
Etoposide Toposar VePesid (B)	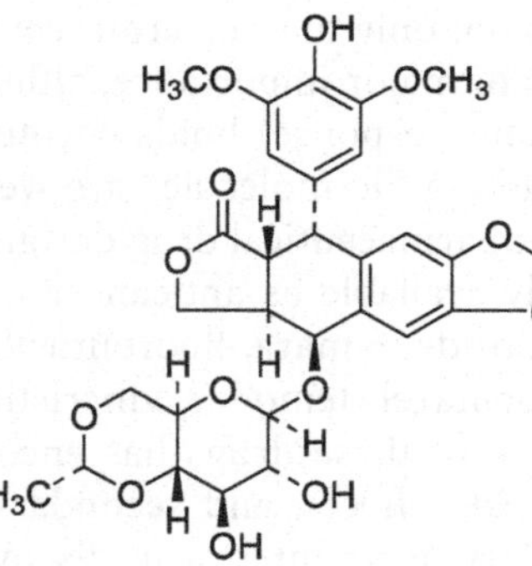	Variety of cancers
Femara Letrozole (G)	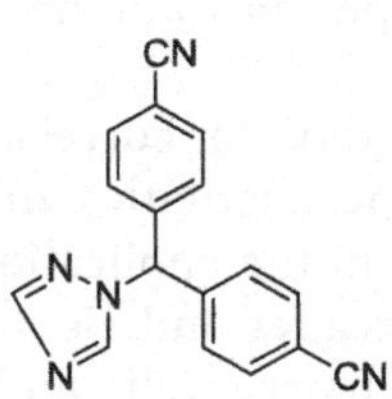	Breast cancers

Table 1 continued

Name of drug	Structure of drug	Type of cancer
Fluorouracil Adrucil Carac, Efudex Fluoroplex (B)	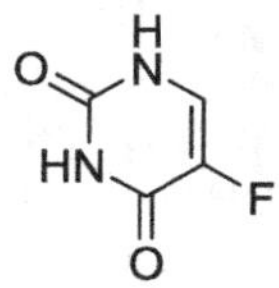	Variety of cancers
Hexalen Altretamine (G)	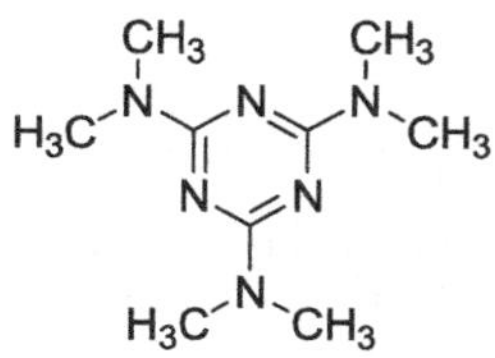	Ovarian cancers
Ifosfamide Ifex (B)	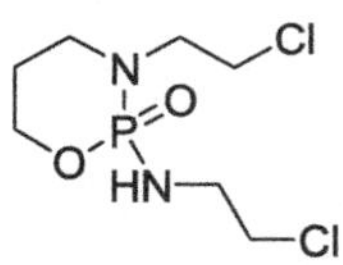	Variety of cancers
Imatinib Gleevec (B)	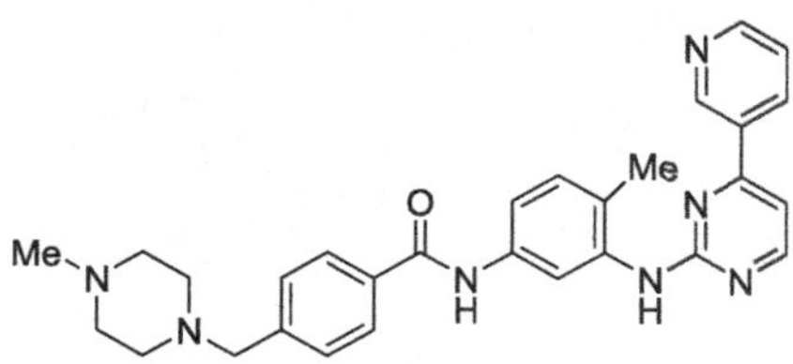	Leukemia and gastrointestinal cancers
Methotrexate Rheumatrex Trexall(B)	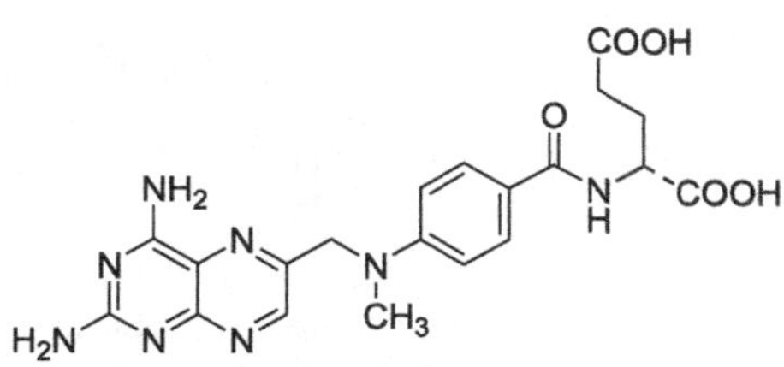	Variety of cancers

Table 1 continued

Name of drug	Structure of drug	Type of cancer
Paclitaxel Onxol Taxol (B)	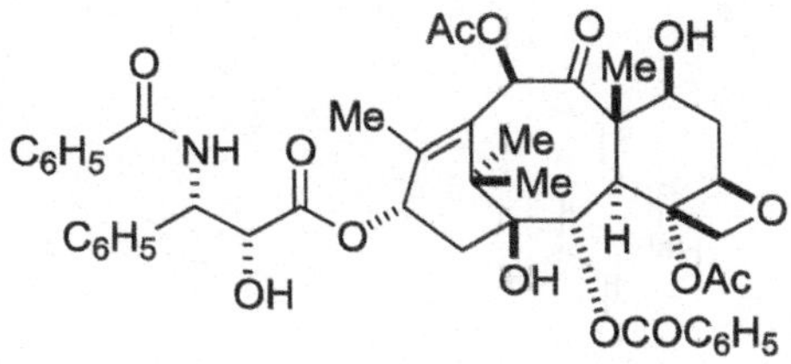	Variety of cancers
Temodar Temozolomide (G)	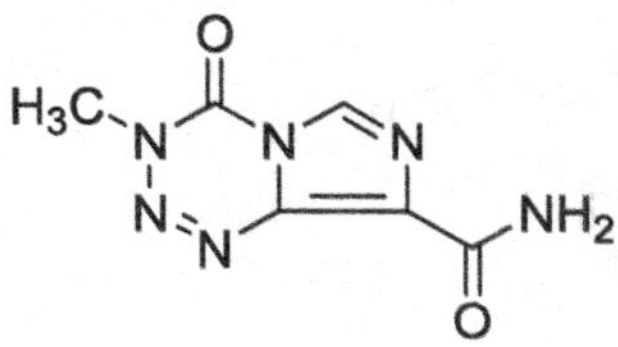	Various brain cancers
Vincristine Oncovin DSC Vincasar PFS (B)	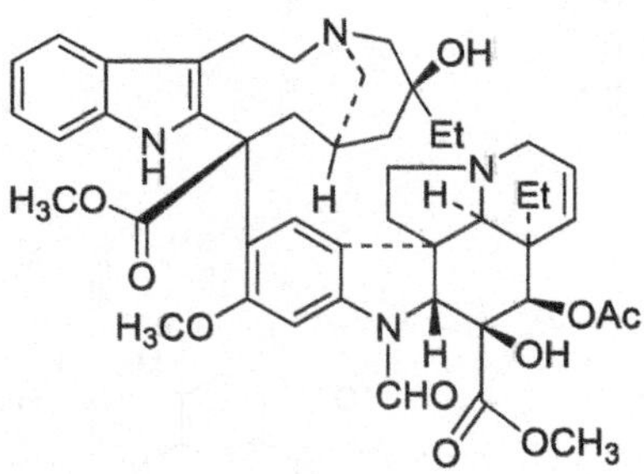	Variety of cancers
Xeloda Capecitabine (G)	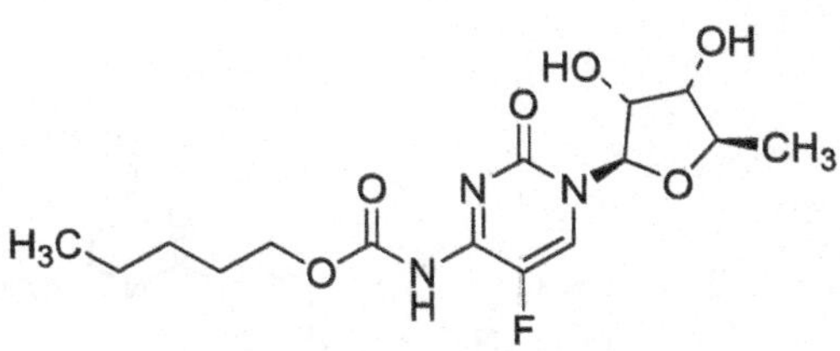	Breast cancers

[a] From BCCA Cancer Drug Manual, which is freely available at http://www.bccancer.bc.ca/HPI/DrugDatabase/DrugIndexPro/default.htm

totoxicity data on 17 different heterocyclic compound series against various cancer cell lines. These are discussed in terms of QSAR to understand the chemical–biological interactions.

In the 44 years since the advent of this methodology [2], the use of QSAR (one of the well-developed areas in computational chemistry) has become increasingly helpful for understanding many aspects of chemical–biological interactions in drug and pesticide research, as well as in toxicology. This method is useful in elucidating the mechanisms of chemical–biological interaction in various biomolecules, particularly enzymes, membranes, organelles, and cells, as well as in humans [3–5]. It has also been utilized for the evaluation of absorption, distribution, metabolism, and excretion (ADME) phenomena in many organisms and whole animal studies [6]. The QSAR approach employs extrathermodynamically derived and computational-based descriptors to correlate biological activity in isolated receptors, cellular systems, and in vivo. Four standard molecular descriptors are routinely used in QSAR analysis: electronic, hydrophobic, steric, and topological indices. These descriptors are invaluable in helping to delineate a large number of receptor–ligand interactions that are critical to biological processes [3, 7]. The quality of a QSAR model depends strictly on the type and quality of the data, and not on the hypotheses, and is valid only for the compound structure analogous to those used to build the model. QSAR models can stand alone to augment other computational approaches or can be examined in tandem with equations of a similar mechanistic genre to establish their authenticity and reliability [7]. Potential use of QSAR models for screening of chemical databases or virtual libraries before their synthesis appears equally attractive to chemical manufacturers, pharmaceutical companies, and government agencies.

2
Materials and Methods

All the data/equations have been collected from the literature (see individual QSAR for respective references). *C* is the molar concentration of a compound and log $1/C$ is the dependent variable that defines the biological parameter for the QSAR equations. Physicochemical descriptors are autoloaded, and multiregression analyses (MRA) used to derive the QSAR are executed with the C-QSAR program [8]. Selection of descriptors is made on the basis of permutation and correlation matrix among the descriptors (to avoid collinearity problems). Details of the C-QSAR program, the search engine, the choice of parameters and their use in the development of QSAR models have already been discussed [9]. The parameters used in this chapter have also been discussed in detail along with their application [3]. Briefly, Clog *P* is the calculated partition coefficient in *n*-octanol/water and is a measure of hydrophobicity, and π is the hydrophobic parameter for substituents. σ, σ^+ and $\sigma^-$are Hammett electronic parameters that apply to substituent effects on aromatic systems.

*B*1, *B*5 and *L* are Verloop's sterimol parameters for substituents [10]. *B*1 is a measure of the minimum width of a substituent, *B*5 is an attempt to define maximum width of the substituent, and *L* is the substituent length. CMR is the calculated molar refractivity for the whole molecule. MR is calculated from the Lorentz–Lorenz equation and is described as follows: $\left[(n^2 - 1)/(n^2 + 2)\right](MW/\delta)$, where n is the refractive index, MW is the molecular weight, and δ is the density of the substance. MR is dependent on volume and polarizability. It can be used for a substituent or for the whole molecule. A new polarizability parameter, NVE, was developed, which is shown to be effective at delineating various chemico-biological interactions [11–14]. NVE represents the total number of valence electrons and is calculated by simply summing up the valence electrons in a molecule, that is, H = 1, C = 4, Si = 4, N = 5, P = 5, O = 6, S = 6, and halogens = 7. It may also be represented as: NVE = $n_\sigma + n_\pi + n_\mathrm{n}$, where n_σ is the number of electrons in the σ orbital, n_π is the number of electrons in π orbitals, and n_n is the number of lone pair electrons. MgVol is the molar volume for the whole molecule [15]. The indicator variable *I* is assigned the value of 1 or 0 for special features with special effects that cannot be parameterized, and has been explained wherever used.

In QSAR equations, n is the number of data points, r is the correlation coefficient between observed values of the dependent and the values predicted from the equation, r^2 is the square of the correlation coefficient and represents the goodness of fit, q^2 is the cross-validated r^2 (a measure of the quality of the QSAR model), and s is the standard deviation. The cross-validated r^2 (q^2) is obtained by using leave-one-out (LOO) procedure [16]. Q is the quality factor (quality ratio), where $Q = r/s$. Chance correlation, due to the excessive number of parameters (which increases the r and s values also), can, thus, be detected by the examination of the Q value. F is the Fischer statistics (Fischer ratio), $F = fr^2/\left[(1 - r^2)m\right]$, where f is the number of degree of freedom $f = n - (m + 1)$, n is the number of data points, and m is the number of variables. The modeling was taken to be optimal when Q reached a maximum together with F, even if slightly non-optimal F values have normally been accepted. A significant decrease in F with the introduction of one additional variable (with increasing Q and decreasing s) could mean that the new descriptor is not as good as expected, that is, its introduction has endangered the statistical quality of the combination. However, the statistical quality could be improved by the introduction of a more useful descriptor [17–19]. Compounds were assigned to be outliers on the basis of their deviation between observed and calculated activities from the equation ($> 2s$) [20–22]. Each regression equation includes 95% confidence limits for each term in parentheses.

3 QSAR Studies

3.1 Acridine

DACA {*N*-[2-(dimethylamino)-ethyl]acridine-4-carboxamide; NSC 601316} (**I**; X = Y = H) is a DNA-intercalating agent and dual topoisomerase (topo) I/II inhibitor currently in clinical trial as an anticancer drug. Substitutions in the acridine ring of DACA have significant effects on biological activity; most 7-substituted DACA analogues had cytotoxicities similar to DACA, whereas most 5-substituted derivatives were more cytotoxic but relatively less effective against JL_A and JL_D cell lines than the wild-type JL_C. Cell line studies showed that the 5,7-disubstituted analogues of DACA retained both the broad-spectrum effectiveness of the 7-substituted derivatives and the higher cytotoxic potency of the 5-substituted derivatives. Spicer et al. [23] published the cytotoxicity data of compound **I** against six cancer cell lines (P388, LLTC, HT-29, JL_A, JL_C and JL_D). From their published cytotoxic data of compound **I** against murine P388 leukaemia cells, we derived Eq. 1 (Table 2):

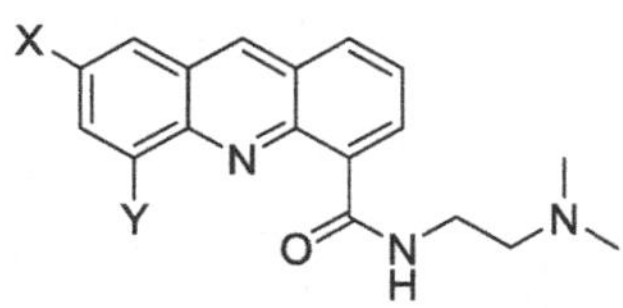

I

$$\log 1/C = 1.43(\pm 0.35)\text{Clog}\,P - 3.30(\pm 0.68)\text{MR}_X + 3.13(\pm 1.19) \quad (1)$$

$$n = 10, \quad r^2 = 0.949, \quad s = 0.146,$$

$$q^2 = 0.876, \quad Q = 6.671, \quad F_{2,7} = 65.127$$

The major conclusion to be drawn from the above QSAR is that Clog *P* (calculated hydrophobicity of the whole molecule) promotes the cytotoxic activity. MR_X refers to the molar refractivity of X-substituents. Since MR_X is primarily a measure of the bulk of X-substituents, a negative coefficient (– 3.30) suggests steric hindrance. We also obtained very similar QSAR models (using the same descriptors Clog *P* and MR_X) from the cytotoxic data of compound **I** against five other cancer cell lines (LLTC, HT-29, JL_A, JL_C, and JL_D), which suggests that compound **I** may target to the involvement of similar kinds of enzymes and/or DNA binding in each of these six cancer cell lines.

Telomerase is an attractive target for the design of new anticancer drugs. Harrison et al. [24] synthesized a series of 3,6-disubstituted acridines (**II**) on the basis that inhibition of telomerase occurs by stabilizing G-quadruplex

Table 2 Biological and physicochemical parameters used to derive QSAR Eq. 1

			log 1/*C* (Eq. 1)				
No.	X	Y	Obsd.	Pred.	Δ	Clog *P*	MR_X
1	H	H	7.01	7.21	- 0.20	3.10	0.10
2	Cl	H	6.60	6.62	- 0.02	3.84	0.60
3	H	Cl	8.19	8.27	- 0.08	3.84	0.10
4	CH_3	H	6.48	6.40	0.08	3.60	0.56
5	H	CH_3	8.19	7.92	0.27	3.60	0.10
6	Cl	Cl	7.57	7.65	- 0.08	4.56	0.60
7	Br	Br	7.14	7.14	0.00	4.86	0.89
8	CH_3	CH_3	7.21	7.11	0.10	4.10	0.56
9	Cl	CH_3	7.25	7.33	- 0.08	4.34	0.60
10	CH_3	Cl	7.47	7.46	0.01	4.34	0.56

structures formed by the folding of telomeric DNA. Equations 2 and 3 were derived from the cytotoxic data of these compounds (**II**) against CH1 (human ovarian carcinoma) and SKOV-3 (human ovarian carcinoma) cells, respectively [21].

Cytotoxic activities of compound **II** against CH1 (human ovarian carcinoma) cells:

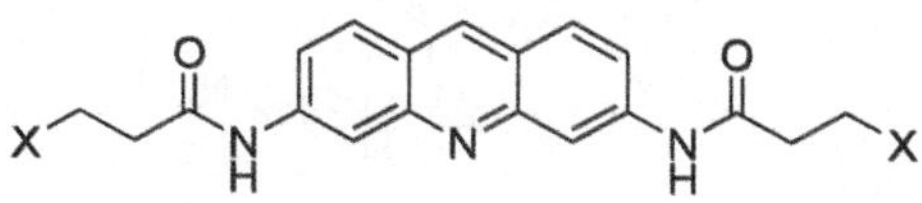

II

$$\log 1/C = 0.19(\pm 0.05)\mathrm{Clog}\,P + 4.33(\pm 0.31) \quad (2)$$
$$n = 15, \quad r^2 = 0.832, \quad s = 0.183,$$
$$q^2 = 0.765, \quad Q = 4.984, \quad F_{1,13} = 64.381$$

Cytotoxic activities of compound **II** against SKOV-3 (human ovarian carcinoma) cells:

$$\log 1/C = 0.16(\pm 0.04)\mathrm{Clog}\,P - 0.57(\pm 0.18)\mathrm{MgVol} + 6.87(\pm 0.63)\,. \quad (3)$$
$$n = 13, \quad r^2 = 0.893, \quad s = 0.095,$$
$$q^2 = 0.775, \quad Q = 9.947, \quad F_{2,10} = 41.729$$

Although the same compounds (**II**) were used in the formulation of QSAR (Eqs. 2 and 3), two different types of equations were obtained. These results suggest that these compounds (**II**) may target an enzyme of one kind in human CH1 and another kind in human SKOV-3 ovarian carcinoma cells, or that a different mechanism is involved.

From the cytotoxic data [25] of another set of acridine derivatives (**III**) against CH1 human ovarian carcinoma cells, Eq. 4 was derived [21]:

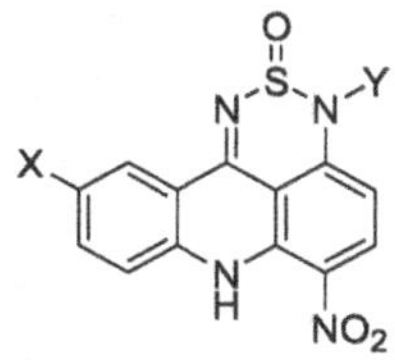

III

$$\log 1/C = 0.69(\pm 0.40)\pi_Y - 1.05(\pm 0.37)\mathrm{CMR} + 16.23(\pm 3.53)\,. \quad (4)$$

$n = 10, \quad r^2 = 0.882, \quad s = 0.140,$

$q^2 = 0.729, \quad Q = 6.708, \quad F_{2,7} = 26.161$

The above QSAR suggests that π_Y (calculated hydrophobicity of Y) promotes the cytotoxic activity. A negative coefficient of CMR (overall molar refractivity) suggests steric hindrance.

3.2 Benzimidazole

Several bibenzimidazoles and tribenzimidazoles have recently been identified as topoisomerase I poisons. Kim et al. [26] synthesized a number of 5-substituted terbenzimidazoles (**IV**) and evaluated as mammalian topoisomerase I poisons and for cytotoxicity against a human lymphoblastoma cell line, RPMI-8402. We used cytotoxic data of these compounds (**IV**) against RPMI-8402 and derived Eq. 5 (Table 3):

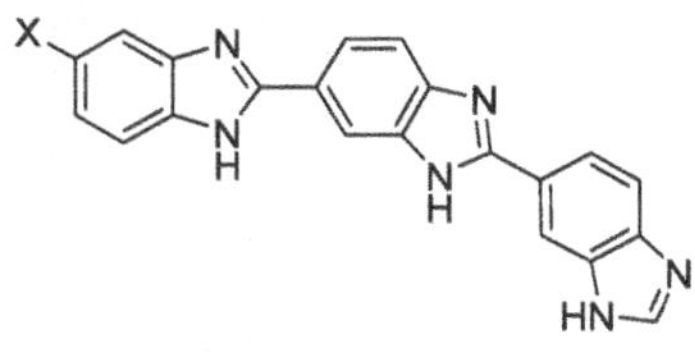

IV

$$\log 1/C = 1.62(\pm 0.71)\pi_X - 0.40(\pm 0.26)\pi_X{}^2 + 5.14(\pm 0.40) \quad (5)$$

$n = 12, \quad r^2 = 0.808, \quad s = 0.515,$

$q^2 = 0.664, \quad Q = 1.746, \quad F_{2,9} = 18.938$

optimum $\pi_X = 2.01(1.61 - 3.42)$

outliers: X = propyl; NH_2

This is a parabolic correlation in terms of π_X, which suggests that the cytotoxic activities of compounds **IV** against RPMI-8402 cells first increase with an increase in hydrophobicity of X-substituents up to an optimum value of

Table 3 Biological and physicochemical parameters used to derive QSAR Eq. 5

No.	X	log 1/*C* (Eq. 5) Obsd.	Pred.	Δ	π_X
1	1-Naphthyl	5.90	6.31	- 0.41	3.07
2	2-Naphthyl	6.80	6.31	0.49	3.07
3	Phenyl	6.72	6.76	- 0.04	1.96
4[a]	Propyl	4.82	6.68	- 1.86	1.55
5	Br	5.79	6.23	- 0.44	0.86
6	Piperidine	6.20	6.22	- 0.02	0.85
7	Cl	5.89	6.08	- 0.19	0.71
8	F	5.77	5.35	0.42	0.14
9	H	5.30	5.14	0.16	0.00
10	OCH_3	6.10	5.10	1.00	- 0.02
11	NO_2	3.94	4.65	- 0.71	- 0.28
12	CN	3.87	4.08	- 0.21	- 0.57
13	OH	3.82	3.87	- 0.05	- 0.67
14[a]	NH_2	4.21	2.53	1.68	- 1.23

[a] Not included in the derivation of QSAR Eq. 5

$\pi_X = 2.01$ and then decrease. The propyl group has a flexible nature in terms of conformation, which may represent its steric flexibility and behavior as an outlier. The other derivative ($X = NH_2$) is much more active than expected by more than three times the standard deviation and is also considered to be an outlier. This anomalous behavior may be attributed to its nature as an aniline. This could result in the hydrogen abstraction or be involved in microsomal N-oxidation [4, 27].

3.3 Benzothiazole

Chung et al. [28] evaluated the cytotoxic activities of 2-[(substituted-1,3-benzothiazole-2-yl)aminomethyl]-5,8-dimethoxy-1,4-naphthoquinones (**V**) against SNU-1 cells. From these data Eq. 6 was derived [29]:

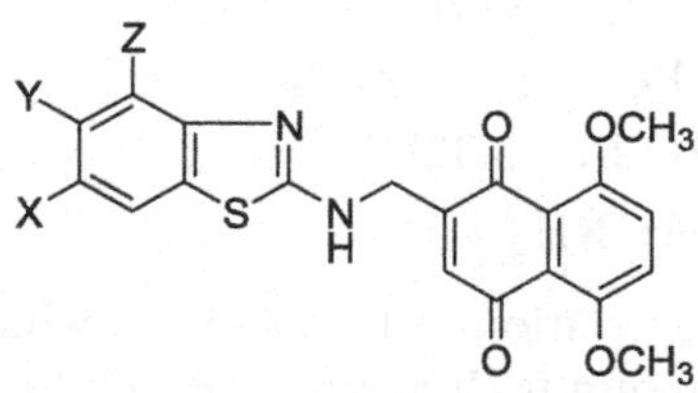

V

$$\log 1/C = 0.32(\pm 0.14)\pi_X + 8.41(\pm 0.05) \quad (6)$$
$$n = 8, \quad r^2 = 0.835, \quad s = 0.047,$$
$$q^2 = 0.727, \quad Q = 19.442, \quad F_{1,6} = 30.364$$

This study shows that only substituents X of the phenyl ring contact the hydrophobic space. The positive coefficient of $\pi_X(0.32)$ suggests that the compounds with highly hydrophobic X-substituents will be more active.

3.4 Camptothecin

In 1985, It was reported by Hsiang et al. [30] that the cytotoxic activity of camptothecin (CPT) was attributed to a novel mechanism of action involving the nuclear enzyme topoisomerase I, and this discovery of unique mechanism of action revived the interest in CPT and its analogues as anticancer agents. From the cytotoxic data [31] of camptothecin derivatives (**VI**) against SKOV-3 human ovarian cancer cells, Eq. 7 was derived [21]:

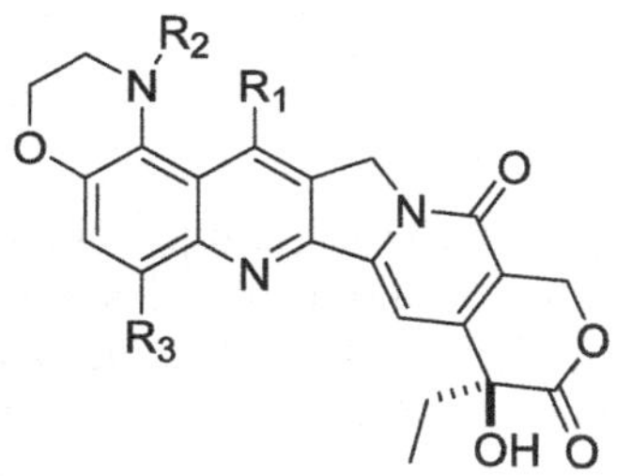

VI

$$\log 1/C = 6.90(\pm 3.25)\text{CMR} - 0.33(\pm 0.14)\text{CMR}^2 - 28.31(\pm 18.62) \quad (7)$$
$$n = 10, \quad r^2 = 0.921, \quad s = 0.278,$$
$$q^2 = 0.723, \quad Q = 3.452, \quad F_{2,7} = 40.804$$
$$\text{optimum} \quad \text{CMR} = 10.50(9.65 - 10.90)$$

This is a parabolic correlation in terms of CMR, which suggests that the cytotoxic activities of camptothecin derivatives (**VI**) against SKOV-3 cells first increase with an increase in molar refractivity up to an optimum CMR value of 10.50 and then decrease. Clog P cannot replace CMR. Substituting log P for CMR in Eq. 7 gives a very poor fit, indicating interaction in non-hydrophobic space (Clog P vs. CMR = 0.036)

3.5 Flavonoid

Flavonoid toxicity to HeLa tumor cells was reported by Mori et al. [32] From these data, Eq. 8 was derived [33]:

$$\log 1/C = 0.34(\pm 0.09)\text{Mlog}P + 3.60(\pm 0.16) \quad (8)$$

$$n = 11, \quad r^2 = 0.899, \quad s = 0.131,$$

$$q^2 = 0.847, \quad Q = 7.237, \quad F_{1,9} = 80.109$$

The above equation predicts the cytotoxic activities of following flavonoids against HeLa cells: galangin, fisetin, quercitin, myricetin, apigenin, eriodictyol, kaempferol, morin, taxifolin, catechin, epicatechin. Mlog *P* is the measured hydrophobicity of the flavonoids. Hydrophobicity is found to be the single important parameter for this data set, which shows the compound's ability to cross the membranes. The linear Mlog *P* model suggests that the highly hydrophobic flavonoids will be more active.

3.6 Indole

Due to interesting biological activities, unique chemical structures, and low availabilities of marine indole alkaloids, indoles have been considered an attractive field in medicinal chemistry for the discovery of new drugs. In an efforts to search for novel antitumor agents, Jiang et al. [34] synthesized a number of indolylpyrimidines and indolylpyrazines and evaluated their cytotoxicities against a panel of 60 human tumor cell lines. QSAR results suggest that the most important determinant for the cytotoxic activities of these compounds (Fig. 1) against different cancer cell lines is the hydrophobic parameters of the whole molecules. We present here two QSAR equations (Eqs. 9 and 10) for the cytotoxic activities of these compounds (**VII–XI**) against SF-539 (CNS) and UO-31 (renal) cancer cell lines, respectively. The biological and physicochemical parameters used to derive these equations are listed in Table 4.

Cytotoxic activities of compounds **VII–XI** against SF-539 (CNS) cancer cells:

$$\log 1/C = 0.44(\pm 0.10)\text{Clog}\,P + 3.26(\pm 0.48) \quad (9)$$

$$n = 8, \quad r^2 = 0.950, \quad s = 0.177,$$

$$q^2 = 0.934, \quad Q = 5.508, \quad F_{1,6} = 114.000$$

Cytotoxic activities of compounds **VII–XI** against UO-31 (renal) cancer cells:

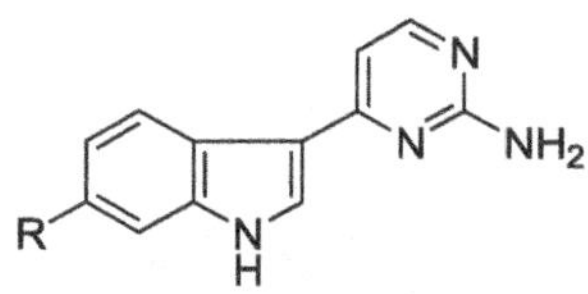

VII

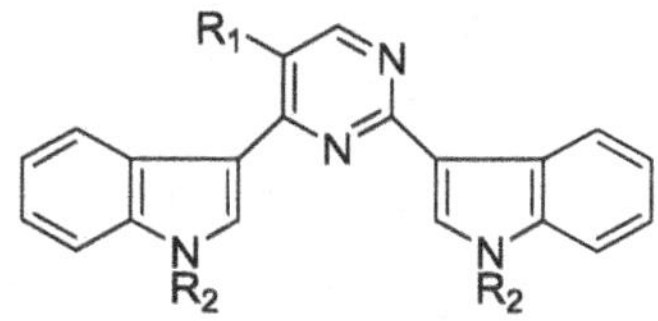

VIIIa-d

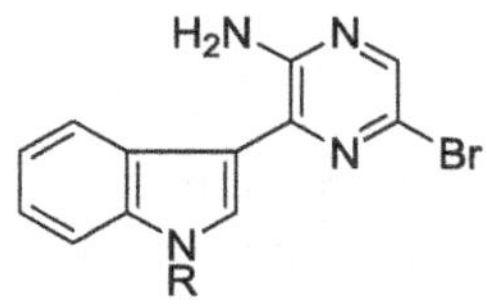

IX

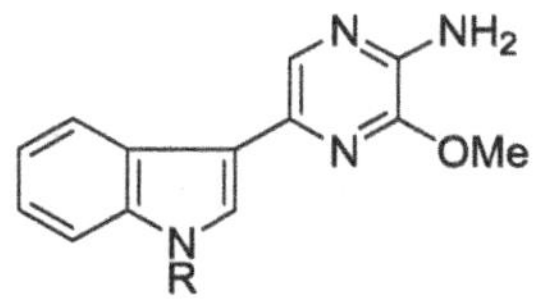

X

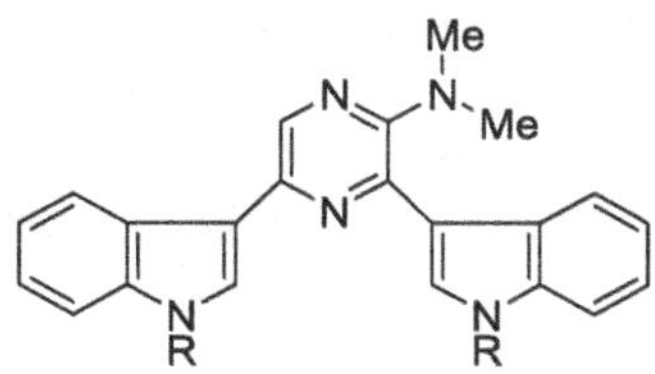

XI

Fig. 1 Structure of compounds used in the development of QSAR Eqs. 9 and 10

Table 4 Biological and physicochemical parameters used to derive QSAR Eqs. 9 and 10

		log 1/*C* (Eq. 9)			log 1/*C* (Eq. 10)			
No.	Substituents	Obsd.	Pred.	Δ	Obsd.	Pred.	Δ	Clog *P*
VII	Meridianin D (R = Br)	4.49	4.51	- 0.02	4.73	4.78	- 0.05	2.86
VIIIa	R_1 = OMe R_2 = Ts	6.80	6.80	0.00	6.24	6.29	- 0.05	8.15
VIIIb	R_1 = R_2 = H	5.36	5.11	0.25	5.40	5.18	0.22	4.25
VIIIc	R_1 = Me R_2 = H	4.93	5.20	- 0.27	5.16	5.24	- 0.08	4.45
VIIId	R_1 = OMe R_2 = H	5.15	5.11	0.04	5.26	5.18	0.08	4.24
IX	R = Ts	5.54	5.36	0.18	5.36	5.34	0.02	4.82
X	R = H	4.47	4.53	- 0.06	4.67	4.80	- 0.13	2.91
XI	R = H	5.38	5.50	- 0.12	5.43	5.44	- 0.01	5.16

Ts $SO_2Ph(4-CH_3)$

$$\log 1/C = 0.29(\pm 0.07)\mathrm{Clog}\,P + 3.96(\pm 0.32) \quad (10)$$
$$n = 8, \quad r^2 = 0.950, \quad s = 0.118,$$
$$q^2 = 0.888, \quad Q = 8.254, \quad F_{1,6} = 114.000$$

The linear Clog *P* models suggest that the highly hydrophobic molecules will be more active.

3.7 Isatin

Isatin (1*H*-indole-2,3-dione) is an endogenous compound identified in humans. This class of compounds possesses a wide range of biological activities [35] that include antiallergic, anticancer, anticonvulsant, antidiuretic, antithrombotic, antitubercular, antiviral, anxiogenic, immunosuppressant, muscle relaxant, and sedative activities. Vine et al. [36] synthesized a variety of isatin derivatives (**XII**) and evaluated their cytotoxic activities against the human monocyte-like histiocytic lymphoma (U937) cell line in vitro. We used these cytotoxic data to develop QSAR Eq. 11 (Table 5):

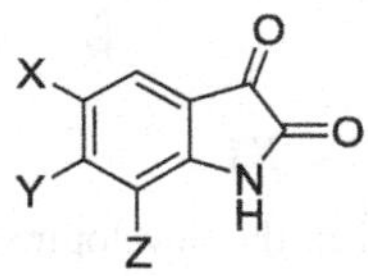

XII

$$\log 1/C = 0.71(\pm 0.20)\mathrm{Clog}\,P + 2.99(\pm 0.39) \quad (11)$$
$$n = 13, \quad r^2 = 0.844, \quad s = 0.267,$$
$$q^2 = 0.774, \quad Q = 3.442, \quad F_{1,11} = 59.513$$
$$\text{outlier}: \quad X = NO_2, \; Y = Br, \; Z = H$$

It is interesting to note here that there is a high mutual correlation between Clog *P* and CMR ($r = 0.899$). Thus, it is very hard to predict for this data set if it is a positive hydrophobic or polarizability effect. We derived Eq. 11a with CMR and finally preferred Eq. 11 on the basis of their statistics, which are better than those of Eq. 11a:

$$\log 1/C = 0.66(\pm 0.32)\mathrm{CMR} + 1.61(\pm 0.94) \quad (11a)$$
$$n = 13, \quad r^2 = 0.654, \quad s = 0.397,$$
$$q^2 = 0.540, \quad Q = 2.038, \quad F_{1,11} = 20.792$$

Table 5 Biological and physicochemical parameters used to derive QSAR Eq. 11

				log 1/C (Eq. 11)			
No.	X	Y	Z	Obsd.	Pred.	Δ	Clog *P*
1	H	H	H	3.25	3.58	- 0.33	0.83
2	Br	H	H	4.19	4.20	- 0.01	1.69
3	H	Br	H	4.13	4.20	- 0.07	1.69
4	H	H	Br	4.08	4.20	- 0.12	1.69
5	F	H	H	4.01	3.68	0.33	0.97
6	I	H	H	4.27	4.38	- 0.11	1.95
7	NO_2	H	H	3.88	3.40	0.48	0.57
8	OCH_3	H	H	3.38	3.52	- 0.14	0.75
9	Br	H	Br	4.98	4.81	0.17	2.55
10	Br	Br	H	4.94	4.67	0.27	2.35
11	I	H	I	5.11	5.18	- 0.07	3.07
12	Br	H	NO_2	3.59	4.01	- 0.42	1.43
13[a]	NO_2	Br	H	4.77	3.87	0.90	1.23
14	Br	Br	Br	5.17	5.14	0.03	3.02

[a] Not included in the derivation of QSAR Eq. 11

3.8 Isoquinoline

Sami et al. [37] synthesized a series of (4, 8, 9, 10, or 11)-X-2-[2′-(dimethylamino)ethyl]-1,2-dihydro-3*H*-dibenz[*de,h*]isoquinoline-1,3-diones (**XIII**) and evaluated their cytotoxic activities against a number of cancer cell lines. From the cytotoxic data of these compounds against OVCAR-3 ovarian and UACC 375 melanoma cancer cells, QSAR Eqs. 12 and 13 were developed respectively [21, 38].

Cytotoxic activities of compound **XIII** against OVCAR-3 ovarian cancer cells:

$$\log 1/C = -0.53(\pm 0.21)\mathrm{Clog}\,P - 1.98(\pm 0.35)I - 0.31(\pm 0.10)B5_{X-4} + 9.84(\pm 0.98) \quad (12)$$

$$n = 18, \quad r^2 = 0.926, \quad s = 0.214,$$
$$q^2 = 0.858, \quad Q = 4.497, \quad F_{3,14} = 58.396$$

Cytotoxic activities of compound **XIII** against UACC 375 melanoma cancer cells:

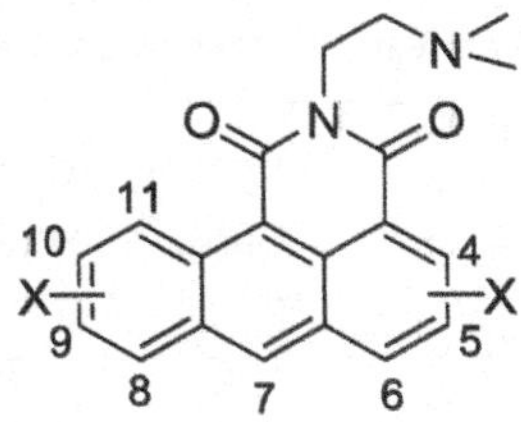

XIII

$$\log 1/C = -0.97(\pm 0.24)\text{Clog}\,P - 1.99(\pm 0.37)I - 0.27(\pm 0.12)B5_{X\text{-}4} + 11.72(\pm 1.12) \quad (13)$$

$n = 20,\quad r^2 = 0.898,\quad s = 0.252,$

$q^2 = 0.803,\quad Q = 3.760,\quad F_{3,16} = 46.954$

I is an indicator variable, which acquired a value of 1 for X = OH and 0 for the others. The negative coefficient of the indicator variable suggests an unfavorable cytotoxic effect for the presence of X = OH against these cancer cell lines. $B5_{X\text{-}4}$ represents the sterimol parameter for the largest width of the groups at position 4, indicating unfavorable steric effect. The negative coefficient with Clog P suggests that highly hydrophobic molecules (**XIII**) will be less active. QSAR Eqs. 12 and 13 are very similar to each other, which suggest that compounds **XIII** may target an enzyme of similar kind in human OVCAR-3 ovarian and UACC 375 melanoma cancer cells, or that a very similar mechanism is involved.

In an another study, the cytotoxicities of a series of 3-arylisoquinolines (**XIV**) against SK-MEL-2 melanoma cells were published by Cho et al. [39]. Using their data Eq. 14 was developed [38]:

$$\log 1/C = 0.43(\pm 0.15)\text{Clog}\,P + 0.40(\pm 0.19)L_{X\text{-}6} + 2.35(\pm 0.74) \quad (14)$$

$n = 12,\quad r^2 = 0.901,\quad s = 0.107,$

$q^2 = 0.805,\quad Q = 8.869,\quad F_{2,9} = 40.955$

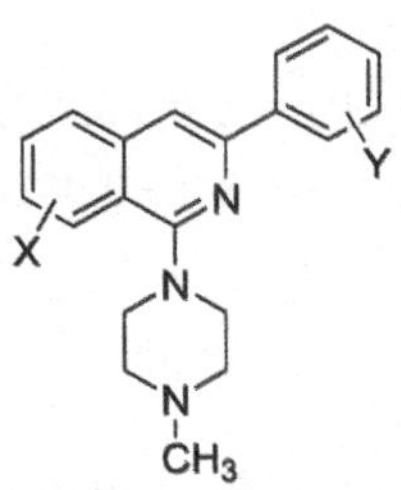

XIV

$L_{X\text{-}6}$ is the sterimol length parameter for X-6 substituents and its positive coefficient suggests that increasing the length of X-6 substituents increases the activity.

3.9 Oncodazole

Kruse et al. [40] synthesized a series of oncodazole analogues (**XV**) and evaluated their cytotoxic activities against monolayer B16 melanoma cells. From these data Eq. 15 was obtained [38]:

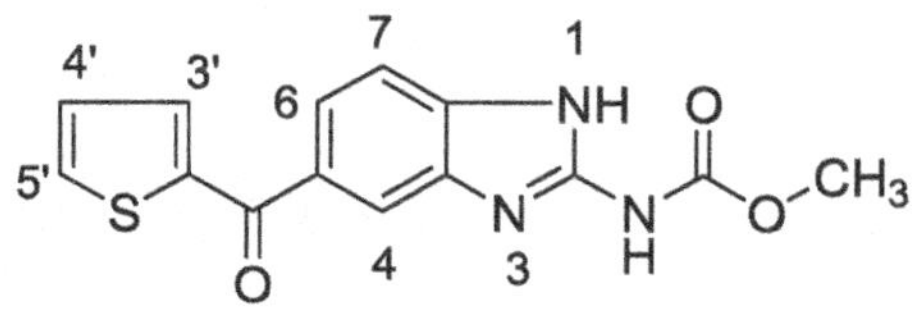

XV

$$\log 1/C = 0.86(\pm 0.38)\mathrm{Clog}\,P + 2.77(\pm 0.65)\mathrm{MR}\text{-}4' - 0.76(\pm 0.23)\mathrm{MR}\text{-}4'^2 - 0.63(\pm 0.42)\pi\text{-}1 + 2.54(\pm 1.09) \quad (15)$$

$n = 18,\quad r^2 = 0.913,\quad s = 0.364,$
$q^2 = 0.767,\quad Q = 2.625,\quad F_{4,13} = 34.106$
optimum $\mathrm{MR}\text{-}4' = 1.82(1.67 - 2.06)$

In this equation, MR-4′ is the calculated molar refractivity of the substituents at position 4′ whereas π-1 is the calculated hydrophobic parameter of the substituents at position 1. This equation contains a positive correlation with Clog P and a negative correlation with π-1, so one should preserve a hydrophilic group at N1 while boosting the molecule's overall hydrophobicity. The parabolic nature of this equation in terms of MR-4′ suggests that the value of MR-4′ should be ≈ 1.82 for the maximum cytotoxicity against B16 melanoma cells.

3.10 Paclitaxel

Paclitaxel (**XVI**; X = C_6H_5, Y = CH_3) is a well-known anticancer drug for the treatment of different kinds of metastatic tumors. Despite its success in the chemotherapy, there are demands to improve its efficacy as well as lower its toxicity. The QSAR of paclitaxel derivatives suggest that the key modification at certain positions may result in the significant improvement of its activity. The cytotoxic activities of a series of paclitaxel derivatives (**XVI**) against prostate cancer (PC3) were evaluated by Baloglu et al. [41]. From their data, we developed Eq. 16 (Table 6):

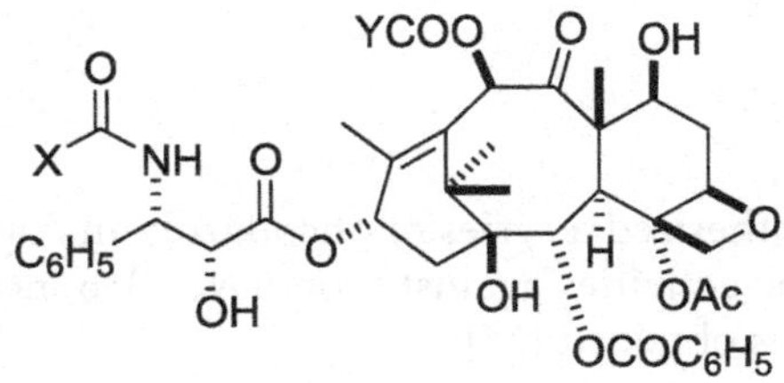

XVI

$$\log 1/C = 10.13(\pm 2.52)\text{Clog}\,P - 0.97(\pm 0.24)\text{Clog}\,P^2 - 19.13(\pm 6.52) \quad (16)$$

$n = 16, \quad r^2 = 0.853, \quad s = 0.207,$

$q^2 = 0.764, \quad Q = 4.459, \quad F_{2,13} = 37.718$

optimum $\text{Clog}\,P = 5.24(5.15 - 5.34)$

outliers : $X = \text{2-Furyl}, Y = CH = CHCH_3; \quad X = \text{2-Furyl}, Y = CH_2CH_3$

Table 6 Biological and physicochemical parameters used to derive QSAR Eq. 16

No.	X	Y	log 1/C (Eq. 16) Obsd.	Pred.	Δ	Clog P
1	C_6H_5	CH_3	7.09	7.16	- 0.07	4.73
2	C_6H_5	$CH = CH(CH_3)_2$	7.25	7.30	- 0.05	5.57
3	C_6H_5	$CH = CHCH_3$	7.25	7.23	0.02	5.67
4	C_6H_5	CH_2CH_3	7.34	7.41	- 0.07	5.26
5	C_6H_5	$CH_2CH_2CH_3$	7.34	7.12	0.22	5.79
6	$(CH_2)_4CH_3$	CH_3	7.23	7.40	- 0.17	5.19
7	2-Furyl	CH_3	5.65	5.69	- 0.04	3.91
8	$(CH_2)_4CH_3$	$CH(CH_3)_2$	6.54	6.82	- 0.28	6.02
9	$(CH_2)_4CH_3$	$CH = CHCH_3$	6.80	6.66	0.14	6.12
10	$(CH_2)_4CH_3$	CH_2CH_3	7.33	7.19	0.14	5.72
11	$(CH_2)_4CH_3$	$CH_2CH_2CH_3$	6.66	6.44	0.22	6.24
12	$O(CH_2)_3CH_3$	$CH = CHCH_3$	6.24	6.73	- 0.49	6.08
13	$O(CH_2)_3CH_3$	CH_2CH_3	7.24	7.23	0.01	5.67
14	$O(CH_2)_3CH_3$	$CH_2CH_2CH_3$	6.64	6.52	0.12	6.20
15	2-Furyl	$CH(CH_3)_2$	7.34	7.17	0.17	4.75
16[a]	2-Furyl	CH = CHCH3	6.40	7.25	- 0.85	4.84
17[a]	2-Furyl	CH_2CH_3	7.33	6.78	0.55	4.44
18	2-Furyl	$CH_2CH_2CH_3$	7.46	7.33	0.13	4.97

[a] Not included in the derivation of QSAR Eq. 16

This is a parabolic correlation in terms of Clog *P*, which suggests that the cytotoxic activities of paclitaxel derivatives (**XVI**) against prostate cancer (PC3) cells first increase with an increase in the hydrophobicity up to an optimum Clog *P* of 5.24 and then decrease.

3.11 Phenanthredin

A series of esters and amides of 2,3-dimethoxy-8,9-methylenedioxy-benzo[*i*] phenanthridine-12-carboxylic acid (**XVII**) was synthesized by Zhu et al. [42] as potent cytotoxic and DNA topoisomerase I-targeting agents. From their cytotoxic data of these compounds against P388/CPT45 (CPT-resistant) cell line, we developed QSAR Eq. 17 (Table 7):

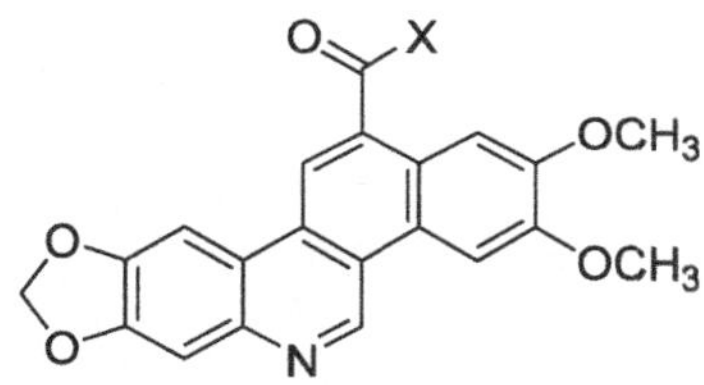

XVII

$$\log 1/C = -0.25(\pm 0.12)\mathrm{MR_X} - 1.08(\pm 0.40)I + 8.07(\pm 0.48) \quad (17)$$

$n = 15, \quad r^2 = 0.861, \quad s = 0.273,$

$q^2 = 0.771, \quad Q = 3.399, \quad F_{2,12} = 37.165$

outliers: $X = OCH_2CH_3$; $OCH_2CH_2CH_2N(CH_3)_2$

$\mathrm{MR_X}$ is the calculated molar refractivity of X-substituents and its negative coefficient suggests steric hindrance. *I* is an indicator variable, which acquired a value of 1 for amides and 0 for the esters. The negative coefficient of the indicator variable suggests an unfavorable cytotoxic effect for the amide derivatives against this cancer cell line. It is interesting to note here that there is a high mutual correlation between π_X and $\mathrm{MR_X}$ ($r = 0.877$). Thus, it is very hard to predict for this data set if it is a positive hydrophobic or polarizability effect of the X-substituents. We derived Eq. 17a with $\mathrm{MR_X}$ and finally preferred Eq. 17 on the basis of their statistics, which are better than those of Eq. 17a:

$$\log 1/C = -0.31(\pm 0.19)\pi_X - 1.35(\pm 0.46)I + 7.57(\pm 0.43) \quad (17a)$$

$n = 15, \quad r^2 = 0.805, \quad s = 0.324,$

$q^2 = 0.715, \quad Q = 2.769, \quad F_{2,12} = 24.769$

Table 7 Biological, physicochemical, and structural parameters used to derive QSAR Eq. 17

No.	X	log 1/*C* (Eq. 17) Obsd.	Pred.	Δ	MR_X	*I*
1[a]	OCH_2CH_3	5.49	7.80	− 2.31	1.08	0
2	$OCH_2CH_2N(CH_3)_2$	7.48	7.47	0.01	2.38	0
3	$OCH(CH_3)CH_2N(CH_3)_2$	7.70	7.36	0.34	2.84	0
4	$OC(CH_3)_2CH_2N(CH_3)_2$	6.89	7.24	− 0.35	3.30	0
5[a]	$OCH_2CH_2CH_2N(CH_3)_2$	6.60	7.36	− 0.76	2.84	0
6	$NHCH_2CH_2N(CH_3)_2$	6.49	6.34	0.15	2.59	1
7	$NHCH(CH_3)CH_2N(CH_3)_2$	6.11	6.22	− 0.11	3.06	1
8	$NHCH_2CH(CH_3)N(CH_3)_2$	6.27	6.22	0.05	3.06	1
9	$N(CH_3)CH_2CH_2N(CH_3)_2$	6.52	6.22	0.30	3.06	1
10	$NHCH_2CH_2N(C_2H_5)_2$	6.43	6.10	0.33	3.52	1
11	$NHCH_2CH_2NCH_3(CH_2Ph)$	5.56	5.71	− 0.15	5.10	1
12	$NHCH_2CH_2N(CH_2Ph)_2$	5.19	5.08	0.11	7.62	1
13	$N(-CH_2CH_2CH_2CH_2-)$	6.48	6.47	0.01	2.05	1
14	$N(-CH_2CH_2CH_2CH_2CH_2-)$	6.00	6.36	− 0.36	2.51	1
15	$N(-CH_2CH_2N(CH_3)CH_2CH_2-)$	5.87	6.26	− 0.39	2.88	1
16	$NHCH_2CH_2CH_2N(CH_3)_2$	6.47	6.22	0.25	3.06	1
17	$N[CH_2CH_2N(CH_3)_2]_2$	5.59	5.78	− 0.19	4.82	1

[a] Not included in the derivation of QSAR Eq. 17

3.12 Phenazine

A series of bis[(phenazine-1-carboxamide)propyl]methylamines (**XVIII**) was synthesized by Spicer et al. [43] and their cytotoxicities evaluated against a panel of tumor cell lines. From the cytotoxic data of these compounds against P388, LL_C, and JL_C cancer cells, QSAR Eqs. 18, 19, and 20 were developed, respectively [44].

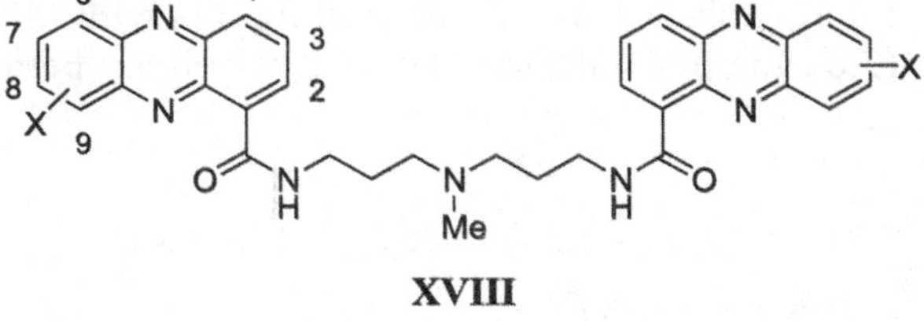

XVIII

Cytotoxic activities of compound **XVIII** against murine P388 leukemia cells:

$$\log 1/C = 6.26(\pm 5.11)\text{Clog}\,P - 0.42(\pm 0.37)\text{Clog}\,P^2 + 0.51(\pm 0.35)L_8 - 1.25(\pm 0.47)\text{Es}_9 - 17.85(\pm 17.73) \quad (18)$$

$n = 17, \quad r^2 = 0.838, \quad s = 0.356,$
$q^2 = 0.762, \quad Q = 2.571, \quad F_{4,12} = 15.519$
optimum Clog $P = 7.37(6.94 - 10.25)$

Cytotoxic activities of compound **XVIII** against murine Lewis lung carcinoma (LL_C) cells:

$$\log 1/C = 11.62(\pm 5.01)\text{Clog}\,P - 0.79(\pm 0.37)\text{Clog}\,P^2 - 1.08(\pm 0.45)\text{Es}_9 - 35.12(\pm 17.71) \quad (19)$$

$n = 17, \quad r^2 = 0.845, \quad s = 0.357,$
$q^2 = 0.617, \quad Q = 2.575, \quad F_{3,13} = 23.624$
optimum Clog $P = 7.32(7.09 - 7.77)$

Cytotoxic activities of compound **XVIII** against human Jurkat leukemia wild-type (JL_C) cells:

$$\log 1/C = 5.86(\pm 3.21)\text{Clog}\,P - 0.40(\pm 0.23)\text{Clog}\,P^2 + 0.43(\pm 0.22)L_8 - 1.08(\pm 0.29)\text{Es}_9 - 15.69(\pm 11.16) \quad (20)$$

$n = 18, \quad r^2 = 0.903, \quad s = 0.226,$
$q^2 = 0.837, \quad Q = 4.205, \quad F_{4,13} = 30.255$
optimum Clog $P = 7.38(7.09 - 8.11)$

The Eqs. 18, 19, and 20 for the different cell lines are very similar to each other, which suggests that similar modes of enzyme and/or DNA binding may be involved. Clog P becomes a descriptor of great importance, with close agreement in these three examples on an optimum value of Clog $P \approx 7.4$. Es_9 is the Taft's steric constant for the 9-substituents while L_8 is the sterimol length parameter for the 8-substituents. The Es_9 confirms a positive steric effect. Bearing in mind that the more sterically hindering the substituent, the more negative its Es-value, the negative coefficient with the 9-substituents shows that substitution at position 9 makes more effective compounds. Positive coefficient of L_8 suggests that increasing the length of 8-substituents increases activity.

3.13 Podophyllotoxin

Duca el al. [45] synthesized a series of 4-β-amino-4′-*O*-demethyl-4-desoxy-podophyllotoxins (**XIX**) as DNA topoisomerase II inhibitors and also evaluated their cytotoxicities against L1210 cell line. From the cytotoxic data of these compounds, we developed Eq. 21 (Table 8):

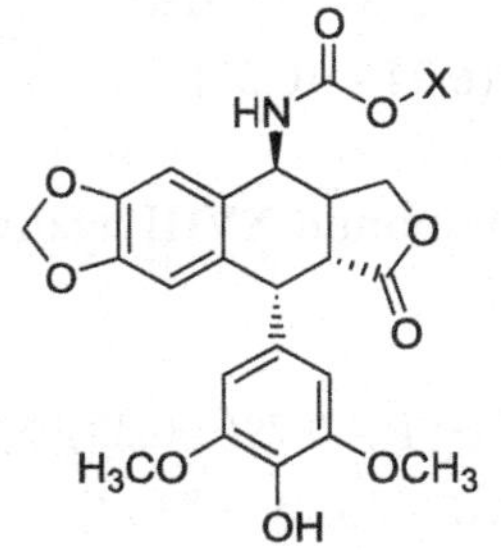

XIX

$$\log 1/C = 1.22(\pm 0.46)\pi_X - 0.32(\pm 0.11)\pi_X^2 + 5.74(\pm 0.45) \quad (21)$$
$$n = 11, \quad r^2 = 0.880, \quad s = 0.101,$$
$$q^2 = 0.676, \quad Q = 9.287, \quad F_{2,8} = 29.333$$
$$\text{optimum} \quad \pi_X = 1.93(1.74 - 2.07)$$
$$\text{outliers}: \quad X = CH_2Ph; \; CH_2Ph(4-F)$$

Table 8 Biological and physicochemical parameters used to derive QSAR Eq. 21

No.	X	log 1/*C* (Eq. 21) Obsd.	Pred.	Δ	π_X
1	CH_3	6.74	6.70	0.04	1.09
2	CH_2CH_3	6.85	6.89	- 0.04	1.62
3	$(CH_2)_3CH_3$	6.64	6.74	- 0.10	2.68
4	$(CH_2)3Cl$	6.82	6.92	- 0.10	2.01
5	$CH_2CH=CH_2$	6.96	6.92	0.04	1.87
6	$CH_2C\equiv -CH$	7.05	6.90	0.15	1.69
7	$(CH_2)_2OCH_3$	6.85	6.77	0.08	1.24
8	$CH_2CH[-O(CH_2)_3CH_2-]$	6.85	6.91	- 0.06	2.11
9	$CH_2CH_2N(-COCH_2CH_2CO-)$	6.36	6.46	- 0.10	0.73
10[a]	CH_2Ph	6.57	6.92	- 0.35	1.90
11[a]	$CH_2Ph(4-F)$	6.44	6.91	- 0.47	2.05
12	$CH_2CH_2Ph(4-F)$	6.41	6.29	0.12	3.33
13	$CH_2Ph(2,4-Cl_2)$	6.24	6.29	- 0.05	3.33

[a] Not included in the derivation of QSAR Eq. 21

This is a parabolic correlation in terms of π_X, which suggests that the cytotoxic activities of compounds **XIX** against L1210 cells first increase with an increase in hydrophobicity of X-substituents up to an optimum value of $\pi_X = 1.93$ and then decrease.

3.14
Pyrrole

Pyrrole-based chemotherapeutic agents have a long drug history, which includes antiinflammatory, antihelminthic, antimycotic, and antibiotic drugs. Furthermore, the pyrrole moiety has been incorporated in several non-nucleoside reverse transcriptase inhibitors and antiproliferative agents, as well as the DNA minor groove binders Distamycin A and Tallimustine. In recent years, the in vitro anticancer activity of diazopyrroles and triazenopyrroles has been reported. A number of agents of this class are under intensive development by research groups throughout the world. Cocco et al. [46] synthesized a series of *N*-phenyl-3-pyrrolecarbothioamides (**XX**) and evaluated their cytotoxic activities against melanoma cell lines. Utilization of the cytotoxic data of these compounds, QSAR 22, 23, and 24 were developed [38].

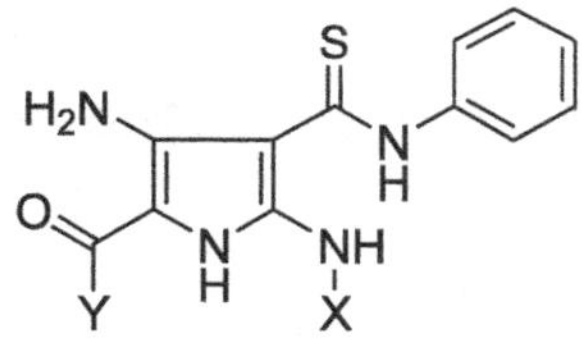

XX

Cytotoxic activities of compound **XX** against MALME-3M melanoma cells:

$$\log 1/C = 0.12(\pm 0.03)\text{Clog}\,P + 4.29(\pm 0.11) \quad (22)$$

$n = 8, \quad r^2 = 0.953, \quad s = 0.039,$

$q^2 = 0.910, \quad Q = 25.031, \quad F_{1,6} = 121.660$

Cytotoxic activities of compound **XX** against UACC-257 melanoma cells:

$$\log 1/C = 0.09(\pm 0.03)\text{Clog}\,P + 4.43(\pm 0.11) \quad (23)$$

$n = 9, \quad r^2 = 0.920, \quad s = 0.042,$

$q^2 = 0.862, \quad Q = 22.837, \quad F_{1,7} = 80.500$

Cytotoxic activities of compound **XX** against SK-MEL-28 melanoma cells:

$$\log 1/C = 0.24(\pm 0.06)\text{Clog}\,P + 3.86(\pm 0.25) \quad (24)$$

$n = 10$, $r^2 = 0.919$, $s = 0.112$,
$q^2 = 0.881$, $Q = 8.559$, $F_{1,8} = 90.765$

Linear Clog *P* correlations are the more significant models.

3.15 Quinocarcin

Cytotoxic activities of quinocarcin derivatives (**XXI**) against HeLa cells were reported by Saito et al. [47]. From their data Eq. 25 was obtained [33]:

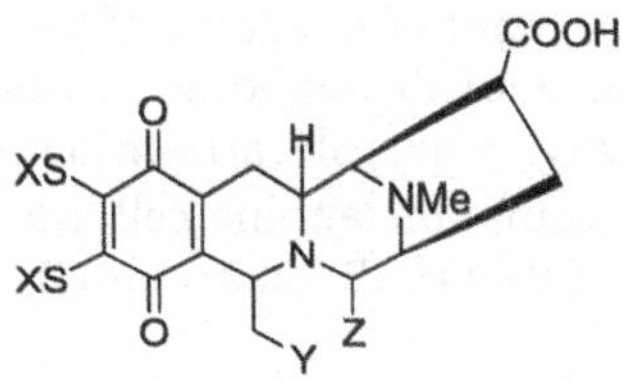

XXI

$$\log 1/C = 0.51(\pm 0.21)\text{Clog}\,P - 0.24(\pm 0.12)B5_X + 8.81(\pm 0.85) \quad (25)$$

$n = 10$, $r^2 = 0.894$, $s = 0.375$,
$q^2 = 0.828$, $Q = 2.521$, $F_{2,7} = 29.519$

$B5_X$ is the sterimol parameter and expresses the largest width of X-substituents. Its negative coefficient (-0.24) suggests a negative steric effect.

3.16 Quinoline

Deady et al. [48] synthesized a series of substituted *N*-[2-(dimethylamino)-ethyl]-11-oxo-11*H*-indeno[1,2-*b*]quinoline-6-carboxamides (**XXII**) and also evaluated their cytotoxic activities against three cancer cell lines. From the

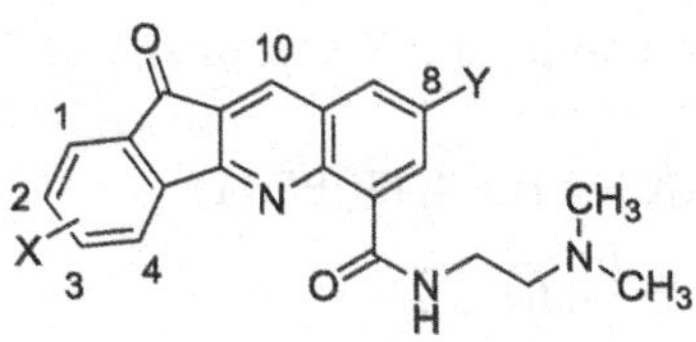

XXII

Table 9 Biological, physicochemical, and structural parameters used to derive QSAR Eqs. 26 and 27

			log 1/*C* (Eq. 26)			log 1/*C* (Eq. 27)					
No.	X	Y	Obsd.	Pred.	Δ	Obsd.	Pred.	Δ	π_X	CMR	*I*
1	H	H	6.74	6.63	0.11	7.04	6.82	0.22	0.00	10.16	0
2	1 – OMe	H	6.97	6.97	0.00	7.47	7.24	0.23	- 0.02	10.78	0
3[a]	2 – OMe	H	7.17	6.97	0.20	7.24	7.24	0.00	- 0.02	10.78	0
4	2 – Cl	H	7.00	7.07	- 0.07	6.94	7.16	- 0.22	0.71	10.66	0
5	3 – OMe	H	6.99	6.97	0.02	7.12	7.24	- 0.12	- 0.02	10.78	0
6	3 – OH	H	6.51	6.56	- 0.05	6.76	6.92	- 0.16	- 0.67	10.32	0
7	3 – Me	H	6.98	7.02	- 0.04	7.23	7.14	0.09	0.56	10.63	0
8[b]	4 – OMe	H	7.15	7.24	- 0.09	7.64	8.05	- 0.41	- 0.02	10.78	1
9	4 – Me	H	7.46	7.29	0.17	7.82	7.95	- 0.13	0.56	10.63	1
10	4 – Cl	H	7.26	7.34	- 0.08	8.09	7.96	0.13	0.71	10.66	1
11	H	OMe	6.92	6.97	- 0.05	6.99	7.24	- 0.25	0.00	10.78	0
12	H	Cl	6.89	6.90	- 0.01	7.28	7.16	0.12	0.00	10.66	0
13	2,3 – Me_2	H	7.38	7.30	0.08	7.77	7.67	0.10	- 0.04	11.40	0

[a] Not included in the derivation of QSAR Eq. 26
[b] Not included in the derivation of QSAR Eq. 27

cytotoxic data of these compound against JL_C and LL cancer cell lines, we developed Eqs. 26 and 27 (Table 9).

Cytotoxic activities of compound **XXII** against Human Jurkat leukemia (JL_C) cells:

$$\log 1/C = 0.23(\pm 0.18)\pi_X + 0.55(\pm 0.23)\mathrm{CMR} + 0.27(\pm 0.16)I + 2.40(\pm 1.02) \quad (26)$$

$n = 12,\quad r^2 = 0.905,\quad s = 0.095,$

$q^2 = 0.700,\quad Q = 10.011,\quad F_{3,8} = 25.404$

outlier : $X = 2 - OCH_3$

Cytotoxic activities of compound **XXII** against murine Lewis lung carcinoma (LL) cells:

$$\log 1/C = 0.69(\pm 0.43)\mathrm{CMR} + 0.81(\pm 0.33)I - 4.63(\pm 0.22) \quad (27)$$

$n = 12,\quad r^2 = 0.822,\quad s = 0.187,$

$q^2 = 0.650,\quad Q = 4.845,\quad F_{2,9} = 20.781$

outlier : $X = 4 - OCH_3$

Clog P vs. CMR : $r = 0.243$

π_X is the sum of the hydrophobicity of X-substituents. A small hydrophobic effect of the X-substituents was observed against human Jurkat leukemia (JL_C) cells but not against murine Lewis lung carcinoma (LL) cells. CMR is the most important determinant for the cytotoxic activities of these compounds (**XXII**) against both cell lines and cannot be replace by Clog P.

3.17 Quinolone

From the cytotoxic data of Soural et al. [49] for 2-oxo-2-phenylethyl-3-hydroxy-4-oxo-2-phenyl-1,4-dihydroquinoline-7-carboxylates (**XXIII**) against A549 and CEM cancer cell lines, we derived QSAR Eq. 28 and Eq. 29 respectively (Table 10).

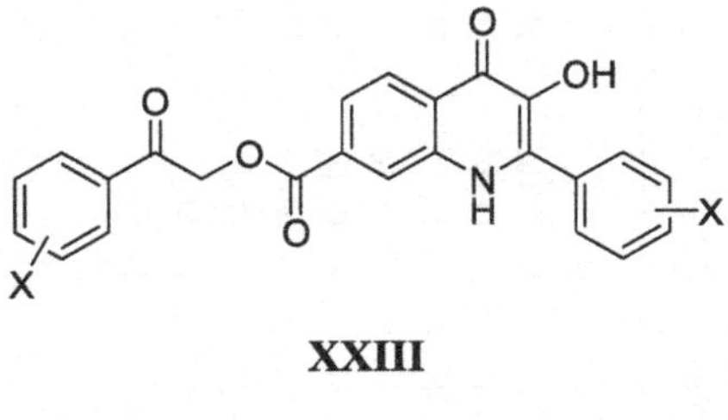

XXIII

Table 10 Biological, physicochemical, and structural parameters used to derive QSAR Eqs. 28 and 29

No.	X	log 1/*C* (Eq. 28) Obsd.	Pred.	Δ	log 1/*C* (Eq. 29) Obsd.	Pred.	Δ	Clog *P*	*I*
1	H	3.92	4.14	-0.22	3.83	4.10	-0.27	2.76	0
2[a]	$2-NO_2$	4.33	2.74	1.59	4.59	3.60	0.99	2.23	1
3	2-I	4.95	5.12	-0.17	5.26	5.56	-0.30	4.33	1
4	2-Cl	4.42	4.47	-0.06	4.76	5.03	-0.27	3.76	1
5	3-Cl	5.35	5.85	-0.51	5.10	5.51	-0.41	4.27	0
6	2-Br	4.98	4.76	0.23	5.31	5.26	0.05	4.01	1
7	$4-CH_3$	5.55	5.27	0.28	5.42	5.03	0.39	3.76	0
8	$3,5-Cl_2$; $4-NH_2$	6.17	5.79	0.38	6.12	5.45	0.67	4.22	0
9	$3-NO_2$	3.79	3.77	0.02	3.99	3.79	0.20	2.43	0
10	$4-NO_2$	3.82	3.77	0.05	3.74	3.79	-0.05	2.43	0

[a] Not included in the derivation of QSAR Eq. 28 and Eq. 29

Cytotoxic activities of compound **XXIII** against A549 cancer cells:

$$\log 1/C = 1.13(\pm 0.39)\mathrm{Clog}\,P - 0.80(\pm 0.62)I + 1.34(\pm 1.02) \tag{28}$$

$n = 9,\quad r^2 = 0.892,\quad s = 0.320,$

$q^2 = 0.738,\quad \mathrm{Q} = 2.950,\quad F_{2,6} = 24.778$

outlier : $\mathrm{X} = 2 - \mathrm{NO_2}$

Cytotoxic activities of compound **XXIII** against CEM cancer cells:

$$\log 1/C = 0.93(\pm 0.41)\mathrm{Clog}\,P + 1.53(\pm 1.48) \tag{29}$$

$n = 9,\quad r^2 = 0.806,\quad s = 0.387,$

$q^2 = 0.695,\quad \mathrm{Q} = 2.320,\quad F_{1,7} = 29.082$

outlier : $\mathrm{X} = 2 - \mathrm{NO_2}$

The major conclusions to be drawn from the above QSARs are that the hydrophobicity of the whole molecules promotes the cytotoxic activities. I is an indicator variable, which acquired a value of 1 for X-substitution at the 2 position and 0 for X-substitution at other positions. The negative coefficient of this indicator variable suggests an unfavorable cytotoxic effect for the presence of substituents at the 2 position against A549 cells, as represented by QSAR Eq. 28. Both equations have one outlier ($\mathrm{X} = 2 - \mathrm{NO_2}$) that is more cytotoxic than expected; it may be possibly due to the hydrogen bonding.

In an another study, the cytotoxic activities of a series of quinolone derivatives (**XXIV**) against OVCAR-3 human ovarian cancer cells were published by Li et al. [50]. Using their data Eq. 30 was obtained [21]:

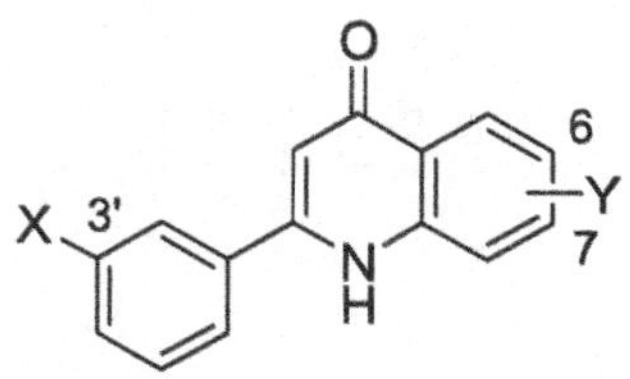

XXIV

$$\log 1/C = 1.52(\pm 0.55)\mathrm{Clog}\,P - 3.06(\pm 0.86)B1_{\mathrm{X}} + 8.55(\pm 1.25) \tag{30}$$

$n = 12,\quad r^2 = 0.891,\quad s = 0.328,$

$q^2 = 0.780,\quad \mathrm{Q} = 2.878,\quad F_{2,9} = 36.784$

$B1_{\mathrm{X}}$ is the sterimol parameter for the smallest width of X-substituents; it has a negative sign indicating that steric interaction at the X-substituents of the phenyl ring is unfavorable.

4
Validation of the QSAR Models

QSAR model validation is an essential task for developing a statistically valid and predictive model, because the real utility of a QSAR model is in its ability to predict accurately the modeled property for new compounds. The following approaches have been used for the validation of QSARs Eqs. 1–30:

- **Fraction of the variance:** The fraction of the variance of an MRA model is expressed by r^2. It is believed that the closer the value of r^2 to unity, the better the QSAR model. The values of r^2 for these QSAR models are from 0.806 to 0.953, which suggests that these QSAR models explain 80.6–95.3% of the variance of the data. According to the literature, the predictive QSAR model must have $r^2 > 0.6$ [51, 52].
- **Cross-validation test:** The values of q^2 for these QSAR models are from 0.617 to 0.934. The high values of q^2 validate these QSAR models. From the literature, it must be greater than 0.50 [51, 52].
- **Standard deviation** (s)**:** s is the standard deviation about the regression line. The smaller the value of s the better the QSAR model. The values of s for these QSAR models are from 0.039 to 0.515.
- **Quality factor or quality ratio(**Q**)**: The high values of Q (1.746–25.031) for QSAR models suggest the high predictive power for these models as well as no over-fitting.
- **Fischer statistics** (F)**:** Fischer statistics (F) is the ratio between explained and unexplained variance for a given number of degree of freedom. The larger the F value the greater the probability that the QSAR equation is significant. The F values obtained for these QSAR models are from 15.519 to 121.660, which are statistically significant at the 95% level.
- All the QSAR models also fulfill the thumb rule condition that (number of data points)/(number of descriptors) ≥ 4.

5
Conclusion

In this chapter, an attempt has been made to present a total of 30 QSAR models on 17 different heterocyclic compound series (acridine, benzimidazole, benzothiazole, camptothecin, flavonoid, indole, isatin, isoquinoline, oncodazole, paclitaxel, phenanthridine, phenazine, podophyllotoxin, pyrrole, quinocarcin, quinoline, and quinolone) for their cytotoxic activities against various cancer cell lines. The QSARs have been found to be well correlated with a number of physicochemical and structural parameters. The most important parameter for these correlations is hydrophobicity, which is one of the most important determinants for the activity.

Out of 30 QSAR, 27 contain a correlation between cytotoxic activity and hydrophobicity. A positive linear correlation is found in 19 equations (Eqs. 1–4, Eq. 6, Eqs. 8–11, Eq. 14, Eq. 15, Eqs. 22–26, and Eqs. 28–30). The coefficient with the hydrophobic parameter varies considerably, from a low value of 0.09 (Eq. 23) to a high value of 1.52 (Eq. 30). These data suggest that the cytotoxic activity might be improved by increasing compound/substituent hydrophobicity. A negative linear correlation is found in three equations (Eq. 12, Eq. 13, and Eq. 15), and the coefficients range from – 0.97 (Eq. 13) to – 0.53 (Eq. 12). Less hydrophobic congeners in these compound families might display enhanced activity (note: Eq. 15 contains a positive correlation with log P and a negative correlation with $\pi - 1$, so one should preserve a hydrophilic group at N1 while boosting the molecule's overall hydrophobicity). Parabolic correlations with the hydrophobic parameter of the substituents are found in two equations (Eq. 5 and Eq. 21), which reflect situations where activity increases with increasing hydrophobicity of the substituents up to an optimal value and then decreases. These are the encouraging examples, where the optimal values of π (hydrophobic parameter of the substituents) are well defined at 2.01 and 1.93, respectively. Parabolic correlations with the molecule's overall hydrophobicity are found in four equations (Eqs. 15–18). The optimal log P for these equations are 5.24, 7.37, 7.32, and 7.38, respectively.

Other parameters (molar refractivity of the molecules/substituents, molar volume, Taft's steric constant, and Verloop's sterimol parameters) also appear in several QSAR. In some cases, these parameters correlate all of the observed variations in activity, but they do not seem to play as important a role as hydrophobicity for the data sets that we have examined.

References

1. Jemal A, Murray T, Ward E, Samuels A, Tiwari RC, Ghafoor A, Feuer EJ, Thun MJ (2005) CA Cancer J Clin 55:10
2. Hansch C, Maloney PP, Fujita T, Muir RM (1962) Nature 194:178
3. Hansch C, Leo A (1995) Exploring QSAR: fundamentals and applications in chemistry and biology. Am Chem Soc, Washington, DC
4. Selassie CD, Garg R, Kapur S, Kurup A, Verma RP, Mekapati SB, Hansch C (2002) Chem Rev 102:2585
5. Verma RP, Kurup A, Mekapati SB, Hansch C (2005) Bioorg Med Chem 13:933
6. Hansch C, Leo A, Mekapati SB, Kurup A (2004) Bioorg Med Chem 12:3391
7. Selassie CD, Mekapati SB, Verma RP (2002) Curr Top Med Chem 2:1357
8. BioByte Corp (2006) C-QSAR program. Claremont, CA, USA, http://www.biobyte.com, last visited: 13 April 2007
9. Hansch C, Hoekman D, Leo A, Weininger D, Selassie CD (2002) Chem Rev 102:783
10. Verloop A (1987) The sterimol approach to drug design. Marcel Dekker, New York
11. Hansch C, Steinmetz WE, Leo AJ, Mekapati SB, Kurup A, Hoekman D (2003) J Chem Inf Comput Sci 43:120

12. Hansch C, Kurup A (2003) J Chem Inf Comput Sci 43:1647
13. Verma RP, Kurup A, Hansch C (2005) Bioorg Med Chem 13:237
14. Verma RP, Hansch C (2005) Bioorg Med Chem 13:2355
15. Abraham MH, McGowan JC (1987) Chromatographia 23:243
16. Cramer RD III, Bunce JD, Patterson DE, Frank IE (1988) Quant Struct-Act Relat 7:18
17. Pogliani L (1996) J Phys Chem 100:18065
18. Pogliani L (2000) Chem Rev 100:3827
19. Agrawal V, Singh J, Khadikar PV, Supuran CT (2006) Bioorg Med Chem Lett 16:2044
20. Selassie CD, Kapur S, Verma RP, Rosario M (2005) J Med Chem 48:7234
21. Verma RP, Hansch C (2006) Mol Pharmaceutics 3:441
22. Verma RP, Hansch C (2006) Virology 359:152
23. Spicer JA, Finlay GJ, Baguley BC, Velea L, Graves DE, Denny WA (1999) Anti-Cancer Drug Design 14:37
24. Harrison RJ, Gowan SM, Kelland LR, Neidle S (1999) Bioorg Med Chem Lett 9:2463
25. Antonini I, Polucci P, Magnano A, Cacciamani D, Konieczny MT, Paradziej-Lukowicz J, Martelli S (2003) Bioorg Med Chem 11:399
26. Kim JS, Sun Q, Yu C. Liu A, Liu LF, LaVoie EJ (1998) Bioorg Med Chem 6:163
27. Kapur S, Shusterman A, Verma RP, Hansch C, Selassie CD (2000) Chemosphere 41:1643
28. Chung Y, Shin YK, Zhan CG, Lee S, Cho H (2004) Arch Pharm Res 27:893
29. Verma RP (2006) Anti-Cancer Agents Med Chem 6:489
30. Hsiang YH, Hertzberg R, Hecht S, Liu LF (1985) J Biol Chem 260:14873
31. Kim DK, Ryu DH, Lee JY, Lee N, Kim YW, Kim JS, Chang K, Im GJ, Kim TK, Choi WS (2001) J Med Chem 44:1594
32. Mori A, Nishino C, Enoki N, Tawata S (1988) Phytochemistry 27:1017
33. Verma RP, Hansch C (2006) Curr Med Chem 13:423
34. Jiang B, Yang CG, Xiong WN, Wang J (2001) Bioorg Med Chem 9:1149
35. Pandeya SN, Smitha S, Jyoti M, Sridhar SK (2005) Acta Pharm 55:27
36. Vine KL, Locke JM, Ranson M, Pyne SG, Bremner JB (2007) Bioorg Med Chem 15:931
37. Sami SM, Dorr RT, Alberts DS, Solyom AM, Remers WA (1996) J Med Chem 39:4978
38. Verma RP, Mekapati SB, Kurup A, Hansch C (2005) Bioorg Med Chem 13:5508
39. Cho W-J, Kim E-K, Park M-J, Choi S-U, Lee C-O, Cheon SH, Choi B-G, Chung B-H (1998) Bioorg Med Chem 6:2449
40. Kruse LI, Ladd DL, Harrsch PB, McCabe FL, Mong SM, Faucette L, Johnson R (1989) J Med Chem 32:409
41. Baloglu E, Hoch JM, Chatterjee SK, Ravindra R, Bane S, Kingston DGI (2003) Bioorg Med Chem 11:1557
42. Zhu S, Ruchelman AL, Zhou N, Liu AA, Liu LF, LaVoie EJ (2005) Bioorg Med Chem 13:6782
43. Spicer JA, Gamage SA, Rewcastle GW, Finlay GJ, Bridewell DJA, Baguley BC, Denny WA (2000) J Med Chem 43:1350
44. Garg R, Denny WA, Hansch C (2000) Bioorg Med Chem 8:1835
45. Duca M, Arimondo PB, Léonce S, Pierré A, Pfeiffer B, Monneret C, Dauzonne D (2005) Org Biomol Chem 3:1074
46. Cocco MT, Congiu C, Onnis V (2003) Bioorg Med Chem 11:495
47. Saito H, Hirata T, Kasai M, Fujimoto K, Ashizawa T, Morimoto M, Sato A (1991) J Med Chem 34:1959
48. Deady LW, Desneves J, Kaye AJ, Thompson M, Finlay GJ, Baguley BC, Denny WA (1999) Bioorg Med Chem 7:2801

49. Soural M, Hlaváč J, Hradil P, Fryšová I, Hajdúch M, Bertolasi V, Maloň M (2006) Eur J Med Chem 41:467
50. Li L, Wang HK, Kuo SC, Wu TS, Mauger A, Lin CM, Hamel E, Lee KH (1994) J Med Chem 37:3400
51. Golbraikh A, Tropsha A (2002) J Mol Graph Modl 20:269
52. Tropsha A, Gramatica P, Gombar VK (2003) QSAR Comb Sci 22:69

Top Heterocycl Chem (2007) 9: 87–118
DOI 10.1007/7081_2007_088

Published online: 21 July 2007

Synthesis, Reactivity and Biological Activity of Benzimidazoles

Mahiuddin Alamgir (✉) · David St. C. Black · Naresh Kumar

School of Chemistry, The University of New South Wales, Sydney, NSW-2052, Australia
m19alamgir@yahoo.com

Abstract Benzimidazole is a biologically important scaffold which displays important biological activities. Recent progress in the synthesis and bioactivity of benzimidazoles is reviewed. New synthetic procedures, including microwave-assisted synthesis, solid phase synthesis, natural product synthesis, and synthesis of bisbenzimidazoles are briefly described. Functionalization and cyclization reactions of benzimidazoles lead to a wide variety of novel benzimidazole structures. Selected bioactivity, such as anti-infective, anti-inflammatory, antitumor and receptor agonist/antagonist activities are presented.

Keywords Benzimidazole · Bioactivity · Bisbenzimidazole · Chemical reactivity · Microwave synthesis

Abbreviations

AIDS	Acquired Immunodeficiency Syndrome
CCl_4	carbon tetrachloride
CNS	central nervous system
DEAD	diethyl azodicarboxylate
DMAP	4-dimethylaminopyridine
DME	dimethyl ether
DMF	*N*,*N*-dimethylformamide
DNA	deoxyribonucleic acid
EtOAc	ethyl acetate
HIV	Human Immunodeficiency Virus
IC_{50}	half-maximal inhibitory concentration
MeOH	methanol
MIC	minimum inhibitory concentration
MW	microwave
NBS	*N*-bromosuccinimide
NMDA	*N*-methyl-D-aspartic acid
PEG	polyethylene glycol
PPA	polyphosphoric acid
PPh_3	triphenylphosphine
TBDMS	*tert*-butyldimethylsilyl

1 Introduction

Benzimidazole is a fused aromatic imidazole ring system where a benzene ring is fused to the 4 and 5 positions of an imidazole ring. Benzimidazoles are also known as benziminazoles and 1,3-benzodiazoles [1, 2]. They possess both acidic and basic characteristics. The NH group present in benzimidazoles is relatively strongly acidic and also weakly basic. Another characteristic of benzimidazoles is that they have the capacity to form salts. Benzimidazoles with unsubstituted NH groups exhibit fast prototropic tautomerism, which leads to equilibrium mixtures of asymmetrically substituted compounds [1].

The benzimidazole scaffold is a useful structural motif for the development of molecules of pharmaceutical or biological interest. Appropriately substituted benzimidazole derivatives have found diverse therapeutic applications such as in antiulcers, antihypertensives, antivirals, antifungals, anticancers, and antihistaminics [3]. The optimization of benzimidazole-based structures has resulted in various drugs that are currently on the market, such as omeprazole **1** (proton pump inhibitor), pimobendan **2** (ionodilator), and mebendazole **3** (anthelmintic) (Fig. 1). The spectrum of pharmacological activity exhibited by benzimidazoles has been reviewed by several authors [3–6].

Since the publications of these reviews, a number of new methods for the synthesis of benzimidazoles have been discovered and reported; such work

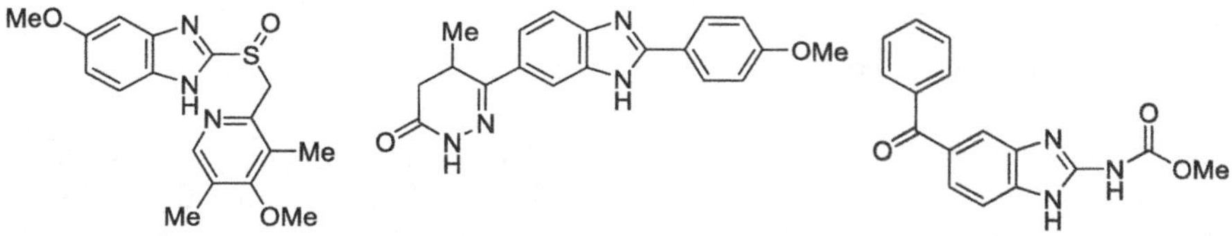

Omeprazole **1** Pimobendan **2** Mebendazole **3**

Fig. 1 Pharmacologically active benzimidazole drugs

continues due to their wide range of pharmacological activities and their industrial and synthetic applications. The present review focuses on the synthetic methodologies and biological activities of the benzimidazoles reported from 2000 to early 2007.

2
General Synthetic Methodologies for Benzimidazoles

Traditionally, benzimidazoles have most commonly been prepared from the reaction of 1,2-diaminobenzenes with carboxylic acids under harsh dehydrating reaction conditions, utilizing strong acids such as polyphosphoric acid, hydrochloric acid, boric acid, or *p*-toluenesulfonic acid [7]. However, the use of milder reagents, particularly Lewis acids [8], inorganic clays [9], or mineral acids [6], has improved both the yield and purity of this reaction [10]. On the other hand, the synthesis of benzimidazoles via the condensation of 1,2-diaminobenzenes with aldehydes requires an oxidative reagent to generate the benzimidazole nucleus. Various oxidative reagents, such as nitrobenzene, benzoquinone, sodium metabisulfite, mercuric oxide, lead tetraacetate, iodine, copper(II) acetate, indium perfluorooctane sulfonates, ytterbium perfluorooctane sulfonates, and even air, have been employed for this purpose [11]. Moreover, a variety of benzimidazoles could also be produced via coupling of 1,2-diaminobenzenes with carboxylic acid derivatives such as nitriles, imidates, orthoesters, anhydrides or lactones [12].

Alternatively, benzimidazoles have also been prepared from 2-nitroanilides, in a two-step process. In the first step, the nitro group is reduced using one of many possible reagents (such as zinc, iron, tin(II) chloride, hydrogen, or Raney nickel). The second step involves the ring closure of the 2-aminoanilide derivative with either a carboxylic acid or an aldehyde [2, 10, 13]. However, this procedure sometimes requires multistep reactions to prepare the starting anilides, resulting in compromised yields and purity. In recent years, some innovative and improved pathways for the synthesis of benzimidazoles have been developed and these are discussed in the following sections.

2.1 Benzimidazole Ring Closure

Various substituted benzimidazoles have been synthesized in very good yields in solvent-free conditions from 1,2-diaminobenzene and aldehydes in the presence of titanium(IV) chloride as a catalyst. The method is applicable to most aromatic, unsaturated and aliphatic aldehydes and to substituted 1,2-diaminobenzenes without significant differences [14]. Several other catalysts, namely iodine [15], hydrogen peroxide [16], zirconyl(IV) chloride [17], boron trifluoride diethyl etherate [18], ytterbium perfluorooctane sulfonates [19, 20], zeolite [11, 21], and L-proline [22], have been effectively used for the synthesis of benzimidazole derivatives.

A palladium-catalyzed *N*-arylation reaction provided a novel synthesis of benzimidazoles **5** from (*o*-bromophenyl)amidine precursors **4** under microwave irradiation. The route was found to be flexible with respect to various substituents and allows for the preparation of highly substituted benzimidazoles, including *N*-substituted examples (Scheme 1) [23]. The method was later improved and optimized to achieve the rapid formation of benzimidazoles in high yield [24]. It has been found that 50% aqueous dimethyl ether (DME) is an optimal solvent for the reaction and that catalyst loading of palladium can be reduced to 1 mol %.

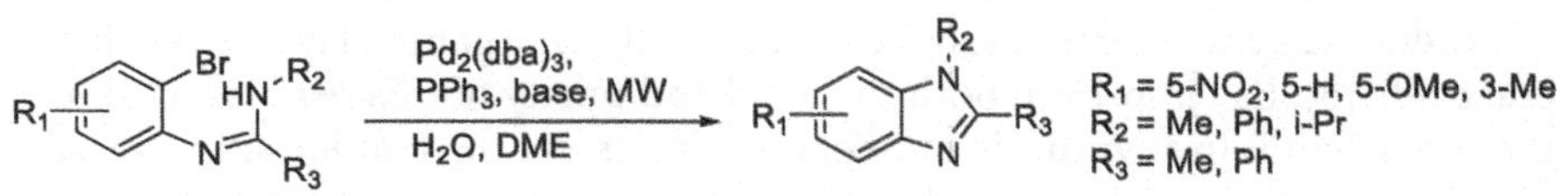

4 **5**

Scheme 1 Palladium-catalyzed synthesis of *N*-arylbenzimidazole

Imidazole *o*-quinodimethane intermediate **8**, synthesized from 2-bromo-4,5-bis(bromomethyl)imidazole derivative **7** via *N*-bromosuccinimide-mediated bromination of imidazole **6** undergoes a Diels–Alder reaction with several symmetrically and asymmetrically substituted dienophiles to yield the benzimidazole derivatives **9–12** in moderate yields [25]. The annulations of the aromatic systems depicted in Scheme 2 illustrate the ability of this reaction to give a variety of benzimidazoles.

2.2 Microwave-Assisted Synthesis of Benzimidazoles

The use of microwave irradiation as a source of heat in synthetic chemistry has been heralded as a promising method of increasing productivity and quality and reducing reaction time since its first use by Gedye et al. in 1986 [26]. It has become a focal point in chemical synthesis in recent years in

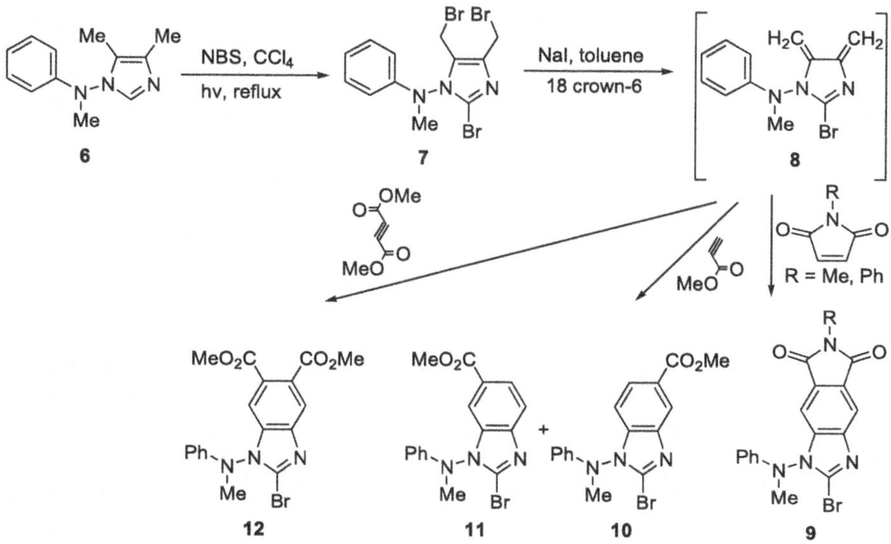

Scheme 2 Synthesis of benzimidazoles by Diels–Alder reaction

terms of sustainable and green chemistry for improved resource management and the need to develop environmentally benign processes. Of particular importance is the reduction in the amounts of solvents and hazardous chemicals required to perform chemical reactions enabled by this approach, and its more efficient use of energy [27]. Since 1995, various substituted benzimidazole derivatives have been synthesized through microwave heating [28]. In this section, selected literature on the synthesis of benzimidazole by microwave technology is discussed.

Recently, 2-alkyl- and 2-aryl-substituted benzimidazole derivatives **15** have been synthesized from 1,2-diaminobenzene dihydrochloride **13** and its corresponding acids **14** in the presence of polyphosphoric acid using microwave-assisted methods (Scheme 3). The reaction time required for the synthesis of benzimidazole derivatives **15** was reduced to minutes by this method compared to conventional synthesis, which required up to four hours of heating to complete the reaction. Furthermore, it was found that the application of microwave irradiation increased yields by 10–50% (Table 1). It has been

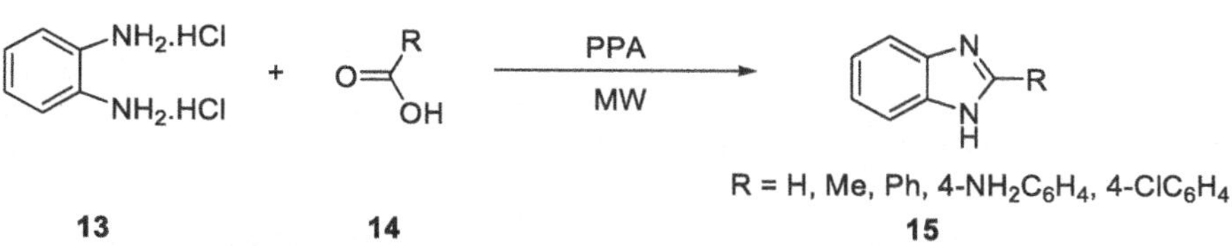

Scheme 3 Synthesis of alkyl and aryl benzimidazoles under microwave conditions

Table 1 Yield and reaction time for benzimidazole synthesis using microwave irradiation [29]

15 R =	RT	Yield (%)	MW	Yield (%)
H	2 h	80	1 min 20 s	92
Me	45 min	48	1 min 20 s	89
Ph	4 h	34	4 min 30 s	84
4-$NH_2C_6H_4$	4 h	57	5 min	95
4-ClC_6H_4	4 h	43	4 min 30 s	89

proposed that microwave heating easily provides the energy of activation required for the chemical reaction [29]. A similar one-pot high-yield procedure for the generation of 2-substituted benzimidazoles from the esters using ethane-1,2-diol as a solvent has been described [7]. Moreover, the single-step synthesis of benzimidazoles from a range of other diamines and carboxylic acids under microwave irradiation conditions has been developed, which provided a practical and efficient method for the high-throughput synthesis of 2-substituted benzimidazoles [12].

In addition, benzimidazoles containing furyl and aryl substituents at the C-2 position have been synthesized from 1,2-diaminobenzene and the corresponding carboxylic acids under microwave irradiation in the presence of artificial zeolites and catalytic amounts of DMF, used as the catalyst and energy transfer medium respectively. With this microwave technique, the reaction time was greatly shortened and the products were obtained in higher yields with easier workup than conventional heating methods [21].

Conventional condensation of 1,2-diaminobenzene **16** with 6-fluoro-3,4-dihydro-2H-chroman-2-carboxylic acid **17** under Phillips' conditions or using Eaton's reagent (1 : 10 mixture of phosphorus pentoxide/methanesulfonic acid) yielded 2-(6-fluorochroman-2-yl)-1*H*-benzimidazole **18** (Scheme 4) [30]. However, irradiating the reaction mixture containing polyphosphoric acid as a catalyst with microwaves afforded the compound **18** in comparable yields in a matter of three minutes [30].

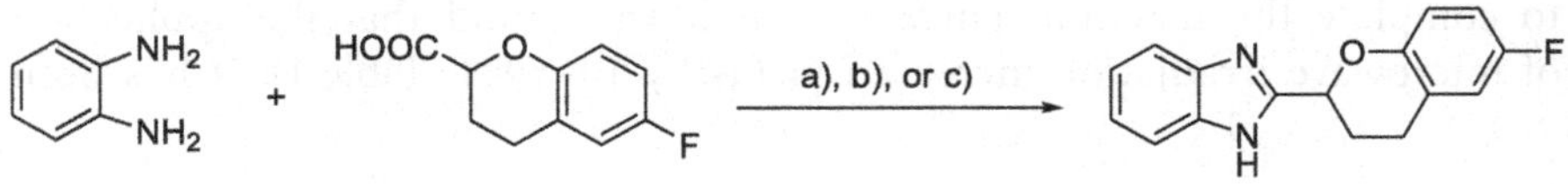

16 **17** **18**

Scheme 4 Reagents and conditions: **a** 4 N HCl, reflux, 6 h, 85%; **b** MW, PPA, 100 W, 170 °C, 3 min, 85%; or **c** Eaton's reagent, 100 °C, 5 h, 80%

Recently, microwave-assisted synthesis of eighteen 2-(alkyloxyaryl)-1H-benzimidazole derivatives **20** related to the natural stilbenoid family has been

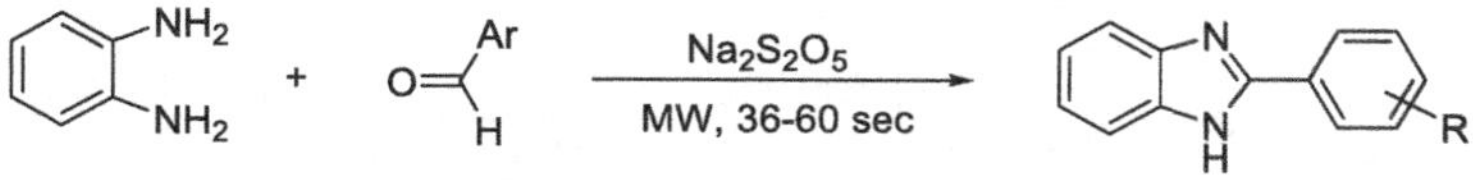

16 19 20 75-94%

R = 2-OH, 2-OMe, 2-OEt, 2-OPr, 2-NO_2, 4-OH, 4-OMe, 4-OH/3-OMe, 3,4-di-OMe, 2,3,4-tri-OMe, 2,4,5-tri-OMe, 2-benzyloxy, 4-Cl-benzyloxy, 4-methyl-benzyloxy, 3,4-methylenedioxy, 4-pyridyl,

Scheme 5 Synthesis of 2-(alkyloxyaryl)-benzimidazole

reported (Scheme 5) [31]. These bioisosteric benzimidazole analogs **20** have been synthesized in high yields through a rapid three-component reaction starting from commercially available aldehydes **19** and 1,2-diaminobenzene **16**, and sodium metabisulfite in the absence of solvent. The in vitro spasmolytic activity of these compounds on the spontaneous contractions of the rat ileum suggests that bioactivity of these compounds depends upon the presence of oxygenated groups attached at C-2 and/or C-4 of the phenyl ring respectively [31].

Recently, a facile, rapid one-pot procedure for the generation of 2-substituted benzimidazoles **23** directly from 2-nitroanilines **21** using a microwave procedure has been demonstrated (Scheme 6). An advantage of this approach is that the intermediate N-acyl derivatives **22** need not be isolated prior to cyclization [10].

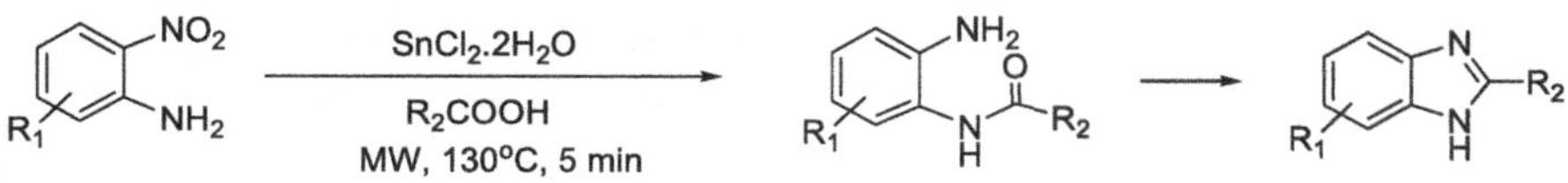

21 22 23

R_1 = H, 4,5-dimethyl, 5-OH, 5-OMe, 5-COOH, 5-CN, 5-CF_3, 4,6-dichloro
R_2 = H, Me, CF_3

Scheme 6 Synthesis of 2-substituted benzimidazoles from 2-nitroanilines

Classical condensation-cyclization reactions using 1,2-diaminobenzenes **24**, 2-mercaptoacetic acid **25** and appropriately substituted aromatic aldehydes **26** in dry benzene under reflux required a long reaction time to afford the thiazobenzimidazoles **27**, which are potent anti-HIV agents, by Scheme 7. On the other hand, the microwave-assisted synthesis of 1*H*,3*H*-thiazolo[3,4-a]benzimidazoles **27** was completed in toluene within 12 minutes [32].

Furthermore, a versatile and efficient microwave-promoted combinatorial library synthesis of two long alkyl chain benzimidazoles from *o*-substituted amines and fatty acids employing either bentonite, alumina or silica gel as solid supports has been developed [33]. Bismuth chloride [34], montmorillonite clay K-10 [35] and silica impregnated with sulfuric acid [36] have also

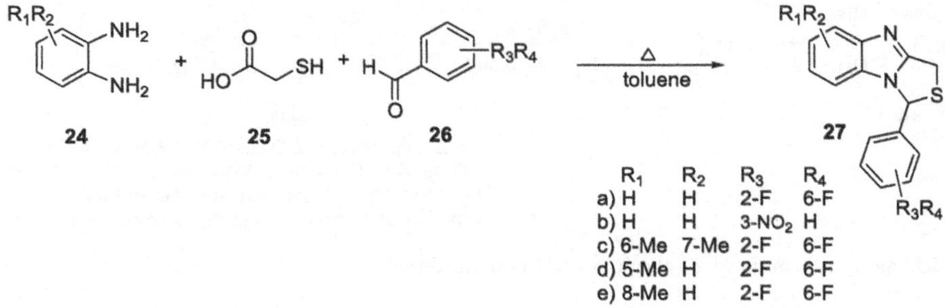

Scheme 7 Synthesis of thiazobenzimidazoles

been reported to act as inorganic catalysts for the benzimidazole ring closure reaction under microwave irradiation conditions.

2.3 Synthesis of Bisbenzimidazoles

Bisbenzimidazole derivatives such as Hoechst 33258 (also known as Pibenzimol) **28** (Fig. 2) is a A/T base pair selective compound that binds in the minor groove of DNA [37]. To investigate its full potential, a number of benzimidazole Hoechst motifs have been synthesized and evaluated for various biological activities [38–40].

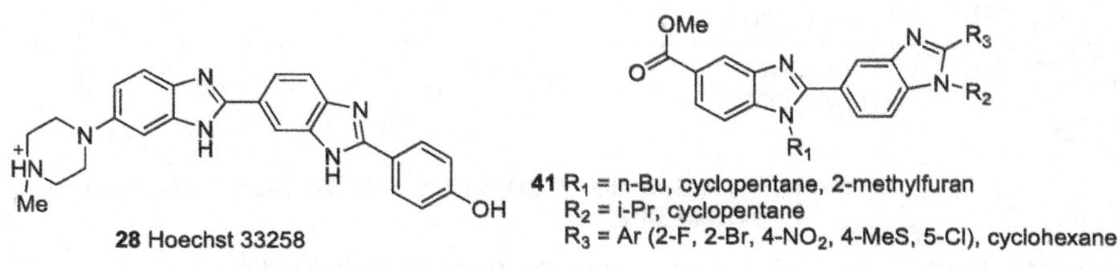

Fig. 2 Examples of some bisbenzimidazoles

Mann and coworkers have synthesized a new class of head-to-head bisbenzimidazoles **31** as DNA minor groove binding agents [41]. This new class of 6,6′-bisbenzimidazoles **31** was synthesized in moderate yields by the condensation of 3,3′,4,4′-tetraaminobiphenyl **30** with requisite aromatic aldehyde in nitrobenzene under reflux for 8–12 hours (Scheme 8).

In order to target the minor groove of a longer sequence in the A/T rich region, the previous group also synthesized a novel dimeric bisbenzimidazole **35** where the two bisbenzimidazole rings are linked together via an appropriate linker [42]. The benzimidazole **33** was first obtained by condensation between 4-methoxybenzoic acid and 3,3′-diaminobenzidine tetrahydrochlo-

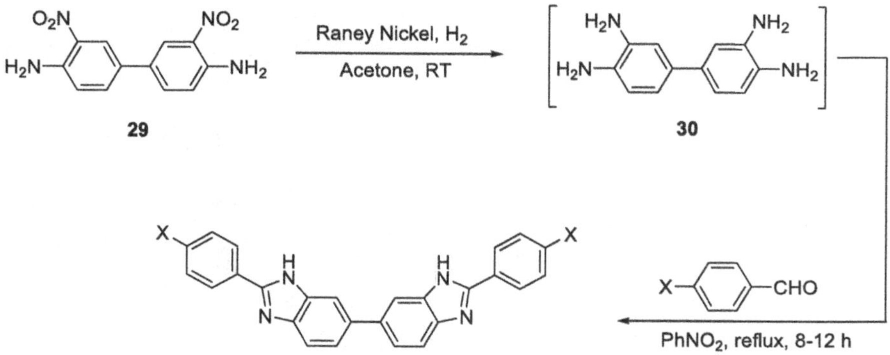

31 X = OH, OMe

Scheme 8 Synthesis of head-to-head bisbenzimidazole

ride **32** (Scheme 9). The compound **33** was then condensed with the diester **34** to provide the desired dimeric bisbenzimidazole **35**.

The microwave-enhanced synthesis of symmetrical and asymmetrical bis(benzimidazol-2-yl)methanes **38** from appropriately substituted benzimidazole 2-acetic acid **36** and substituted-1,2-diaminobenzene **37** under solvent-free conditions without any catalyst has been performed in good yields (Scheme 10) [43]. The symmetrical bis(benzimidazol-2-yl)methanes **40** have similarly been prepared by one-step condensations of malon-

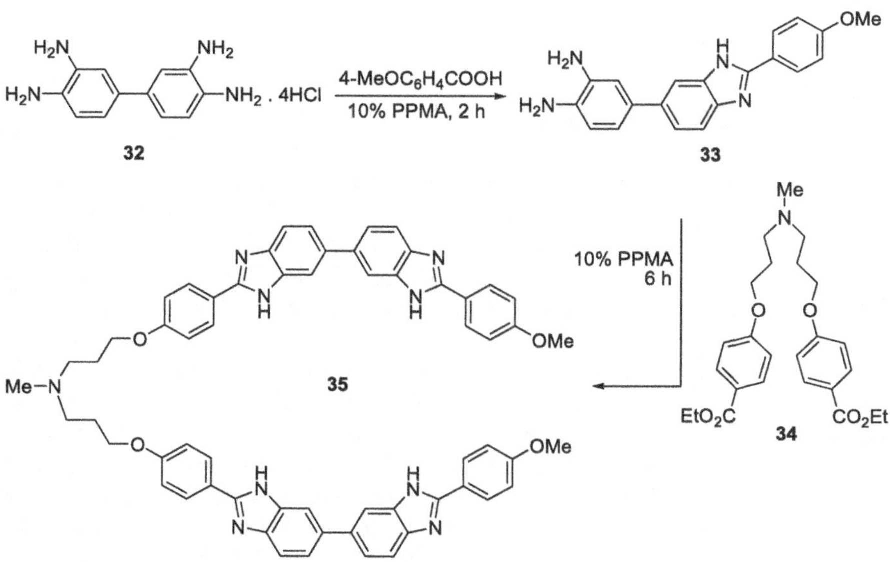

Scheme 9 Synthesis of a dimeric bisbenzimidazole **35**

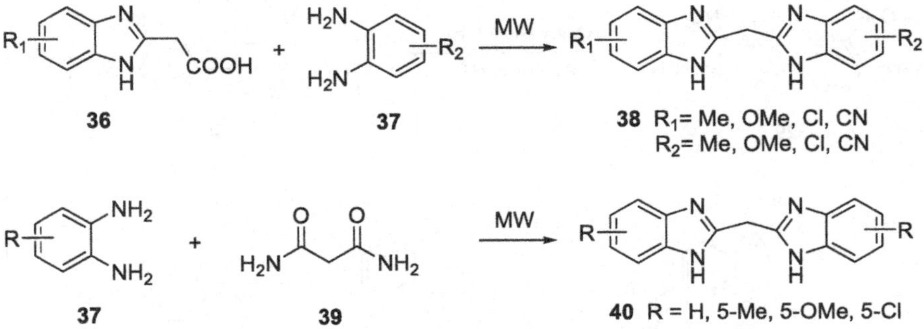

Scheme 10 Synthesis of 2,2′-bisbenzimidazolylmethanes

amide **39** with appropriately substituted-1,2-diaminobenzene [44]. A similar microwave-assisted method for the synthesis of 2,2′-bisbenzimidazoles using 1,2-diaminobenzene **16** and oxalic acid or malonic acid in polyphosphoric acid has also been reported [45].

Combinatorial parallel synthesis of head-to-tail bisbenzimidazoles **41** has been performed using polymer-immobilized 1,2-diaminobenzenes (Fig. 2). The PEG-bound diamines were *N*-acylated at the primary aromatic amino group with 4-fluoro-3-nitrobenzoic acid. The substituted amides were cyclized to benzimidazoles under acidic conditions. Successive reduction and cyclization with various aldehydes yielded 5-(benzimidazol-2-yl)benzimidazoles. Finally, the desired products **41** were released from the polymer support to afford the bisbenzimidazoles in good yields and with high purity [46].

3 Benzimidazole Natural Products

Benzimidazole-derived alkaloids are rare in nature, and only a few examples of these natural products can be found in the literature. On the other hand, the occurrence of the imidazole skeleton in various natural sources is quite common [47–49]. The benzimidazole alkaloid kealiiquinone (Fig. 3) has been isolated from a yellow button-like Micronesian sponge species of *Leucetta* [49].

Recently, Nakamura et al. successfully synthesized a regioisomer of kealiiquinone (Scheme 11) [50]. 1-Methyl-2-phenylthio-1*H*-imidazole **44** was first converted into the 5-substituted imidazole **45**, then the benzylic hydroxyl group in **45** was protected by a *tert*-butyldimethylsilyl (TBDMS) group, and bromination with *N*-bromosuccinimide gave the bromide **46**. Lithiation by *tert*-butyllithium at the 4-position of **46** followed by trapping with 3,4-dimethoxy-2-(methoxymethoxy)benzaldehyde gave the tetrasubstituted imidazole **47** as a diastereomeric mixture. Acetylation of the hydroxy group of **47**

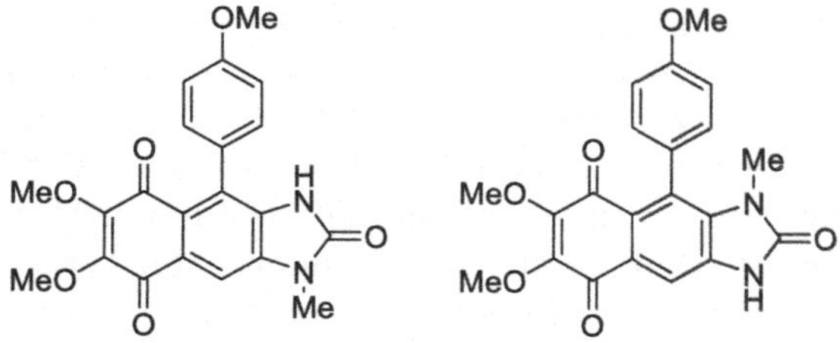

42 Kealiiquinone **43** Regioisomer

Fig. 3 Benzimidazole alkaloid kealiiquinone **42** and its regioisomer **43**

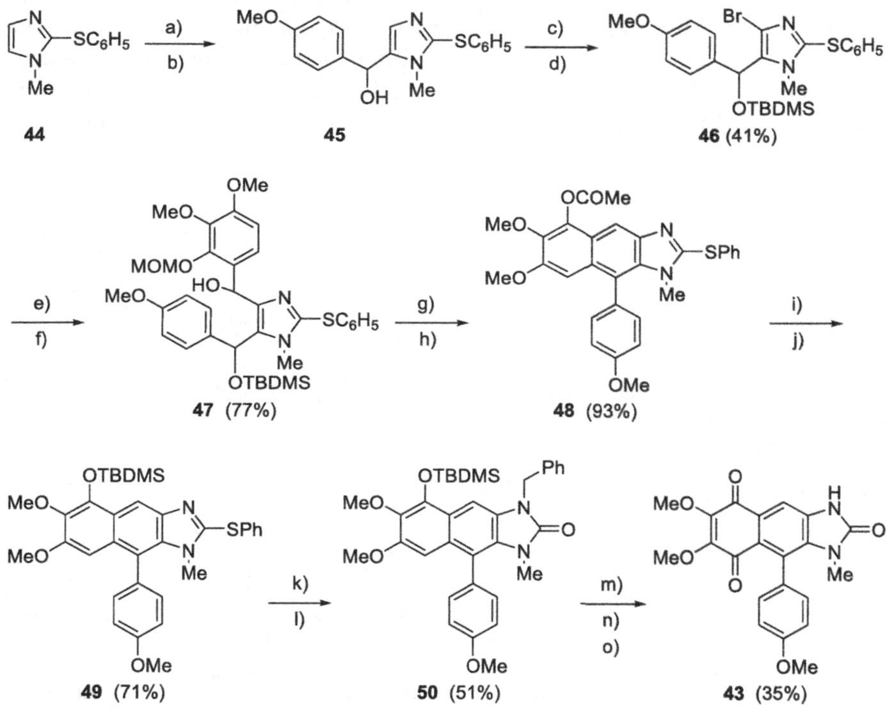

Scheme 11 Reagents and conditions: (**a**) LTMP, DME-THF; (**b**) *p*-anisaldehyde; (**c**) TBDM-SCl, DMF, 12 h, 60 °C; (**d**) NBS, THF, 7 h, 0 °C; (**e**) *tert*-BuLi, *n*-pentane, Et_2O, 1 h, –78 °C; (**f**) 3,4-dimethoxy-2-(methoxymethoxy)benzaldehyde, Et_2O, 3 h, –78 °C; (**g**) Ac_2O, Et_3N, $CHCl_3$, 3 h, 0 °C; (**h**) PPA, Ac_2O, 12 h, 0 °C; (**i**) K_2CO_3, MeOH-H_2O, 3 H, r.t.; (**j**) TBDMSCl, DMF, 6 h, 60 °C; (**k**) benzyl bromide, EtOAc, 6 h, reflux; (**l**) aq. K_2CO_3, 1 h, 80 °C; (**m**) $Pd(OH)_2/C$, H_2 ($4.2\ kg/cm^{-2}$), 48 h, r.t.; (**n**) TBAF, THF, 5 min, r.t.; (**o**) O_2, salcomin, 1 h, r.t.

and cyclization with polyphosphoric acid in the presence of acetic anhydride gave the tricyclic compound **48**. Alkaline hydrolysis of the ester group of **48** followed by conversion of the phenolic hydroxy group into a TBDMS group afforded the silyl ether **49**. Quaternization of **49** with benzyl bromide followed by heating in aqueous potassium carbonate successfully afforded the 2-oxo compound **50**. The benzyl group and the TBDMS group of **50** were removed by Pd/C-catalyzed hydrogenation followed by treatment with TBAF, and the product was auto-oxidized in the presence of salcomin in THF to give the desired regioisomer of kealiiquinone compound **43** [50]. The kealiiquinone **42** and its synthetic regioisomer **43** both have relatively weak activities against a panel of 39 human cancer cell lines but are considered to have a unique mechanism of action [50].

Makaluvamines (pyrroloiminoquinones) **51** (Fig. 4) isolated from a Fijian sponge in the early 1990s display in vitro cytotoxicity against human colon tumor cell lines and also inhibit human topoisomerase II in vitro. The benzimidazole analog of this indole-based marine natural product, imidazoquinoxalinone **52**, has been synthesized starting from *p*-methoxydiacetanilide (Scheme 12) [51]. Treatment of the dinitration product **54** of *p*-methoxy-

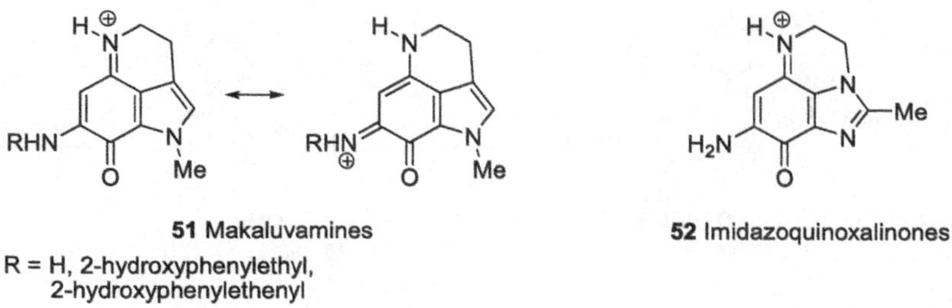

Fig. 4 Makaluvamine and its benzimidazole analog

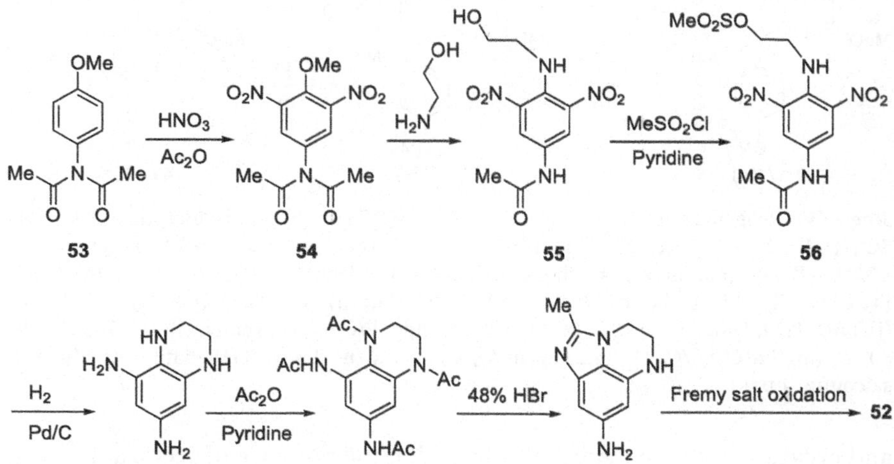

Scheme 12 Synthesis of imidazoquinoxalinone

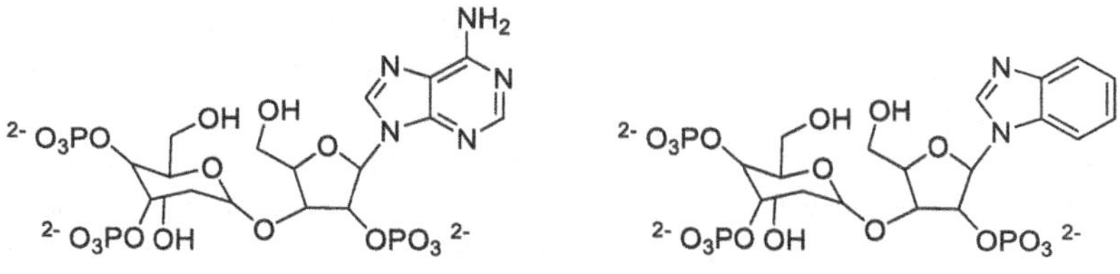

60 Adenophostin **61** Benzimidazole analog

Fig. 5 Adenophostin and its benzimidazole analog

diacetanilide **53** with ethanolamine resulted in the formation of aminoethanol compound **55**. Conversion of the alcohol **55** to mesylate **56** followed by catalytic reduction of the nitro group afforded the tetrahydroquinoxoline **57**. Peracetylation of quinoxoline **57** afforded a stable amide **58**, which upon acid-catalyzed cyclization yielded the precursor benzimidazole **59**. Finally, Fremy salt oxidation of benzimidazole **59** afforded the iminoquinone **52**.

In comparison with the natural inositol 1,4,5-triphosphate, the adenophostins **60** (Fig. 5) exhibit higher receptor binding activity and Ca^{2+} mobilizing potencies and thus have significant biological importance. A total synthesis of a benzimidazole analog of adenophostin A **61** has been described by Shuto et al. [52].

4 Functionalization of the Benzimidazole Molecule

4.1 Substitution at the N-1 Position

N-substituted benzimidazoles **62** and **63** (Fig. 6) have been reported to show anti-hepatitis B virus activity, and thus several derivatives of novel benzimidazoles **62** and **63** have been prepared by Li et al. [53]. The precursor benzimidazoles readily undergo *N*-substitution reactions with sulfonyl chlorides in dichloromethane using DMAP as a base, whereas methylation can

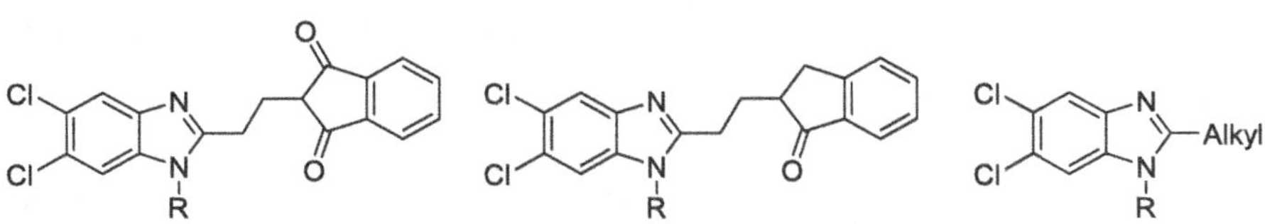

R = $MeSO_2$, Ts, Me, 4-Me-benzyl R = $MeSO_2$, Ts, Me, 4-Me-benzyl R = Me, Et

62 **63** **64**

Fig. 6 *N*-substituted benzimidazoles

be achieved by potassium carbonate in DMF or sodium hydroxide in acetonitrile. Several N-alkyl benzimidazole derivatives **64** behave as selective androgen receptor antagonists [54].

Ionic liquids have been reported to accelerate slow *N*-benzylation reactions of benzimidazole **65** utilizing dibenzyl carbonate **66** as an alkylating reagent (Scheme 13) [55]. An additional rate enhancement was observed when microwave irradiation was applied in this reaction to afford the *N*-benzylbenzimidazole **67**.

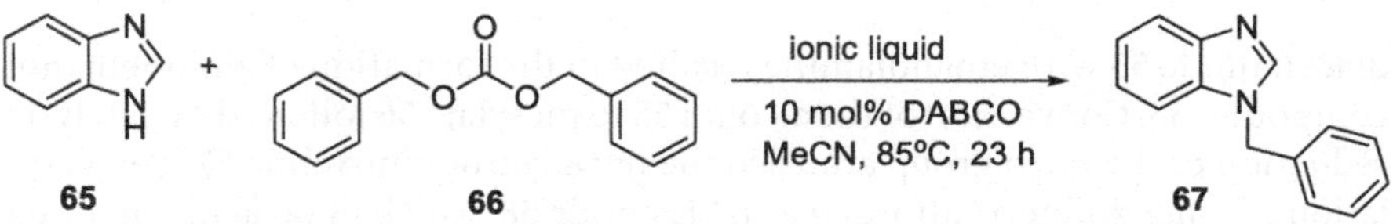

Scheme 13 *N*-benzylation reaction via dibenzyl carbonate

4.2 Direct Coupling at the C-2 Position

Functionalization of C – H bonds of heterocycles to C-arylation is an important synthetic reaction that is used to build important bioactive structures. Recently palladium and copper-mediated C-2 arylations of benzimidazole with aryl iodides under ligandless and base-free conditions have been described (Scheme 14) [56]. These reactions show complete selectivity under these conditions and allow for the use of substrates containing base-sensitive groups without their prior protection, such the NH group of benzimidazoles [56, 57]. The aryl-substituted benzimidazole compounds were obtained in high purities and yields within 48 hours, and the scheme was also applicable to other azoles.

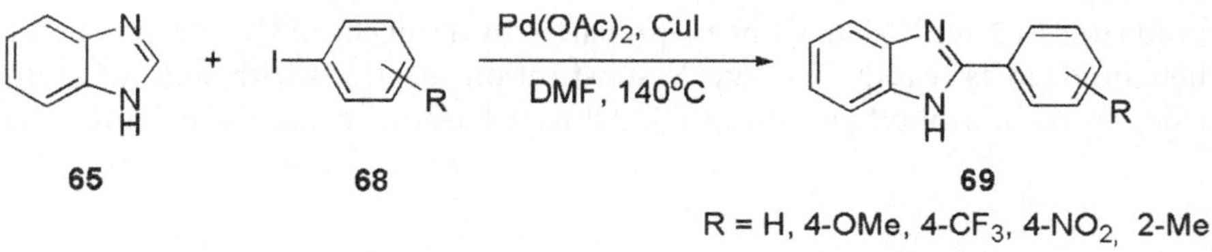

Scheme 14 Palladium-catalyzed direct C-2 coupling

A similar general method for the rhodium-catalyzed direct coupling of benzimidazoles with aryl bromides or iodides has also been developed under microwave-assisted conditions, and it has shown to provide rapid access to medicinally relevant compounds. Both electron-rich and electron-poor aryl iodides were observed to couple with tricyclohexylphosphine. The desired arylated products were obtained in good yields by conventionally heating the reaction mixtures in sealed tubes. N-Heterocyclic carbene/rhodium complexes were considered to be intermediates during C – C bond formation [58].

4.3
Doebner–Von Miller Reaction at the C-7 Position

Imidazoquinolines **71** have been synthesized in good yield by coupling benzimidazole **70** with dimethyl *trans*-2-ketoglutaconate under Doebner–Von Miller reaction conditions in dichloromethane (Scheme 15). The compound **71** was demethylated using hydrobromic acid in glacial acetic acid, re-esterifed with methanolic hydrochloric acid, and oxidized to afford benzimidazole quinone **72** [59].

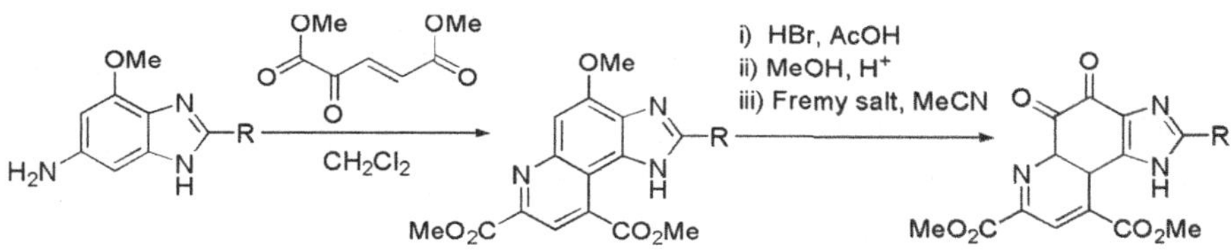

70 **71** R = H, Me **72** R = H, Me

Scheme 15 Doebner–Von Miller reaction at C-7 position

4.4
Cyclization Between the N-1 and C-2 Positions

Various benzimidazole derivatives can be fused between the C-2 and N-1 positions in order to build novel heterocyclic ring systems. For example, the reaction of 2-cyanomethylbenzimidazole **73** with hydrazonoyl halides **74** in the presence of triethylamine led to the formation of pyrrolo[1,2-a]benzimidazoles **76** (Scheme 16) [60]. It has been suggested that the reaction starts with the nucleophilic substitution of the halogen by the benzimidazole carbanion to give intermediate **75**, which upon cyclization via elimination of water gives the desired cyclic pyrrolobenzimidazoles **76**. On the other hand, the reaction of hydrazonoyl chlorides **77** with 2-cyanomethylbenzimidazole **73** in sodium ethoxide afforded pyrazole-3-carboxylate **80**, which upon treatment under triethylamine yielded the pyrazolopyrrolobenzimidazole **81**. The product was also obtained by the direct reaction of 2-cyanomethylbenzimidazole **73** with hydrazonoyl chlorides **77** in the presence of triethylamine.

Dzvinchuk has synthesized several pyrido[1,2-*a*]benzimidazoles (**84**, **86** and **88**) via reactions of 2-acylmethylbenzimidazole **82** (Scheme 17). Treatment of **82** with malononitrile led to the formation of the dicyanomethylene-

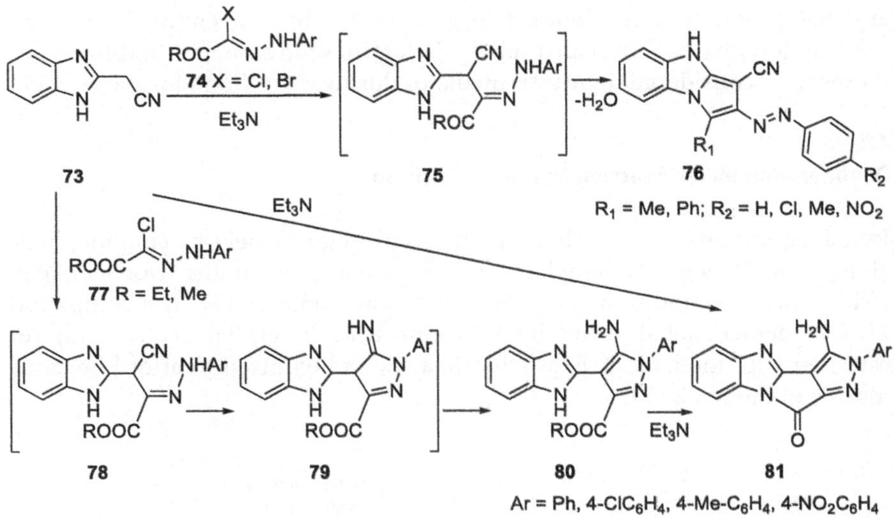

Scheme 16 Synthesis of pyrrolobenzimidazoles

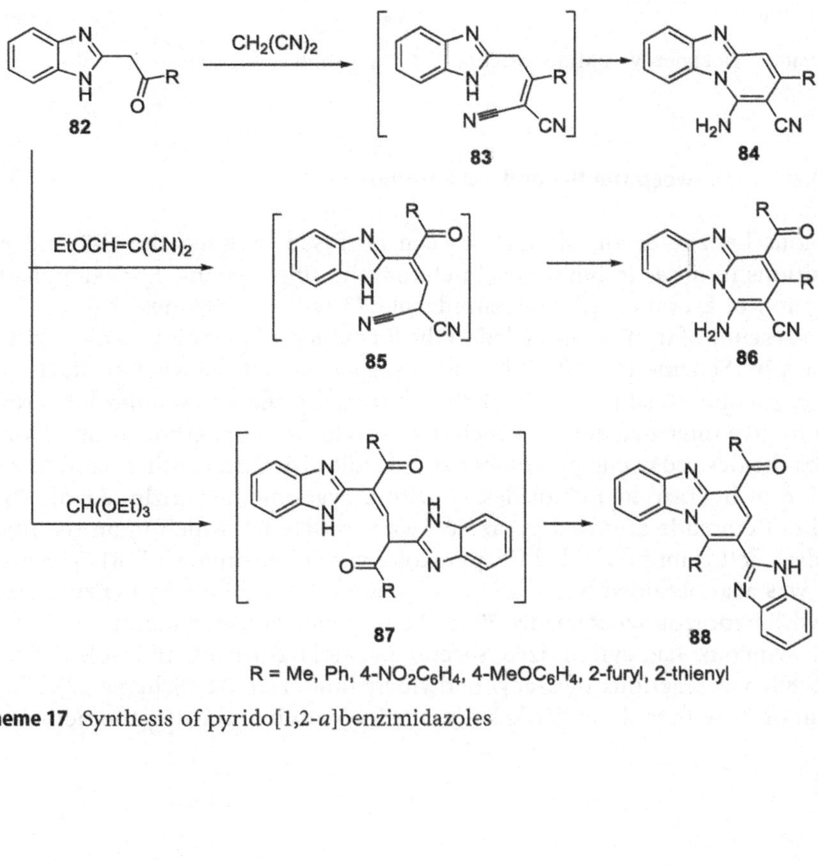

Scheme 17 Synthesis of pyrido[1,2-*a*]benzimidazoles

substituted compounds **83**, which by intramolecular addition of the benzimidazole imino group to the nitrile gave the pyridobenzimidazoles **84**. The similar reaction of 2-acylmethylbenzimidazole **82** with ethoxymethylenemalononitrile yielded the pyridobenzimidazoles **86** by the cyclization of the intermediate compound **85**. On the other hand, attachment of the 2-acylmethylbenzimidazole **82** to triethylorthoformate joins two benzimidazole molecules of the starting keto compound at the active methylene group to give the intermediate **87**, which then undergoes cyclocondensation of a benzimidazole imino group at the keto group to give the substituted pyridobenzimidazoles **88** [61].

Polyheterocyclic structures such as benzimidazoquinazolines **91** made up of two fused heterocyclic rings often possess potent biological activity, like antiproliferative and DNA-intercalator activity [62], antifertility activity [63], anticonvulsant activity, and myorelaxant activity [64]. These benzimidazoquinazoline compounds **91** have been obtained by the condensation of 2-cyanobenzothiazoles **89** or benzoxazoles **89** with 2-(2-aminophenyl)benzimidazole **90** under microwave conditions in the presence of graphite as a catalyst [65].

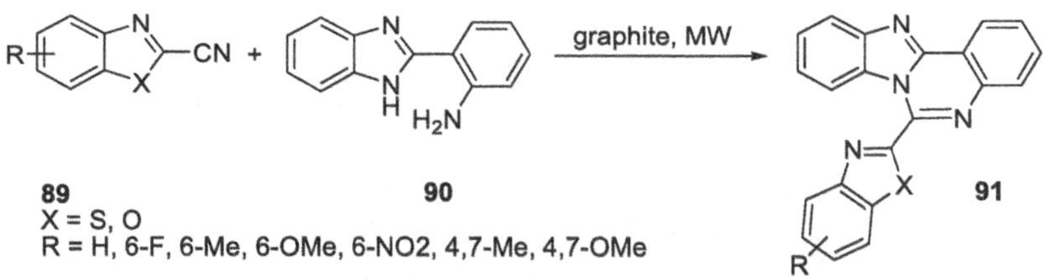

Scheme 18 Synthesis of polyheterocyclic benzimidazoquinazolines

The intermediate 1,3-dipolar nitrile imines **93**, generated in situ from hydrazonyl chloride **92**, have been reacted with 2-chloromethylbenzimidazole **94** in the presence of triethylamine and silica gel under microwave irradiation for four minutes to afford the synthesis of the novel tricyclic benzimidazole system **95** [66].

Radical cyclizations of nucleophilic N-alkyl radicals **96** onto the benzimidazole 2-position, mediated by tributyltin hydride and activated by quaternizing the pyridine-like N-3 of imidazole with camphorsulfonic acid, have recently been reported (Scheme 20) [67]. These new five-, six- and seven-membered homolytic aromatic substitutions of nucleophilic N-alkyl radicals onto the benzimidazole-2-position occurred upon the use of large excesses of the azo-initiator, 1,1′-azobis(cyclohexanecarbonitrile), to supplement the non-chain reaction. The intermediate **97** aromatizes in high yields to the cyclized benzimidazoles **98**.

Microwave irradiation has been shown to strongly accelerate the rhodium-catalyzed intramolecular C – H bond coupling of benzimidazole alkenes **99** to

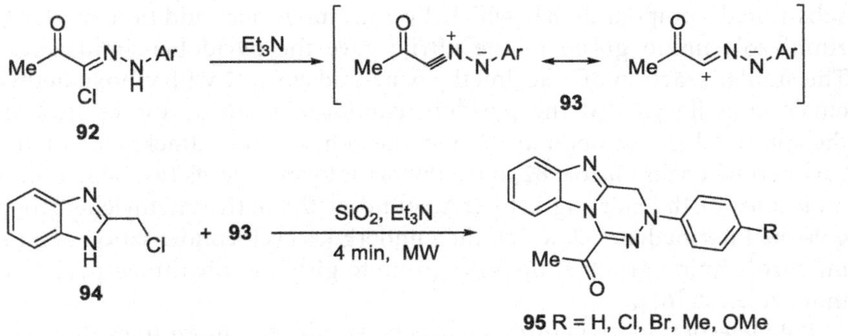

Scheme 19 Synthesis of novel tricyclic benzimidazole

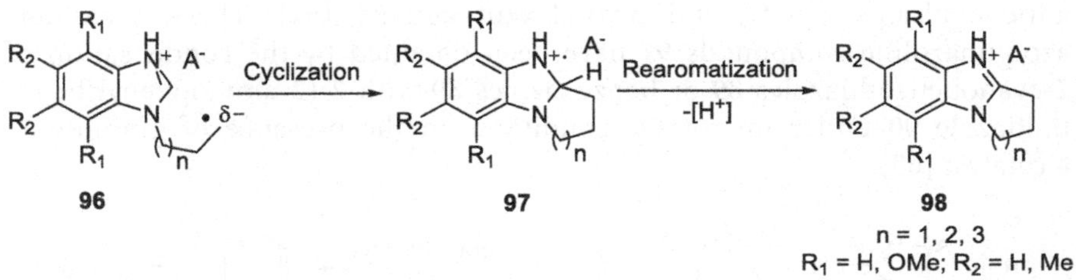

Scheme 20 Intramolecular homolytic aromatic substitution mechanism

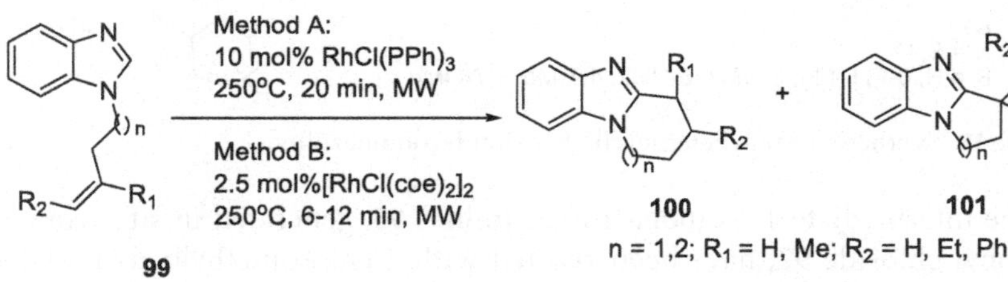

Scheme 21 Microwave-assisted intramolecular coupling reaction

cyclic benzimidazoles **100** and **101**. These products were formed in moderate to excellent yields with reaction times of less than 20 minutes. Additionally, the use of microwave irradiation allowed the reactions to be performed without any solvent and purification and with minimal precautions to exclude air [68].

4.5 Solid Phase Synthesis

The use of solid phase synthesis has been directed towards high-speed synthesis and biological screening of diverse libraries as part of the drug discovery process. The widespread use of solid phase synthesis is due to its advantages of inexpensive apparatus, easy workup and filtration. Thus, it has become an increasingly popular tool for combinatorial synthesis as well as green chemistry in recent years [69, 70]. From a combinatorial chemistry perspective, the benzimidazole scaffold allows the stepwise incorporation of diverse functionality with control over regiochemistry, making it a suitable target for library synthesis using solid and solution phase approaches as well as parallel polymer-assisted synthesis [10, 71–75].

Resin-bound iminophosphoranes **103** derived from the reaction of resin-bound 2-aminobenzimidazole **102** with triphenylphosphine oxide were reacted with aryl isocyanates in an abnormal aza-Wittig reaction with a chemoselectivity that depends on the reaction temperature and the nature of the aryl isocyanate (Scheme 22). The mechanism considered for the solid phase synthesis reaction involves the loss of triphenylphosphinimide instead of triphenylphosphine oxide, resulting in the formation of isocyanates instead of carbodiimides as intermediates. Optimization studies revealed that employing electron-poor aryl isocyanates at high temperature leads to 95% of the abnormal aza-Wittig products 3-aryl 2,4-dioxo-1,3,5,-triazino[1,2-*a*]benzimidazoles **104** [76].

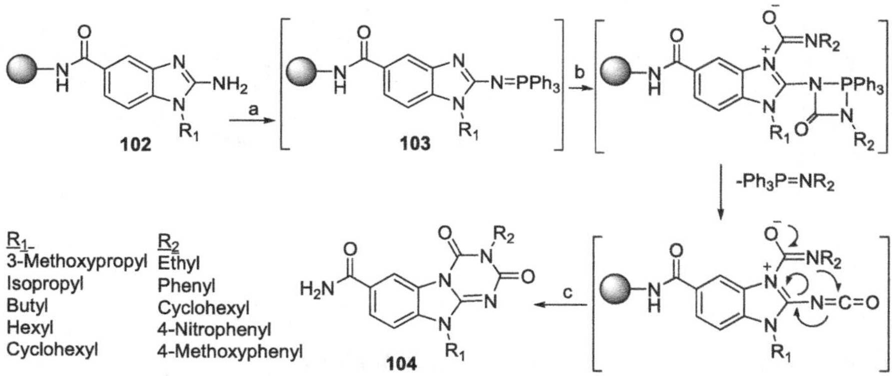

R_1	R_2
3-Methoxypropyl	Ethyl
Isopropyl	Phenyl
Butyl	Cyclohexyl
Hexyl	4-Nitrophenyl
Cyclohexyl	4-Methoxyphenyl

Scheme 22 Solid phase synthesis of triazinobenzimidazoles (reagents and conditions: (**a**) PPh_3, DEAD, THF, 25 °C, 3 days; (**b**) R_2NO_2, toluene, 25 °C, 2 days; (**c**) HF, anisole, 0 °C, 1.5 hour)

Using polymer-immobilized liquid phase synthesis and controlled microwave irradiation, trisubstituted bisbenzimidazoles have been prepared and released with good yield and purity [77]. Furthermore, a wide range of benzimidazole derivatives have been synthesized with excellent yields and purities by simple washing and filtration using liquid phase synthesis on a soluble polymer support [78].

4.6 Benzimidazole-Derived Metal Complexes

Recently, neutral dimeric zinc(II) complexes have been constructed from phenolic benzimidazole derivatives containing N and O donor atoms. These complexes exhibited a trigonal-bypyramidal geometry [79]. Pal recently reported a benzimidazole N-donor dinuclear palladacycle complex **105** [80] (Fig. 7). The X-ray crystal structure revealed that the two bisbenzimidazole ligands assembled through complexation to two palladium (II) ions to give a compressed rectangular metallamacrocycle. The complex effectively catalyzed Suzuki cross-coupling reactions in methanol at room temperature [80]. Moreover, a series of nickel(II) complexes **106** and **107** ligated by 2-(2-benzimidazole)-pyridine derivatives and nickel dichloride hexahydrate have been prepared [81]. Interestingly, benzimidazole-derived copper(II) and nickel(II) complexes have revealed antibacterial, antifungal and DNA intercalator activities [82], whereas lanthanide(III) complexes exhibited seed germination inhibition activity [83].

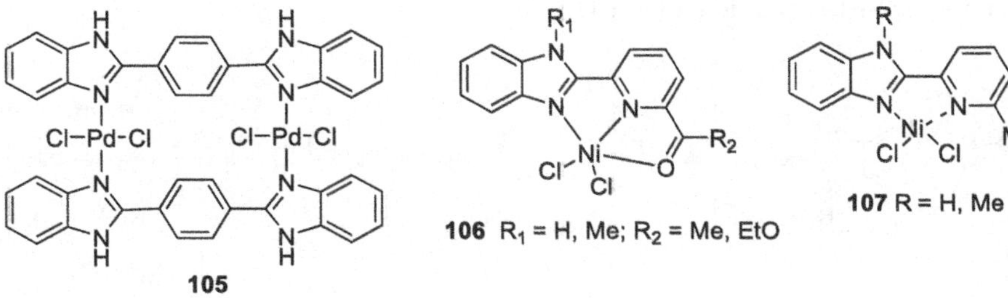

Fig. 7 Some benzimidazole–metal complexes

5 Biological Activities of the Benzimidazole Analogs

5.1 Anti-infective Agents

5.1.1 Antibacterial and Antifungal Agents

The search for compounds with antibacterial activity has gained increasing importance in recent times, due to growing worldwide concern over the alarming increase in the rate of infection by antibiotic-resistant microorganisms [84]. Owing to the current importance of developing novel antimicrobials and the varied bioactivities exhibited by benzimidazoles, several reseachers have investigated the antimicrobial activities of benzimidazole derivatives.

2-Mercaptobenzimidazole derivatives are known to possess varied biological activities [85]. Recently, an efficient and rapid synthesis of novel benzimidazole azetidin-2-ones **108** has been established [86], and antibacterial screening revealed that all newly synthesized azetidin-2-ones **108** exhibited potent antibacterial activity against *Bacillus subtilis*, *Staphylococcus aureus* and *Escherichia coli*. In general, compounds **108a**, **108i** and **108j** exhibited more pronounced antibacterial activity than compounds **108b–h**, with better activity against both Gram-positive and Gram-negative bacteria (Fig. 8). Among all of the compounds investigated, **108i** and **108j** exhibited the greatest antibacterial activity against Gram-negative *E. coli* as compared to the antibiotic streptomycin [86]. Benzimidazole benzyl ethers **109** have exhibited good antibacterial activity against *S. aureus* and antifungal activity against *Candida albicans* and *Candida krusei*. In general the dichlorophenyl-substituted benzimidazoles **109e**, **109f**, and **109h** showed the best antibacterial (MIC 3.12 μg/mL) and antifungal (MIC 12.5 μg/mL) activity [87]. In addition, 5-fluoro benzimidazole carboxamide derivatives **110** [88] and benzimidazole isoxazolines **111** [89] have been reported to show antibacterial and antifungal activities. N-alkylated or acylated derivatives of benzimidazole **18** also exhibited good antibacterial activities [30]. Numerous other reports of benzimidazole derivatives with antimicrobial activities have been published [90–99].

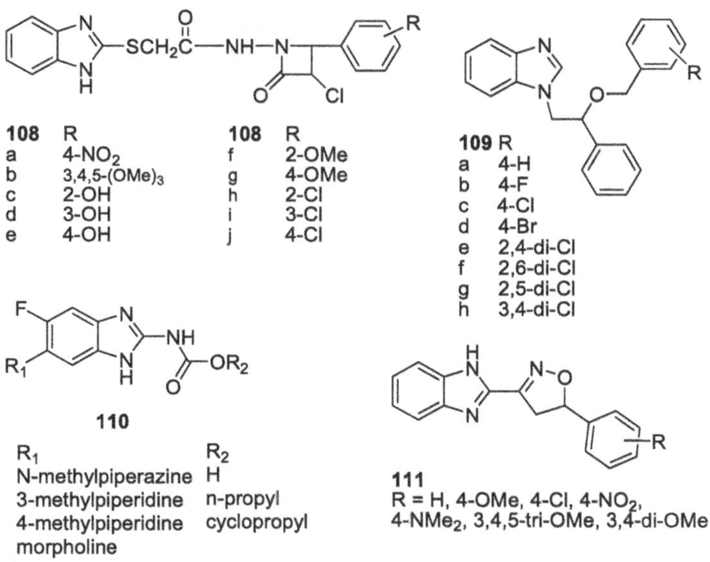

Fig. 8 Some antimicrobial benzimidazole derivatives

5.1.2 Anthelmintic Agents

Anthelmintic resistance is almost cosmopolitan in distribution and it has been reported in almost all species of domestic animals and even in some parasites of human beings. All of the major groups of anthelmintics have encountered variable degrees of resistance from different species of gastrointestinal nematodes [100]. Bearing in mind previous benzimidazole anthelmintics (e.g., albendazole, mebendazole), the search for new anthelmintic drugs is being actively pursued. Synthetic benzimidazole piperazine derivatives exhibited 50% anthelmintic activity in mice infected with *Syphacia obvelata* [101]. Furthermore, piperazine derivatives of 5(6)-substituted-(1*H*-benzimidazol-2-ylthio) acetic acids **112–114** [102] and benzimidazolyl crotonic acid anilide **115** have shown good anthelmintic activity [103] (Fig. 9).

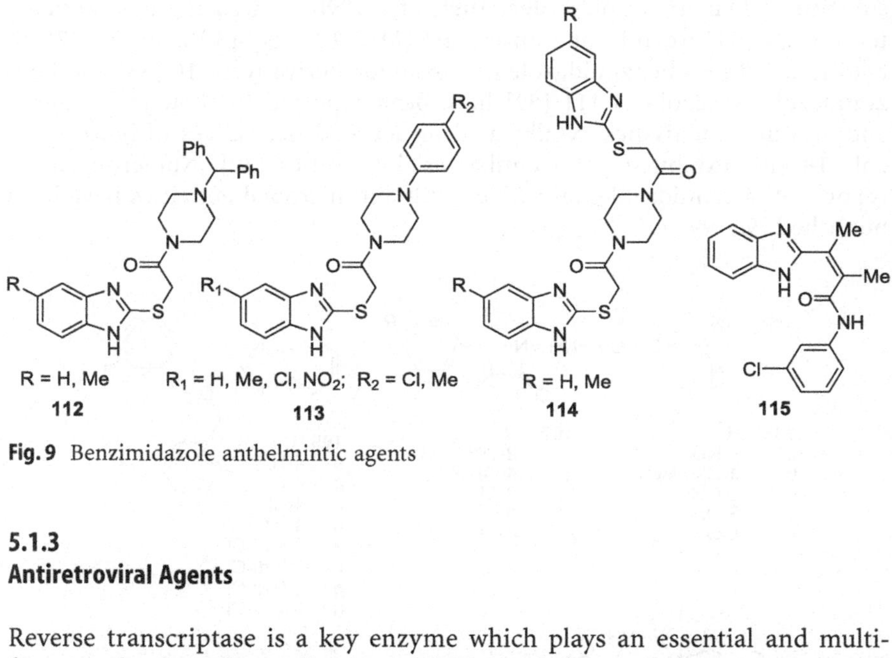

Fig. 9 Benzimidazole anthelmintic agents

5.1.3 Antiretroviral Agents

Reverse transcriptase is a key enzyme which plays an essential and multifunctional role in the replication of HIV-1 and thus constitutes an attractive target for the development of new drugs that could be used in AIDS therapy. A combination of reverse transcriptase and protease inhibitors is an effective approach to the treatment of AIDS [32]. However, side effects and the clinical emergence of resistant mutants suggests an increasing need for novel antiviral drugs.

Thiazolobenzimidazoles **27** proved to be a highly potent inhibitor of HIV-1-induced cytopathic effects. Structure–activity relationship studies showed that the C-1 substituents in benzimidazole greatly influence the interaction of the active compound with the receptor. Substitution on the benzene-fused ring influences the inhibitory potency depending on the nature and position of the substituent; the presence of a methyl group at C-3 is favorable to the pharmacological profile [104].

5.2 Anti-inflammatory and Antiulcer Agents

Structure–activity relationship studies of the 5,6-dialkoxy-2-thiobenzimidazole derivatives **116** have revealed that compounds **116a–116k** possess pronounced anti-inflammatory properties [105] (Fig. 10). Using the carrageenan model, the most significant anti-inflammatory effects were observed for com-

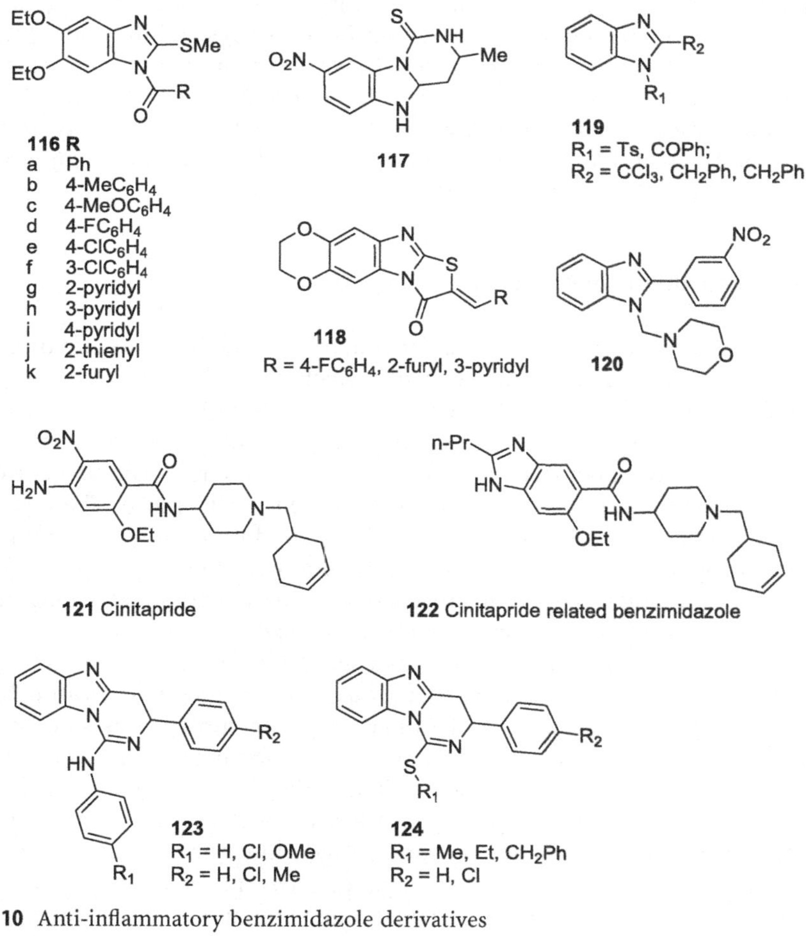

Fig. 10 Anti-inflammatory benzimidazole derivatives

pounds **116a, 116d, 116h, 116i,** and **116j**. While using the bentonite model, the maximum activities were observed for compounds **116e** and **116h**. These results indicated that benzimidazoles are promising leads for the development of new anti-inflammatory agents.

Pyrimidobenzimidazole **117** [106] and dioxinobenzimidazothiazol-9-ones **118** [107] exhibited anti-inflammatory and analgesic activity, as evaluated by carrageenan-induced rat paw edema and phenylquinone-induced writhing tests. In addition, N-benzoyl and N-tosyl benzimidazole compounds **119** showed significant anti-inflammatory activity, as indicated by ear swelling induced by xylene in mice, and their ulcer indices were all lower than those of aspirin [108]. Furthermore, N-morpholinomethylbenzimidazole **120** and its derivatives have been recently reported to show significant anti-inflammatory activity [99].

Despite the success of several commercial benzimidazole proton pump inhibitors for the treatment of ulcer disease, work is still in progress to discover new benzimidazole-derived antiulcer drugs. Cinitapride (**121**)-related benzimidazole derivatives **122** have been prepared and studied for their antiulcerative activity [109]. In addition, 1,3-disubstituted 3,4-dihydropyrimido[1,6-a]benzimidazoles and 3-substituted 3,4-dihydropyrimido[1,6-a]benzimidazol-1(2*H*)-thiones exhibited good gastric antisecretory activity (> 50% inhibition) [110].

5.3 Cytotoxic and Antitumor Agents

In cancer chemotherapy there is currently much interest in the design of small molecules that bind to DNA with sequence selectivity and noncovalent interactions [37]. A possible lead for this new class of compounds is Hoechst 33258 **28** (Fig. 2), which recognizes A/T sequences in human DNA and is also an effective inhibitor of mammalian DNA topoisomerase [37]. Several structure–activity relationship studies have been performed on the Hoechst motif. The replacement of the terminal piperazine ring with an amidinium, an imidazoline or a tetrahydropyridinium group significantly reinforces the affinity of the drug for the A/T stretches [111]. The corresponding trisbenzimidazole derivative prepared by the addition of one more benzimidazole unit to the structure of Hoechst 33258 exhibits high A/T-base pair selectivity [112].

Novel bisbenzimidazoles with general formula **125–128** incorporating benzimidazole, pyridoimidazole, and imidazoquinone moieties as one of the units of bisbenzimidazole with a piperazinyl functional group have been synthesized (Fig. 11) [113]. The series of bisbenzimidazoles contains different leaving groups along with *p*-methoxy substituents. The latter may be expected to have some influence on the nitrogen lone pair and consequently on the binding characteristics of the ligand. These novel bisbenzimidazoles are found to be actively cytotoxic against many human cancer cell lines, with

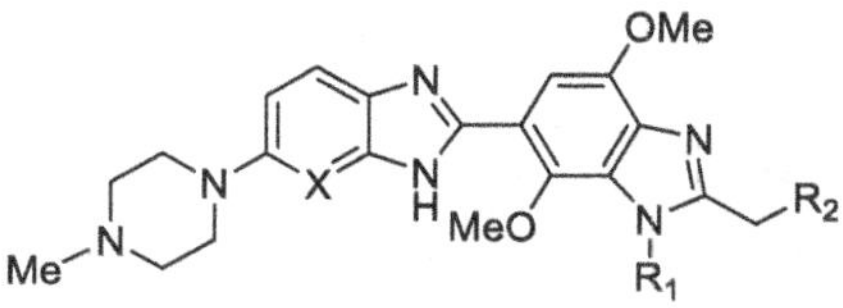

125 X = N, R_1 = H
126 X = CH, R_1 = H
127 X = N, R_1 = Me
R_2 = OMe, OEt, OAc, OH

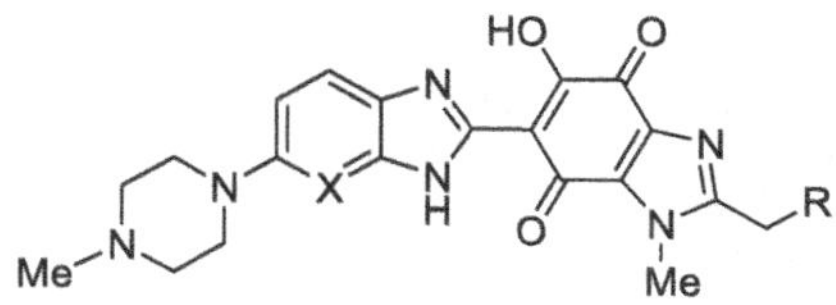

128 X = N, CH; R = OH, Cl

Fig. 11 Cytotoxic benzimidazole derivatives

GI_{50} values of between 0.01 and 100 μM, especially in the cases of renal cancer, CNS cancer, colon cancer, melanoma, and breast cancer cell lines. The pyridoimidazole compounds **125** and **127** are generally more potent. The derivative **128**, characterized by the presence of a *p*-quinone moiety, a characteristic feature found in the bioreductively activated alkylating agent mitomycin C, exhibits enhanced cytotoxic activity. This biological result suggests that the modification of the bisbenzimidazole structure by the incorporation of a quinone moiety might have significant potential for the development of bioreductive quinone-based drugs [37].

Furthermore, novel head-to-head bisbenzimidazole compound **31** binds with high affinity to the minor groove of double-stranded B-DNA with a strong preference for A/T-rich regions. The bisbenzimidazole **31** showed potent growth inhibition in human ovarian carcinoma cell lines (IC_{50} = 200–300 nM), with no significant cross-resistance in two acquired cisplatin-resistant cell lines and a low level of cross-resistance in the *p*-glycoprotein overexpressing doxorubicin-resistant cell line. In addition, compound **31** was found to have significant in vivo activity in the allowed fiber assay and tumor xenografts (CH1 cells) [37, 41, 114].

The bioactive benzimidazolequinone **131** has been synthesized by demethylation of the dimethoxybenzimidazole **129** followed by facile oxidation of the intermediate dihydroxy compound **130** by ferric chloride to yield the quinone **131** in excellent yield (Scheme 23). Synthesis of the related benzimidazolequinones **134** was achieved by dinitration of **132** followed by the reduction of **133** and oxidation as above. The benzimidazole-6,9-dione **134** has been found to be 300 times more cytotoxic towards the human skin fibroblast cell line in the MTT assay than the clinically used bioreductive drug, mitomycin C. Attaching methyl substituents onto the quinone moiety increased reductive potential and decreased cytotoxicity and selectivity towards hypoxia [67].

In addition, the alkyl-linked bisbenzimidazole **135** [115] and thiazolylbenzimidazole-4,7-diones **136** [116] exhibited cytotoxic activity against tumor cell lines (Fig. 12).

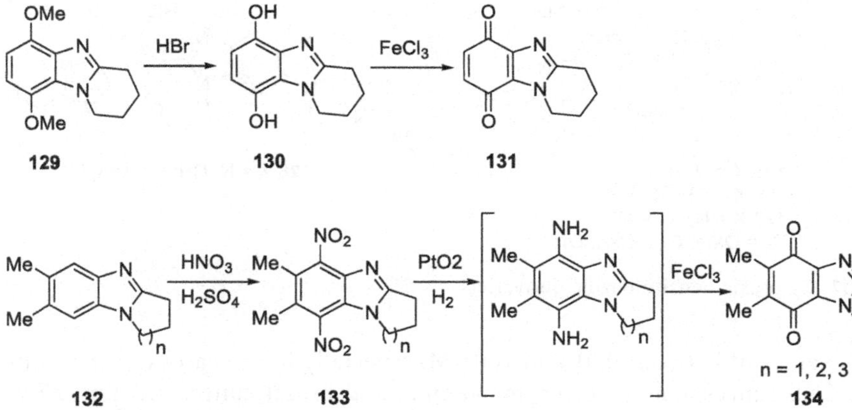

Scheme 23 Synthesis of benzimidazole diones

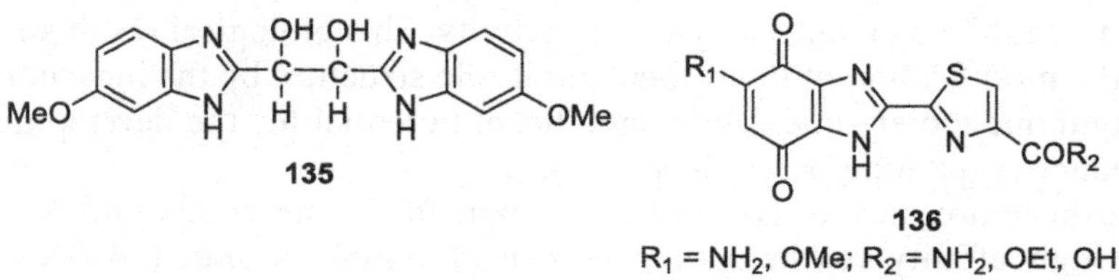

Fig. 12 Cytotoxic benzimidazole derivatives

5.4 Enzyme and Receptor Agonists/Antagonists

Several benzimidazole derivatives have been reported to act on various enzymes and receptors. Some examples of benzimidazoles acting as agonists or antagonists of various receptors and enzymes are listed in Table 2.

6 Conclusions

Conventional synthetic methods remain the mainstream routes for the synthesis of benzimidazoles. However, a few novel synthetic methodologies for the synthesis of benzimidazole have been reported in the time frame selected for this review. The popularity of microwave-assisted synthesis has been increasing rapidly since it enables the effective synthesis of benzimidazoles and its analogs. A wide variety of benzimidazole analogs have been synthesized, and several of these look promising for further drug discovery efforts.

Table 2 Benzimidazole derivatives that act on enzymes/receptors

Compound	Enzyme/receptor	Activity	Refs.
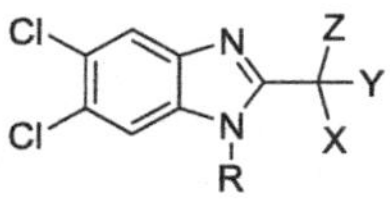 R = H, Me, Et X = H, OH Y = Me, Et, n-Pr Z = Me, Et	Androgen receptor	Antagonist	[54, 117, 118]
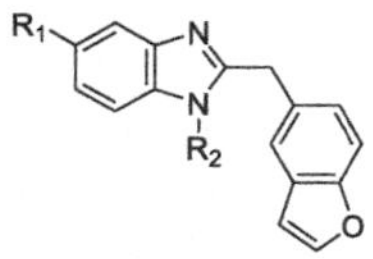 R_1 = Et_2NCO, t-BuCONMe R_2 = CH_2Ph, CH_2CH_2Ph	Cannabinoid 2 (CB2) receptor	Agonist	[119]
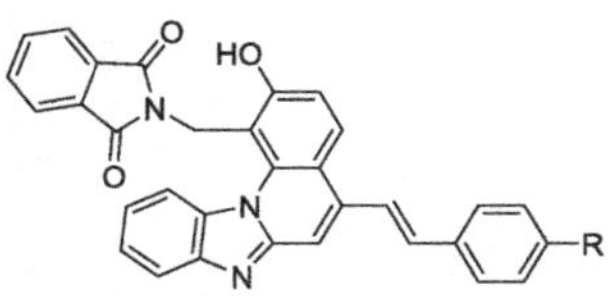 R = H, OH	Cholecystokinin B receptor	Antagonist	[120]
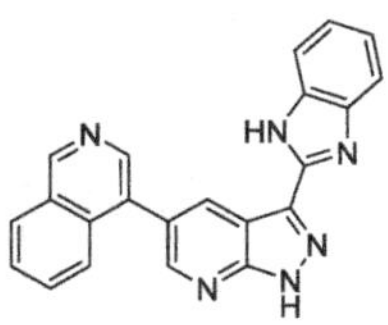	Cyclin-dependent kinase 1 (CDK1)	Inhibitory	[121]
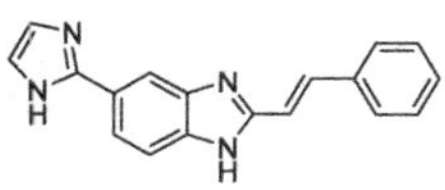	Enkephalinase B (DPP III)	Antagonist	[122]
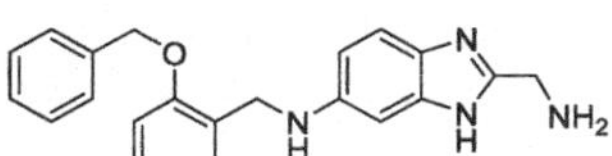	Gelatinase B	Inhibitory	[75]
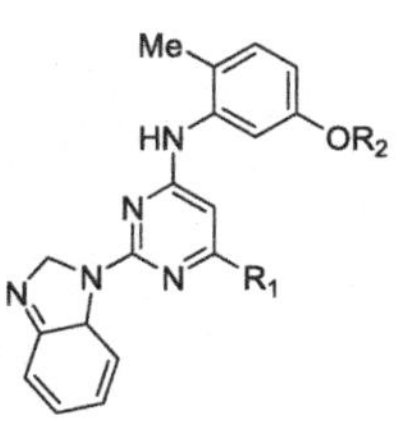 R_1: Cl, NH_2, $NHCH_3$, OMe, Etc R_2: H, Me	Lymphocyte specific kinase	Inhibitory	[123]

Table 2 (continued)

Compound	Enzyme/receptor	Activity	Refs.
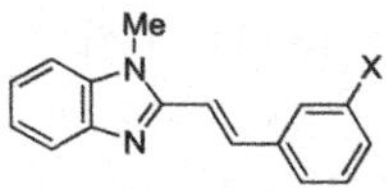 X = H, Cl, F	Monoamine oxidases B receptor	Antagonist	[124]
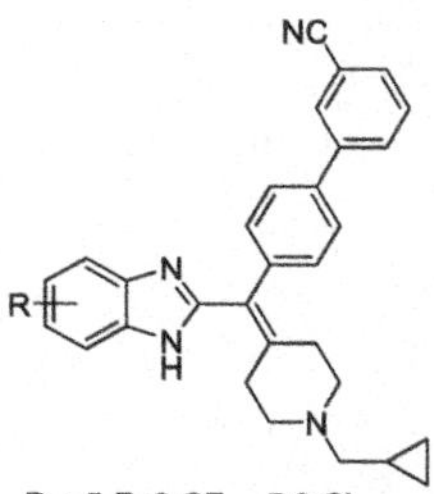 R = 5-F, 6-CF_3, 5,6-Cl_2	Melanin-concentrating hormone receptor 1 (MCH R1)	Antagonist	[125]
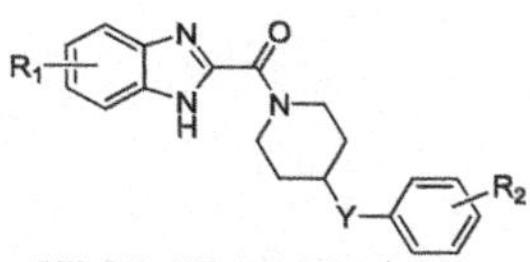 R_1 = 5(6)-OH, 4(7)-OH, 5(6)-NH_2 Y = O, CH_2 R_2 = H, 2-Me, 3-Me, 4-Me, 3-F, 4-F,4-Cl, 3-OMe	NMDA receptor	Antagonist	[126]
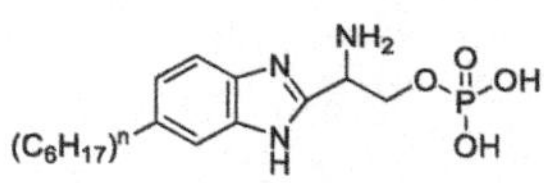	Sphingosine-1-phosphate receptor	Agonist	[127]
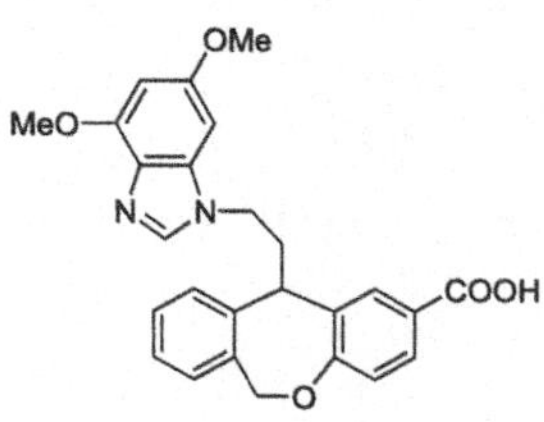	Thromboxane A_2 (TXA_2) receptor	Antagonist	[128]
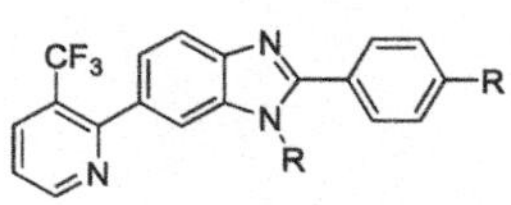 R = CF_3, *tert*-butyl, Me, F	Transient receptor potential vanilloid 1 (TPRV1)	Antagonist	[129]

Table 2 (continued)

Compound	Enzyme/receptor	Activity	Refs.
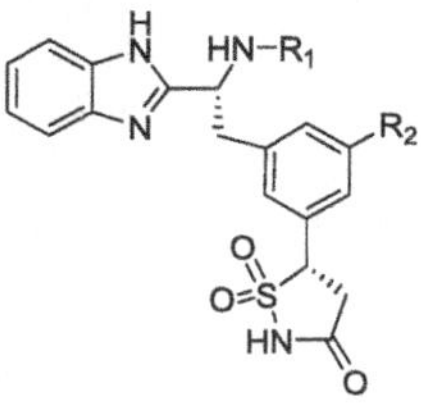 R_1 = 2-(benzothiazolyl), 2-(phenylsulfonyl) R_2 = H, Me, Cl, F, Br	Tyrosine phosphatase 1B	Inhibitory	[53, 130]

References

1. Elderfield RC (ed) (1957) Heterocyclic compounds, vol 5. Wiley, New York
2. Katritzky AR, Rees CW (eds) (1984) Comprehensive heterocyclic chemistry, vol 5. Pergamon, Oxford
3. Spasov AA, Yozhitsa IN, Bugaeva LI, Anisimova VA (1999) Pharm Chem J 33:232
4. Weber J, Antonietti M, Thomas A (2007) Macromolecules 40:1299
5. Soula C, Luu-Duc C (1986) Lyon Pharm 37:297
6. Rastogi R, Sharma S (1983) Synthesis 861
7. Jing X, Zhu Q, Xu F, Ren X, Li D, Yan C (2006) Synth Commun 36:2597
8. Tandon VK, Kumar M (2004) Tetrahedron Lett 45:4185
9. Bougrin K, Loupy A, Petit A, Daou B, Soufiaoui M (2001) Tetrahedron 57:163
10. VanVliet DS, Gillespie P, Scicinski JJ (2005) Tetrahedron Lett 46:6741
11. Hegedus A, Hell Z, Potor A (2006) Synth Commun 36:3625
12. Lin S-Y, Isome Y, Stewart E, Liu J-F, Yohannes D, Yu L (2006) Tetrahedron Lett 47:2883
13. Ibrahim MN (2007) Asian J Chem 19:2419
14. Nagawade RR, Shinde DB (2007) Indian J Chem B 46B:349
15. Sun P, Hu Z (2006) J Heterocycl Chem 43:773
16. Bahrami K, Khodaei MM, Kavianinia I (2007) Synthesis 547
17. Nagawade RR, Shinde DB (2006) Russian J Org Chem 42:453
18. Nagawade RR, Shinde DB (2006) Chin Chem Lett 17:453
19. Shen M-G, Cai C (2007) J Fluorine Chem 128:232
20. Wang L, Sheng J, Tian H, Qian C (2004) Synth Commun 34:4265
21. Peng Y, Chen Z, Liu Y, Mu Q, Chen S (2005) J Sichuan Univ 42:1054
22. Varala R, Nasreen A, Enugala R, Adapa SR (2006) Tetrahedron Lett 48:69
23. Brain CT, Brunton SA (2002) Tetrahedron Lett 43:1893
24. Brain CT, Steer JT (2003) J Org Chem 68:6814
25. Neochoritis C, Livadiotou D, Stephanidou-Stephanatou J, Tsoleridis CA (2007) Tetrahedron Lett 48:2275
26. Gedye R, Smith F, Westaway K, Ali H, Baldisera L, Laberge L, Rousell J (1986) Tetrahedron Lett 27:279
27. Kuhnert N (2002) Angew Chem Int Ed 41:1863

28. Bougrin K, Soufiaoui M (1995) Tetrahedron Lett 36:3683
29. Dubey R, Moorthy NSHN (2007) Chem Pharm Bull 55:115
30. Kumar BVS, Vaidya SD, Kumar RV, Bhirud SB, Mane RB (2006) Eur J Med Chem 41:599
31. Navarrete-Vazquez G, Moreno-Diaz H, Aguirre-Crespo F, Leon-Rivera I, Villalobos-Molina R, Munoz-Muniz O, Estrada-Soto S (2006) Bioorg Med Chem Lett 16:4169
32. Rao A, Chimirri A, Ferro S, Monforte AM, Monforte P, Zappala M (2004) ARKIVOC 147
33. Martinez-Palou R, Zepeda LG, Hoepfl H, Montoya A, Guzman-Lucero DJ, Guzman J (2005) Mol Divers 9:361
34. Su Y-S, Sun C-M (2005) Synlett, p 1243
35. Perumal S, Mariappan S, Selvaraj S (2004) ARKIVOC 46
36. Montazeri N, Rad-Moghadam K (2006) Asian J Chem 18:1557
37. Baraldi PG, Bovero A, Fruttarolo F, Preti D, Tabrizi MA, Pavani MG, Romagnoli R (2004) Med Res Rev 24:475
38. Gromyko AV, Streltsov SA, Zhuze AL (2004) Russian J Bioorg Chem 30:400
39. Tawar U, Jain AK, Dwarakanath BS, Chandra R, Singh Y, Chaudhury NK, Khaitan D, Tandon V (2003) J Med Chem 46:3785
40. Gravatt GL, Baguley BC, Wilson WR, Denny WA (1994) J Med Chem 37:4338
41. Mann J, Baron A, Opoku-Boahen Y, Johansson E, Parkinson G, Kelland LR, Neidle S (2001) J Med Chem 44:138
42. Sun X-W, Neidle S, Mann J (2002) Tetrahedron Lett 43:7239
43. Sun Y-W, Duan G-Y, Liu J-Z, Wang J-W (2006) Chin J Org Chem 26:942
44. Duan G-Y, Sun Y-W, Liu J-Z, Zhao G-L, Zhang D-T, Wang J-W (2006) J Chin Chem Soc 53:455
45. Wang L-Y, Han J, Li X-J, Shi Z (2005) Huaxue Shiji 27:317
46. Lin M-J, Sun C-M (2004) Synlett, p 663
47. Lewis JR (1992) Nat Prod Rep 9:81
48. He HY, Faulkner DJ, Lee AY, Clardy J (1992) J Org Chem 57:2176
49. Faulkner DJ (1992) Nat Prod Rep 9:323
50. Nakamura S, Tsuno N, Yamashita M, Kawasaki I, Ohta S, Ohishi Y (2001) J Chem Soc Perkin Trans 1:429
51. LaBarbera DV, Skibo EB (2004) Bioorg Med Chem 13:387
52. Shuto S, Horne G, Marwood RD, Potter BVL (2001) Chem Eur J 7:4937
53. Combs AP, Zhu W, Crawley ML, Glass B, Polam P, Sparks RB, Modi D, Takvorian A, McLaughlin E, Yue EW, Wasserman Z, Bower M, Wei M, Rupar M, Ala PJ, Reid BM, Ellis D, Gonneville L, Emm T, Taylor N, Yeleswaram S, Li Y, Wynn R, Burn TC, Hollis G, Liu PCC, Metcalf B (2006) J Med Chem 49:3774
54. Ng RA, Guan J, Alford VC, Lanter JC, Allan GF, Sbriscia T, Linton O, Lundeen SG, Sui Z (2007) Bioorg Med Chem Lett 17:784
55. Shieh W-C, Lozanov M, Repic O (2003) Tetrahedron Lett 44:6943
56. Bellina F, Calandri C, Cauteruccio S, Rossi R (2007) Tetrahedron 63:1970
57. Bellina F, Cauteruccio S, Rossi R (2006) Eur J Org Chem 1379
58. Lewis JC, Wu JY, Bergman RG, Ellman JA (2006) Angew Chem Int Ed 45:1589
59. Fouchard DMD, Tillekeratne LMV, Hudson RA (2004) J Org Chem 69:2626
60. Elwan NM (2004) Tetrahedron 60:1161
61. Dzvinchuk IB, Turov AV, Lozinskii MO (2006) Chem Heterocycl Comp 42:929
62. Dalla Via L, Gia O, Magno SM, Da Settimo A, Marini AM, Primofiore G, Da Settimo F, Salerno S (2001) Farmaco 56:159
63. Pandey VK, Raj N, Srivastava UK (1986) Acta Pharm Jugoslavica 36:281

64. Vostrova LN, Voronina TA, Karaseva TL, Gernega SA, Ivanov EI, Kirichenko AM, Totrova MY (1986) Khim-Farm Zh 20:690
65. Frere S, Thiery V, Bailly C, Besson T (2003) Tetrahedron 59:773
66. Abdel-Jalil RJ, Voelter W, Stoll R (2005) Tetrahedron Lett 46:1725
67. Lynch M, Hehir S, Kavanagh P, Leech D, O'Shaughnessy J, Carty MP, Aldabbagh F (2007) Chem Eur J 13:3218
68. Tan KL, Vasudevan A, Bergman RG, Ellman JA, Souers AJ (2003) Organic Lett 5:2131
69. Krchnak V, Holladay MW (2002) Chem Rev 102:61
70. Jung N, Encinas A, Braese S (2006) Current Opin Drug Disc Dev 9:713
71. Akamatsu H, Fukase K, Kusumoto S (2002) J Comb Chem 4:475
72. Yun YK, Porco JA, Jr., Labadie J (2002) Synlett, p 739
73. Yeh C-M, Tung C-L, Sun C-M (2000) J Comb Chem 2:341
74. Yeh CM, Tung CL, Sun CM (2001) J Comb Chem 3:229
75. Wang X, Choe Y, Craik CS, Ellman JA (2002) Bioorg Med Chem Lett 12:2201
76. Hoesl CE, Nefzi A, Houghten RA (2003) Tetrahedron Lett 44:3705
77. Yeh W-B, Lin M-J, Sun C-M (2004) Comb Chem High T Scr 7:251
78. Huang K-T, Sun C-M (2002) Bioorg Med Chem Lett 12:1001
79. Tong Y-P, Zheng S-L, Chen X-M (2007) J Mol Struct 826:104
80. Pal S, Hwang W-S, Lin IJB, Lee C-S (2007) J Mol Catal A 269:197
81. Hao P, Zhang S, Sun W-H, Shi Q, Adewuyi S, Lu X, Li P (2007) Organometallics 26:2439
82. Arjmand F, Mohani B, Ahmad S (2005) Eur J Med Chem 40:1103
83. Gudasi KB, Shenoy RV, Vadavi RS, Patil MS, Patil SA, Hanchinal RR, Desai SA, Lohithaswa H (2006) Bioinorg Chem Appl 9:1
84. Zinner SH (2005) Exp Rev Anti-Infective Ther 3:907
85. Guru N, Srivastava SD (2001) J Sci Ind Res 60:601
86. Desai KG, Desai KR (2006) Bioorg Med Chem 14:8271
87. Guven OO, Erdogan T, Goeker H, Yildiz S (2007) Bioorg Med Chem Lett 17:2233
88. Kus C, Goker H, Altanlar N (2001) Arch Pharmazie 334:361
89. Nyati M, Rao NS, Shrivastav YK, Verma BL (2006) Indian J Heterocycl Chem 15:295
90. Bhatt AK, Shah PR, Karadla H, Patel HD (2004) Orient J Chem 20:385
91. Bhatt AK, Shah PR, Karadiya HG, Patel HD (2003) Orient J Chem 19:643
92. Bhatt AK, Karadiya H, Shah PR, Parmar MP, Patel HD (2003) Indian J Heterocycl Chem 13:187
93. Bhatt AK, Karadiya HG, Shah PR, Parmar MP, Patel HD (2003) Orient J Chem 19:493
94. Ayhan KG, Altanlar N (2006) Turk J Chem 30:223
95. Ayhan KG, Altanlar N (2003) Farmaco 58:1345
96. Kus C, Altanlar N (2003) Turk J Chem 27:35
97. Goker H, Kus C, Boykin DW, Yildiz S, Altanlar N (2002) Bioorg Med Chem 10:2589
98. Ates-Alagoz Z, Yildiz S, Buyukbingol E (2007) Chemotherapy 53:110
99. Leonard JT, Rajesh OS, Jeyaseeli L, Murugesh K, Sivakumar R, Gunasekaran V (2007) Asian J Chem 19:116
100. Jabbar A, Iqbal Z, Kerboeuf D, Muhammad G, Khan MN, Afaq M (2006) Life Sci 79:2413
101. Anichina KK, Vuchev DI, Mavrova AT (2006) Probl Infect Parasit Dis 34:19
102. Mavrova AT, Anichina KK, Vuchev DI, Tsenov JA, Denkova PS, Kondeva MS, Micheva MK (2006) Eur J Med Chem 41:1412
103. Gaur NM, Patil SV, Mourya VK, Wagh SB (2000) Indian J Heterocycl Chem 9:227
104. Barreca ML, Chimirri A, De Clercq E, De Luca L, Monforte A-M, Monforte P, Rao A, Zappala M (2003) Il Farmaco 58:259

105. Labanauskas LK, Brukstus AB, Gaidelis PG, Buchinskaite VA, Udrenaite EB, Dauksas VK (2000) Pharm Chem J 34:353
106. Sondhi SM, Rajvanshi S, Johar M, Bharti N, Azam A, Singh AK (2002) Eur J Med Chem 37:835
107. Labanauskas L, Brukstus A, Udrenaite E, Gaidelis P, Bucinskaite V, Dauksas V (2000) Chemija 11:211
108. Puratchikody A, Sivajothi V, Jaswanth A, Ruckmani K, Nallu M (2002) Indian J Heterocycl Chem 11:241
109. Srinivasulu G, Reddy PP, Hegde P, Chakrabart R (2005) Heterocycl Commun 11:23
110. Shafik RM, El-Din SAS, Eshba NH, El-Hawash SAM, Desheesh MA, Abdel-Aty AS, Ashour HM (2004) Pharmazie 59:899
111. Reddy BSP, Sharma SK, Lown JW (2001) Curr Med Chem 8:475
112. Ji Y-H, Bur D, Hasler W, Schmitt VR, Dorn A, Bailly C, Waring MJ, Hochstrasser R, Leupin W (2001) Bioorg Med Chem 9:2905
113. Singh AK, Lown JW (2000) Anticancer Drug Des 15:265
114. Joubert A, Sun X-W, Johansson E, Bailly C, Mann J, Neidle S (2003) Biochemistry 42:5984
115. Kubota Y, Fujii H, Liu J, Tani S (2001) Nucleic Acids Res Suppl 101
116. Garuti L, Roberti M, Pession A, Leoncini E, Hrelia S (2001) Bioorg Med Chem Lett 11:3147
117. Ng RA, Guan J, Alford VC, Lanter JC, Allan GF, Sbriscia T, Lundeen SG, Sui Z (2007) Bioorg Med Chem Lett 17:955
118. Ng RA, Lanter JC, Alford VC, Allan GF, Sbriscia T, Lundeen SG, Sui Z (2007) Bioorg Med Chem Lett 17:1784
119. Page D, Brochu MC, Yang H, Brown W, St-Onge S, Martin E, Salois D (2006) Lett Drug Des Discov 3:298
120. Tiwari AK, Mishra AK, Bajpai A, Mishra P, Singh S, Sinha D, Singh VK (2007) Bioorg Med Chem Lett 17:2749
121. Huang S, Lin R, Yu Y, Lu Y, Connolly PJ, Chiu G, Li S, Emanuel SL, Middleton SA (2007) Bioorg Med Chem Lett 17:1243
122. Agic D, Hranjec M, Jajcanin N, Starcevic K, Karminski-Zamola G, Abramic M (2007) Bioorg Chem 35:153
123. Sabat M, VanRens JC, Laufersweiler MJ, Brugel TA, Maier J, Golebiowski A, De B, Easwaran V, Hsieh LC, Walter RL, Mekel MJ, Evdokimov A, Janusz MJ (2006) Bioorg Med Chem Lett 16:5973
124. Petzer JP, Steyn S, Castagnoli KP, Chen J-F, Schwarzschild MA, Van der Schyf CJ, Castagnoli N (2003) Bioorg Med Chem 11:1299
125. Wu W-L, Burnett DA, Caplen MA, Domalski MS, Bennett C, Greenlee WJ, Hawes BE, O'Neill K, Weig B, Weston D, Spar B, Kowalski T (2006) Bioorg Med Chem Lett 16:3674
126. Borza I, Kolok S, Gere A, Nagy J, Fodor L, Galgoczy K, Fetter J, Bertha F, Agai B, Horvath C, Farkas S, Domany G (2006) Bioorg Med Chem Lett 16:4638
127. Clemens JJ, Davis MD, Lynch KR, Macdonald TL (2004) Bioorg Med Chem Lett 14:4903
128. Ohshima E, Takami H, Sato H, Mohri S, Obase H, Miki I, Ishii A, Shirakura S, Karasawa A, Kubo K (1992) J Med Chem 35:3402
129. Fletcher SR, McIver E, Lewis S, Burkamp F, Leech C, Mason G, Boyce S, Morrison D, Richards G, Sutton K, Jones AB (2006) Bioorg Med Chem Lett 16:2872
130. Sparks RB, Polam P, Zhu W, Crawley ML, Takvorian A, McLaughlin E, Wei M, Ala PJ, Gonneville L, Taylor N, Li Y, Wynn R, Burn TC, Liu PCC, Combs AP (2007) Bioorg Med Chem Lett 17:763

Top Heterocycl Chem (2007) 9: 119–138
DOI 10.1007/7081_2007_077

Published online: 27 June 2007

Heterocyclic Compounds Against the Enzyme Tyrosinase Essential for Melanin Production: Biochemical Features of Inhibition

Mahmud Tareq Hassan Khan

PhD School of Molecular and Structural Biology, and Department of Pharmacology, Institute of Medical Biology, Faculty of Medicine,
University of Tromsø, Tromsø, Norway
mahmud.khan@fagmed.uit.no

Abstract Tyrosinase is a copper-containing bifunctional metalloenzyme, widely distributed around the phylogeny. This enzyme is involved in the production of melanin and some other pigments in humans, animals, etc. Abnormal accumulation of melanin, which is due to the overexpression of the enzyme, is called hyperpigmentation and underexpression is called vitiligo, which is a major skin problem around the world. The inhibitors of this enzyme have been utilized in cosmetics, especially as depigmenting agents in the case of hyperpigmentation. They are also involved in several other disease conditions. In the last few decades a large number of tyrosinase inhibitors have been discovered and reported by several groups including ours. This chapter principally emphasizes the discovery of some interesting inhibitors, mainly of heterocyclic origin, and their impacts on drug discovery; some of the inhibitors might not be heterocyclic but their chemistry is quite interesting in terms of the inhibition.

Keywords Tyrosinase inhibitors · *Agaricus bisporus* · Polyphenol oxidase · Hyperpigmentation · Vitiligo · Melanogenesis · Alkaloid

Abbreviations

CO Catechol oxidases
EC Enzyme Commission
PO Phenol oxidases
PPO Polyphenol oxidases

1 Introduction

Tyrosinase or polyphenol oxidase (EC 1.14.18.1) is a bifunctional, copper-containing enzyme widely distributed on the phylogenetic tree. This enzyme uses molecular oxygen to catalyze the oxidation of monophenols to their corresponding *o*-diphenols (cresolase activity) as well as their subsequent oxidation to *o*-quinones (catecholase activity). The *o*-quinones thus generated polymerize to form melanin, through a series of subsequent enzymatic and nonenzymatic reactions [1–3].

This enzyme is involved in many biological processes, such as defense, mimetism, protection from UV light, hardening of cell walls in fungi or exoskeleton in Arthropods, and in general the production of melanins [4]. This process is involved in abnormal accumulation of melanin pigments (hyperpigmentation, melasma, freckles, ephelide, senile lentigines, etc.). Therefore, tyrosinase inhibitors have been established as important constituents of cosmetic materials, as well as depigmenting agents for hyperpigmentation [5].

Tyrosinase may also play an important role in neuromelanin formation in the human brain, particularly in the substantia nigra, and could be central to dopamine neurotoxicity as well as contributing to the neurodegeneration associated with Parkinson's disease [6]. Melanoma-specific anticarcinogenic activity is known to be linked with tyrosinase activity [7].

Tyrosinases are often referred to as phenolases, phenol oxidases (PO), polyphenol oxidases (PPO), or catechol oxidases (CO), etc., depending on the particular source and also on the authors who have described any particular enzyme. The term tyrosinase is usually adopted for the animal and human enzymes, and refers to the "typical" substrate, tyrosine. PPO is perhaps the most suitable general definition [8]. The enzyme extracted from the edible (champignon) mushroom *Agaricus bisporus* is usually referred to as tyrosinase, and its high homology with the mammalian ones renders it well suited as a model for studies on melanogenesis. This enzyme has been thoroughly characterized [8]. Nowadays, mushroom tyrosinase has become popular because it is readily available and useful in a number of applications [9].

Additionally, the enzyme is also responsible for the detrimental enzymatic browning of fruits and vegetables [10] that takes place during senescence or damage at the time of postharvest handling, which makes the identification of novel tyrosinase inhibitors extremely important. Nevertheless, besides this role in undesired browning, the activity of tyrosinase is needed in other cases (raisins, cocoa, fermented tea leaves) where it produces distinct organoleptic properties [9].

Tyrosinase oxidizes phenol in two steps [11]:

- Phenol is oxidized to catechol (*o*-benzenediol).
- This catechol is subsequently oxidized (by tyrosinase) to *o*-quinone.

Tyrosinase shows no activity for the oxidation of *p*- and *m*-benzenediols. Laccase, which catalyzes the oxidation of *o*-, *m*-, and *p*-benzenediols to the corresponding *o*-, *m*-, and *p*-quinones, is used for the detection of these benzenediols. Thus, coimmobilization of tyrosinase and laccase allows the detection of several phenolic compounds [11]. Several authors have presented a large number of research and review papers on the structural and kinetic aspects of the enzyme tyrosinase [12–14].

2 Biosynthesis of Melanin

The biosynthetic pathway for melanin formation, operating in insects, animals, and plants, has largely been elucidated by Raper [15], Mason [16], and Lerner et al. [17]. The first two steps in the pathway are the hydroxylation of monophenol to *o*-diphenol (monophenolase or cresolase activity) and the oxidation of diphenol to *o*-quinones (diphenolase or catecholase activity), both using molecular oxygen followed by a series of nonenzymatic steps resulting in the formation of melanin [15, 18, 19]. The whole pathway for melanin biosynthesis is shown in Scheme 1.

3 Tyrosinase Inhibitors

Tyrosinase inhibition may be a potential approach to prevent and control the enzymatic browning reactions and improve the quality and nutritional value of food products [20]. Tyrosinase also plays a major key role in the developmental and defensive functions of insects. Tyrosinase is involved in melanogenesis, wound healing, parasite encapsulation, and sclerotization in insects [21–23]. For these reasons, in recent years the development of tyrosinase inhibitors has become an active alternative approach to control insect pests [20]. Additionally, it is now well-recognized that tyrosinase inhibitors are important for their potential applications in medical and cosmetic products [24–26].

Furthermore, the inhibitors may be clinically used for the treatment of some skin disorders associated with melanin hyperpigmentation and are also important in cosmetics for skin whitening effects [27–42], so there is a need

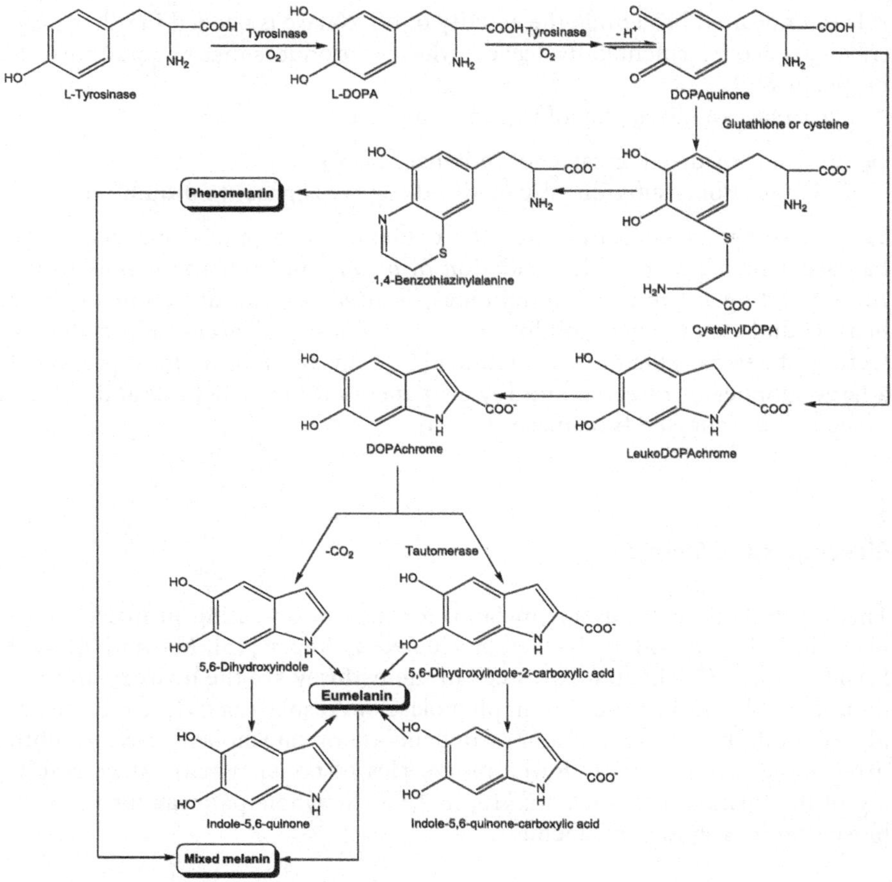

Scheme 1 Melanin biosynthetic pathway [15, 18, 19]

to identify the compounds that inhibit mushroom tyrosinase activity [9]. The molecular structures of some mushroom tyrosinase inhibitors are shown in Fig. 1.

In the next sections some of the promising classes of tyrosinase inhibitors are discussed with especial emphasis on heterocyclic origin; a few of them may not be heterocyclic but their inhibition pattern and chemistry is highly interesting.

3.1 Xanthates

Very recently Saboury et al. (2007) reported the tyrosinase inhibitory potentials of four sodium salts of *N*-alkyl xanthates [43] (see Fig. 2). The xanthates

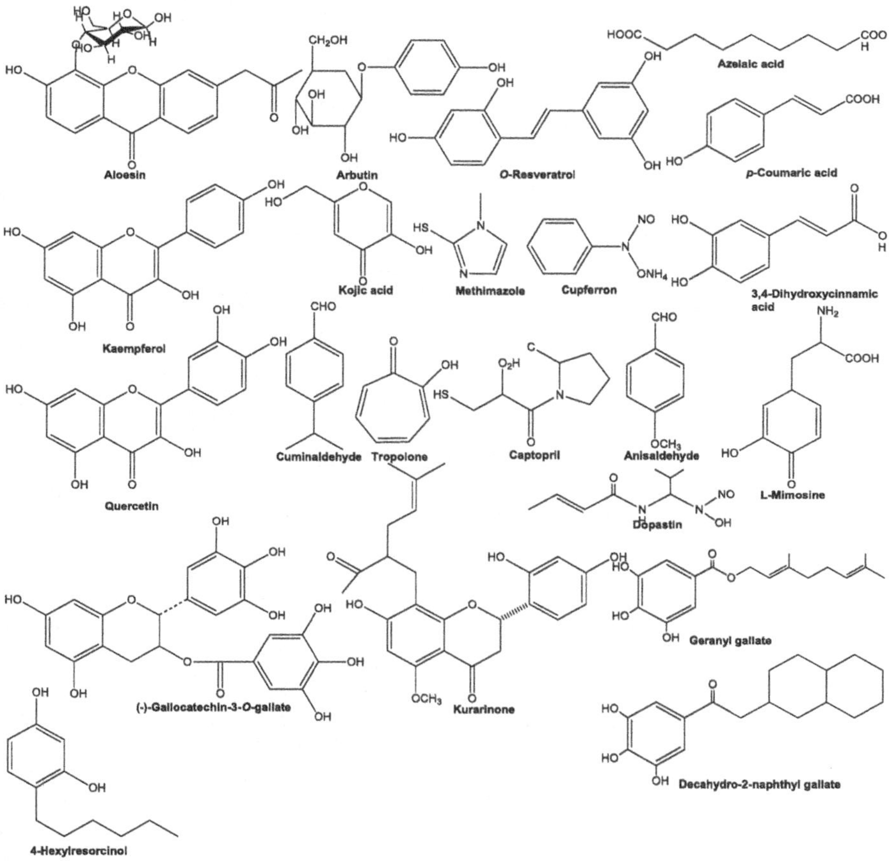

Fig. 1 Molecular structures of some mushroom tyrosinase inhibitors

have been synthesized and examined for their inhibition of both cresolase and catecholase activities against mushroom tyrosinase, taking 4-[(4-methylbenzo)azo]-1,2-benzenediol (MeBACat) and 4-[(4-methylphenyl)azo]-phenol (MePAPh) as substrates (for structures see Fig. 3) [43].

By using Lineweaver–Burk plots the authors found that four xanthates exhibited different patterns of mixed, competitive, or uncompetitive inhibition. For the cresolase activity, **1** and **2** demonstrated uncompetitive inhibition but **3** and **4** exhibited competitive inhibition [43]. For the catecholase activity, **1** and **2** showed mixed inhibition but **3** and **4** showed competitive inhibition against tyrosinase [43]. The xanthates (compounds **1**, **2**, **3** and **4**) have been classified as potent inhibitors against tyrosinase due to their K_i values of 13.8, 11.0, 8.0, and 5.0 μM, respectively, for the cresolase activities, and 1.4, 5.0, 13.0, and 25.0 μM, respectively, for the catecholase activities [43]. The authors concluded that, for the catecholase activity, both substrate and inhibitor can

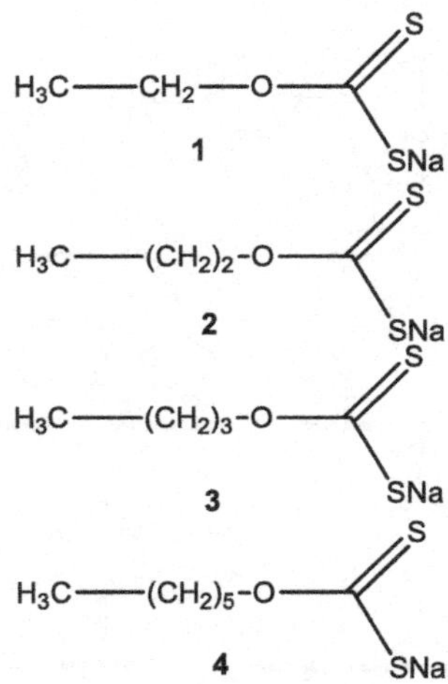

Fig. 2 Molecular structures of four *N*-alkyl xanthates reported by Saboury et al. [43]

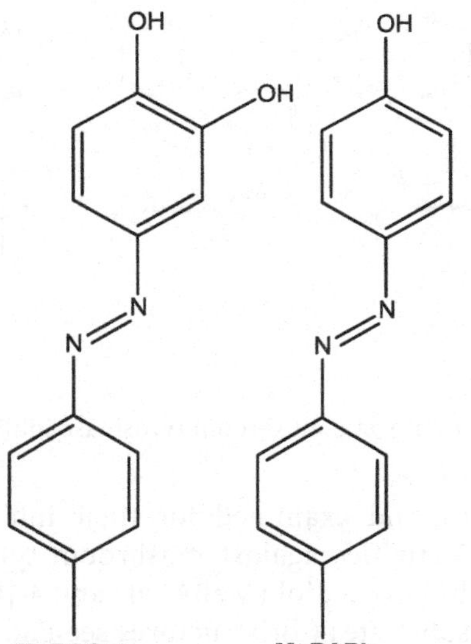

Fig. 3 Structures of 4-[(4-methylbenzo)azo]-1,2-benzenediol (MeBACat) and 4-[(4-methylphenyl)azo]-phenol (MePAPh) used by Saboury et al. as synthetic substrate in their studies against mushroom tyrosinase [43]

be bound to the enzyme with negative cooperativity between the binding sites, and this negative cooperativity increases with increasing length of the aliphatic tail of these compounds. The length of the hydrophobic tail of the xanthates has a stronger effect on the K_i values for catecholase inhibition than for cresolase inhibition. Increasing the length of the hydrophobic tail leads to a decrease of the K_i values for cresolase inhibition and an increase of the K_i values for catecholase inhibition [43].

3.2
N,N-Unsubstituted Selenourea Derivatives

Recently, Ha et al. (2005) reported the inhibitory potentials of *N,N*-unsubstituted selenourea derivatives **5–8** on tyrosinase. Three types of *N,N*-unsubstituted selenourea derivatives exhibited an inhibitory effect on the DOPAoxidase activity of mushroom tyrosinase. For the structures of these selenourea derivatives (**5–8**), see Fig. 4. Compound **8** exhibited 55.5% inhibition at a concentration of 200 μM (IC_{50} = 170 μM). This inhibitory effect was higher than that of reference compound kojic acid (39.4%, for structure see Fig. 1) [44]. Interestingly, this compound (**8**) was identified as a noncompetitive inhibitor by Lineweaver–Burk plot analysis. In addition, **8** also inhibited melanin production in melan-a cells [44].

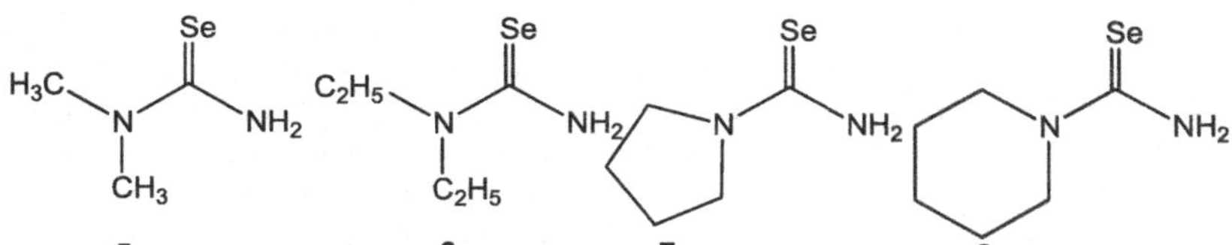

5 **6** **7** **8**

Fig. 4 Molecular structures of *N,N*-unsubstituted selenourea derivatives exhibiting tyrosinase inhibitory activities [44]

3.3
Selenium-Containing Carbohydrates

Ahn and coworkers recently (2006) reported the potency of selenium-containing carbohydrates on depigmentation, based on the direct inhibition of mushroom tyrosinase [45]. The structures of these selenium-containing compounds are shown in Fig. 5. Among them two selenoglycosides, **11** (bis(2,3,4-tri-*O*-acetyl-*β*-D-arabinopyranosyl) selenide) and **16** (4′-methylbenzoyl 2,3,4,6-tetra-*O*-acetyl-D-selenomanopyranoside), have been found to be effective depigmenting compounds against mushroom tyrosinase [45]. In enzyme kinetic studies, compound **11** exhibited a competitive inhibition effect which was found to be similar to that of kojic acid [45]. At 100 and 150 μM concentration, **16** exhibited an uncompetitive inhibition pattern [45].

The authors also performed studies of the same compounds in melan-a cell-originated tyrosinase inhibition assays, which showed that **16** was a less potent inhibitor than the kojic acid [45]. Compound **11** showed a similar kind of inhibitory effect as kojic acid in the melan-a cell-originated tyrosinase inhibitory assay [45].

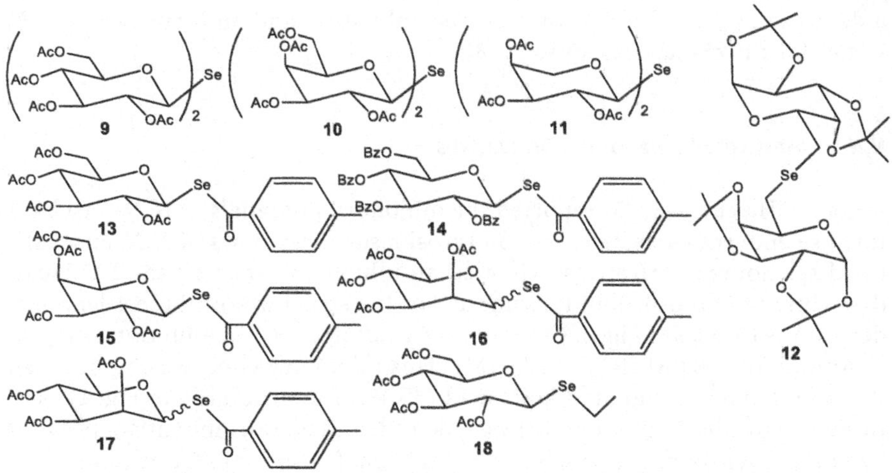

Fig. 5 Structures of the selenium-containing carbohydrates: **9**, bis(2,3,4,6-tetra-*O*-acetyl-*β*-D-glucopyranosyl) selenide; **10**, bis(2,3,4,6-tetra-*O*-acetyl-*β*-D-galactopyranosyl) selenide; **11**, bis(2,3,4-tri-*O*-acetyl-*β*-D-arabinopyranosyl) selenide; **12**, bis(1,2 : 3,4-di-*O*-isopropylidene-6-deoxy-*β*-D-galactopyranosyl) selenide; **13**, 4′-methylbenzoyl 2,3,4,6-tetra-*O*-acetyl-D-selenoglucopyranoside; **14**, 4′-methylbenzoyl 2,3,4,6-tetra-*O*-benzoyl-D-selenoglucopyranoside; **15**, 4′-methylbenzoyl 2,3,4,6-tetra-*O*-acetyl-*β*-D-selenogalactopyranoside; **16**, 4′-methylbenzoyl 2,3,4,6-tetra-*O*-acetyl-D-selenomanopyranoside; **17**, 4′-methylbenzoyl 2,3,4,6-tetra-*O*-acetyl-D-selenorhamnopyranoside; **18**, ethyl 2,3,4,6-tetra-*O*-acetyl-*β*-D-selenoglucopyranoside [45]

Compound **16** showed dose-dependent cytotoxicity in a study of inhibition of melanin synthesis by melan-a cell lines. The cellular survival rate was found to be low after treatment with 20 μM of **16** [45]. Compound **11** inhibited melanin synthesis in the melan-a cells at a concentration of 10 μM. The inhibition of melanin synthesis by **11** was found to be similar to that of phenylthiourea, which is a well-known melanin synthesis inhibitor [45]. The authors concluded that compound **11** is a new candidate for the development of depigmenting agents [45].

3.4
1,3-Selenazol-4-one Derivatives

Koketsu et al. (2002) reported the DOPAoxidase activities of the 1,3-selenazol-4-one derivatives **19–24** against mushroom tyrosinase [46]. All of these compounds exhibited 33.4–62.1% inhibition of DOPAoxidase activity at a concentration of 500 μM. Their inhibitory effects were higher than that of kojic acid (31.7%) [46]. 2-(4-Methylphenyl)-1,3-selenazol-4-one (**19**) exhibited the most potent inhibitory effect among them in a dose-dependent manner. Enzyme kinetic studies showed that compound **19** showed competitive in-

Table 1 The structure–activity relationships of the 1,3-selenazol-4-one derivatives **19–24** against mushroom tyrosinase [46]

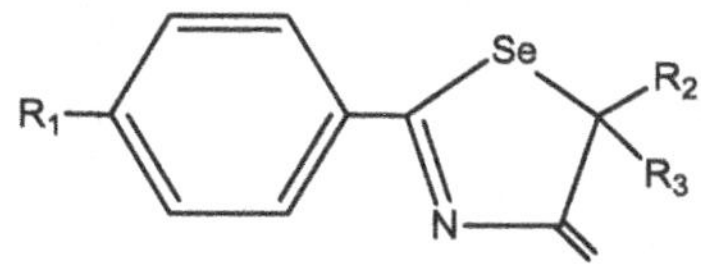

Compound	Substitutions R1	R2	R3	IC_{50} (in μM)
19	$-CH_3$	$-H$	$-H$	333.2
20	$-CH_3$	$-CH_2CH_3$	$-H$	384.3
21	$-CH_3$	$-CH_3$	$-CH_3$	> 500
22	$-H$	$-H$	$-H$	478.1
23	$-Cl$	$-H$	$-H$	498.0
24	$-OCH_3$	$-H$	$-H$	> 500

hibition against tyrosinase [46]. The structure–activity relationships of the 1,3-selenazol-4-one derivatives **19–24** are shown in Table 1.

3.5 Oxadiazole Derivatives

Khan et al. (2005) performed and reported tyrosinase inhibition studies of a combinatorial library of 2,5-disubstituted-1,3,4-oxadiazoles (**25–43**) [47]. The library of oxadiazoles was synthesized under microwave irradiation [47]. The synthetic steps involved for these compounds are shown in Scheme 2. Among the compounds from the library, **29** (30-[5-(40-bromophenyl)-1,3,4-oxadiazol-2-yl]pyridine, for structure see Fig. 6) exhibited the most potent (IC_{50} = 2.18 μM) inhibition against tyrosinase, which has found to be more potent than the standard potent inhibitor L-mimosine (IC_{50} = 3.68 μM, for structure see Fig. 1) [47].

Table 2 shows the structure–activity relationships of the compound library of the 2,5-disubstituted 1,3,4-oxadiazoles **25–43** against the enzyme

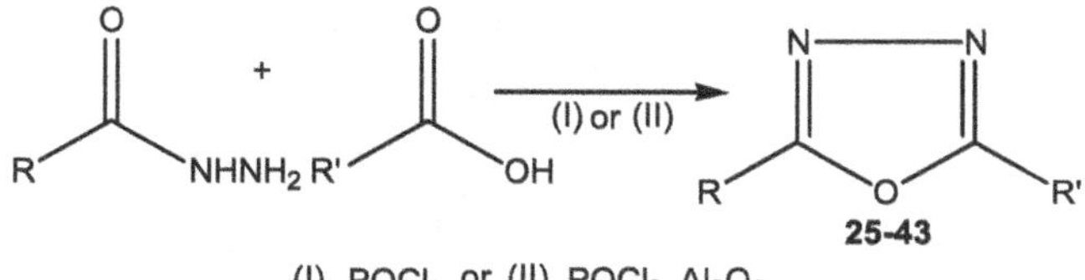

Scheme 2 The steps involved in the synthetic process of the library of 2,5-disubstituted 1,3,4-oxadiazoles **25–43** [47]

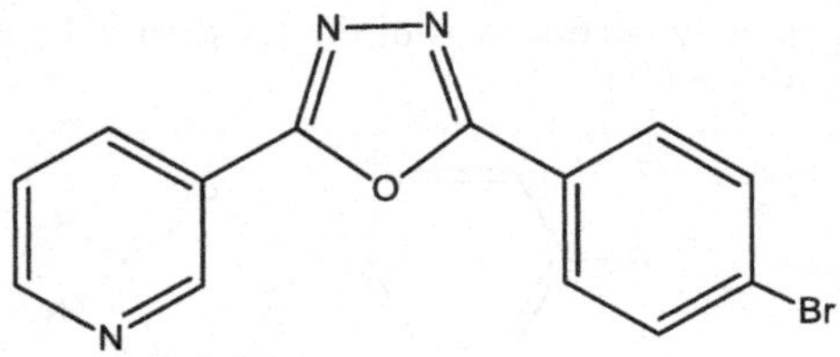

Fig. 6 Structural features of compound **29** (30-[5-(40-bromophenyl)-1,3,4-oxadiazol-2-yl]pyridine) [47]

Table 2 The structure–activity relationships of the compound library of 2,5-disubstituted 1,3,4-oxadiazoles (**25–43**) [47]

Compound	R′	Structure of the compound	IC_{50} (in μM)
25	C_6H_5	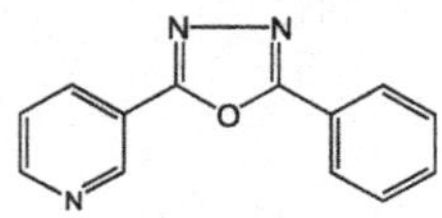	5.15
26	*o*-$NO_2C_6H_4$	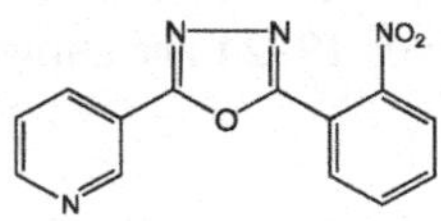	3.18
27	*o*-BrC_6H_4	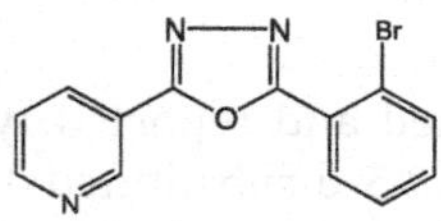	5.23
28	*m*-BrC_6H_4	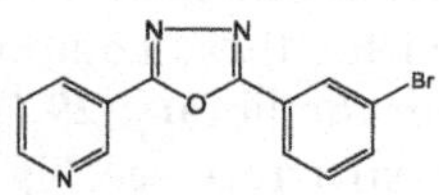	6.04
29	*p*-BrC_6H_4	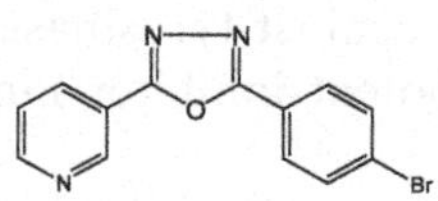	2.18
30	3-Pyridinyl	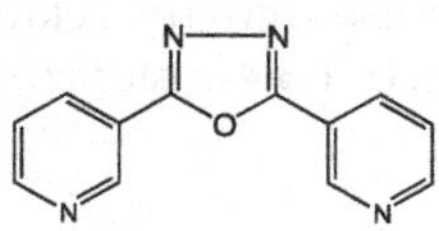	3.29
31	CH_2Cl	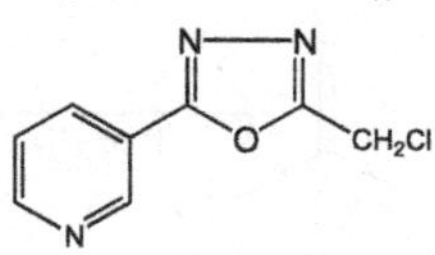	4.18
32	$CHCl_2$	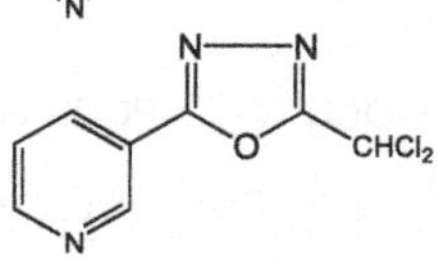	4.01

Table 2 (continued)

Compound	R′	Structure of the compound	IC_{50} (in μM)
33	CCl_3	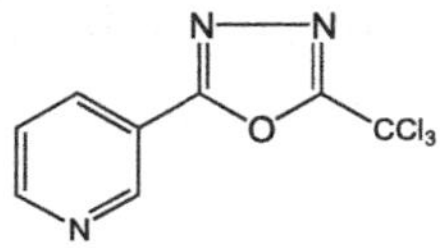	3.98
34	p-$CH_3C_6H_4$	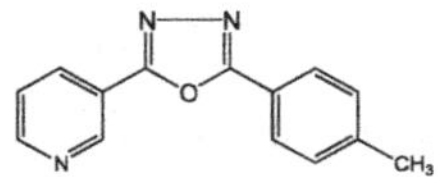	10.40
35	1-$C_{10}H_7$	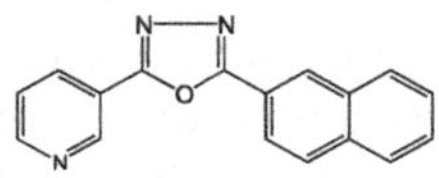	3.23
36	C_6H_5	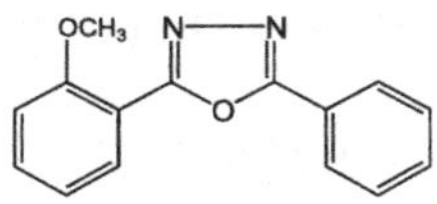	8.71
37	o-BrC_6H_4	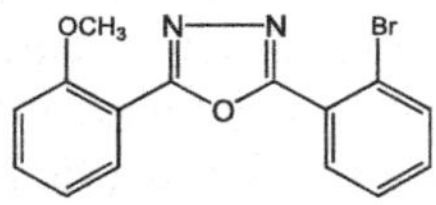	5.16
38	m-BrC_6H_4	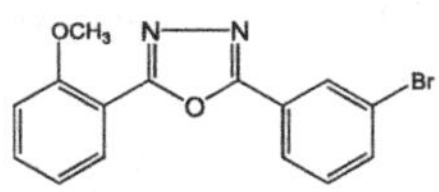	7.18
39	p-BrC_6H_4	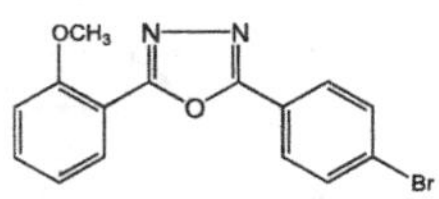	7.82
40	$CHCl_2$	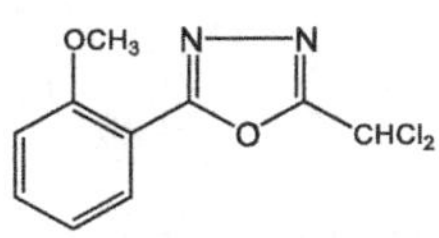	7.28
41	CCl_3	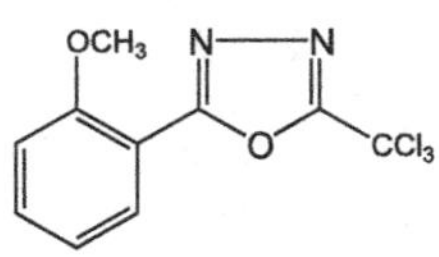	6.21
42	p-$CH_3C_6H_4$	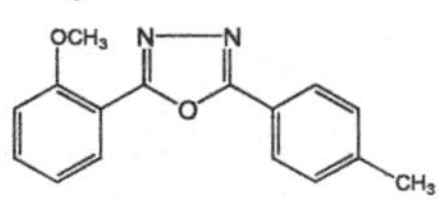	6.43
43	$2C_{10}H_7$	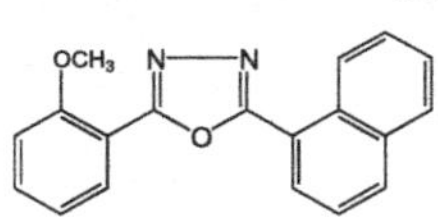	7.81

tyrosinase. The authors deduced that for a better inhibition, electronegative substitution is essential, as most probably the active site(s) of the enzyme contains some hydrophobic site and the position of the substitution also plays a very important role in inhibition, maybe due to the conformational space. The electronegativities of the compounds have been found to be proportional to the inhibitory activity [47].

3.6 Diterpenoid Alkaloids

Shaheen et al. (2005) reported lycoctonine-type norditerpenoid alkaloids isolated from the aerial parts of *Aconitum laeve* Royle, swatinine, delphatine, lappaconitine, puberanine, and *N*-acetylsepaconitine [48]. They performed and reported the anti-inflammatory, antioxidant, and tyrosinase inhibition studies of all these compounds, in which lappaconitine (IC_{50} = 93.33 μM) and puberanine (IC_{50} = 205.21 μM) were found to be active against the enzyme tyrosinase [48].

In another report Sultankhodzhaev et al. discussed the tyrosinase inhibitory potentials and structure–activity relationships of 15 diterpenoid alkaloids with the lycoctonine skeleton, and their semisynthetic derivatives [49]. At least three of them, lappaconitine hydrobromide (**44**, IC_{50} = 13.3 μM), methyllycaconitine perchlorate (**45**, IC_{50} = 477.84 μM), and aconine (**46**, IC_{50} = 220.7 μM), were found to be active against tyrosinase [49]. Their structures are shown in Fig. 7.

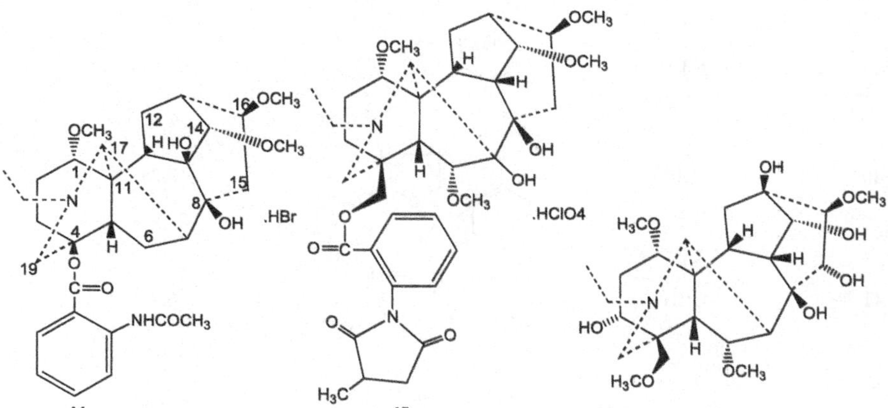

44 **45** **46**

Fig. 7 Structures of the tyrosinase inhibitory diterpenoid alkaloids: lappaconitine hydrobromide (**44**), methyllycaconitine perchlorate (**45**), and aconine (**46**) [49]

3.7 Napelline-Type Alkaloids

Sultankhodzhaev et al. reported the tyrosinase inhibitory pattern of six napelline-type alkaloids [49], and only two compounds exhibited some inhibition against tyrosinase. Napelline (**47**) exhibited moderate inhibition ($IC_{50} = 167.66\ \mu M$) of the enzyme. The presence of a benzoyloxy group at C-1, as in compound **48** (1-*O*-benzoylnapelline), maybe potentiated the inhibitory activity ($IC_{50} = 33.10\ \mu M$) against tyrosinase [49]. It is interesting to note that free hydroxyl groups at C-12 and C-15 were crucial for better potency. Acylation at C-12 and C-15 of other compounds (not shown here), or replacement of the hydroxy function (not shown here), results in total inactivation against tyrosinase [49]. The structures of the compounds are shown Fig. 8.

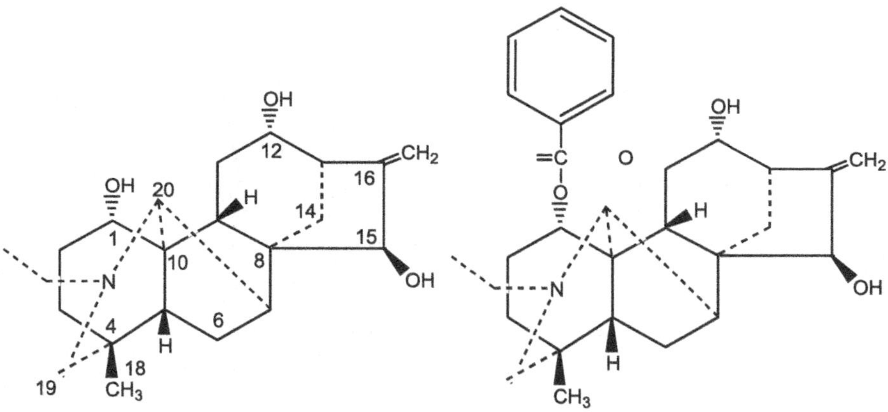

47 **48**

Fig. 8 Structures of the two tyrosinase inhibitory napelline-type alkaloids (napelline, **47**, and 1-*O*-benzoylnapelline, **48**) [49]. The structures of the inactive compounds are not shown

Recently (2006), the quantitative structure–activity relationship (QSAR) modeling of the same compounds based on their atomic linear indices, for finding functions that discriminate between the tyrosinase inhibitor compounds and inactive ones, has been reported by the same group [50].

3.8 Coumarinolignoids

In 2004, Ahmad and coworkers reported a new coumarinolignoid, 8′-epi-cleomiscosin A (**49**), together with the new glycoside 8-*O*-*β*-D-glucopyranosyl-6-hydroxy-2-methyl-4*H*-1-benzopyran-4-one (**50**), isolated from the aerial parts of *Rhododendron collettianum* [51]. These authors also reported the isolation of cleomiscosin A (**51**), aquillochin (**52**), and 5,6,7-trimethoxycoumarin (**53**) from the same plant. Their structures are shown in Fig. 9.

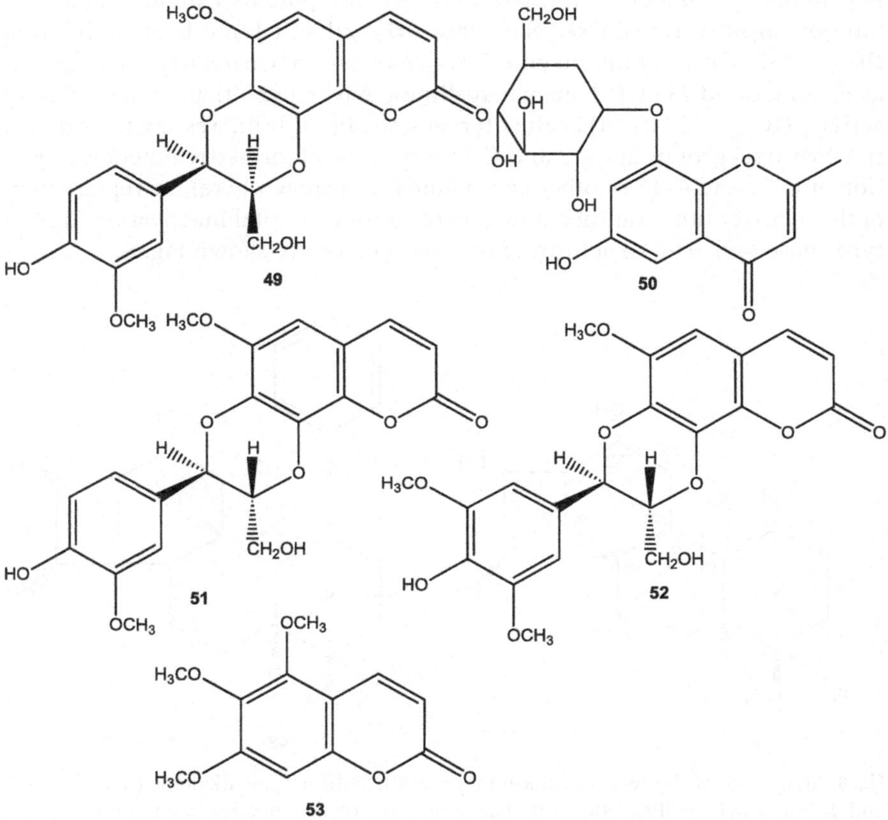

Fig. 9 Molecular structures of 8′-epi-cleomiscosin A (**49**), 8-*O*-*β*-D-glucopyranosyl-6-hydroxy-2-methyl-4*H*-1-benzopyran-4-one (**50**), cleomiscosin A (**51**), aquillochin (**52**), and 5,6,7-trimethoxycoumarin (**53**), isolated from the aerial parts of *Rhododendron collettianum* [51]

Tyrosinase inhibition studies of the same compounds and their structure–activity relationships have also been investigated and reported. The compounds exhibited potent to mild inhibition activity against the enzyme. Especially, compound **49** showed strong inhibition (IC_{50} = 1.33 μM) against the enzyme tyrosinase, as compared to the standard tyrosinase inhibitors kojic acid (IC_{50} = 16.67 μM) and L-mimosine (IC_{50} = 3.68 μM), indicating its potential use for the treatment of hyperpigmentation associated with the overexpression of melanocytes.

3.9
Lignans

Recently, Haq and coworkers (2006) isolated eight lignans from the methanol extract of *Vitex negundo* [52]: negundin A (**54**), negundin B (**55**), 6-hydroxy-4-(4-hydroxy-3-methoxy)-3-hydroxymethyl-7-methoxy-3,4-dihydro-2-naphthaldehyde (**56**), vitrofolal E (**57**), (+)-lyoniresinol (**58**), (+)-lyoniresinol-3α-*O*-β-D-glucoside (**59**), (+)-(−)-pinoresinol (**60**), and (+)-diasyringaresinol (**61**). The structures of these compounds (shown in Fig. 10) were elucidated unambiguously by spectroscopic methods including 1D and 2D NMR analysis, and also by comparing experimental results with literature data [52].

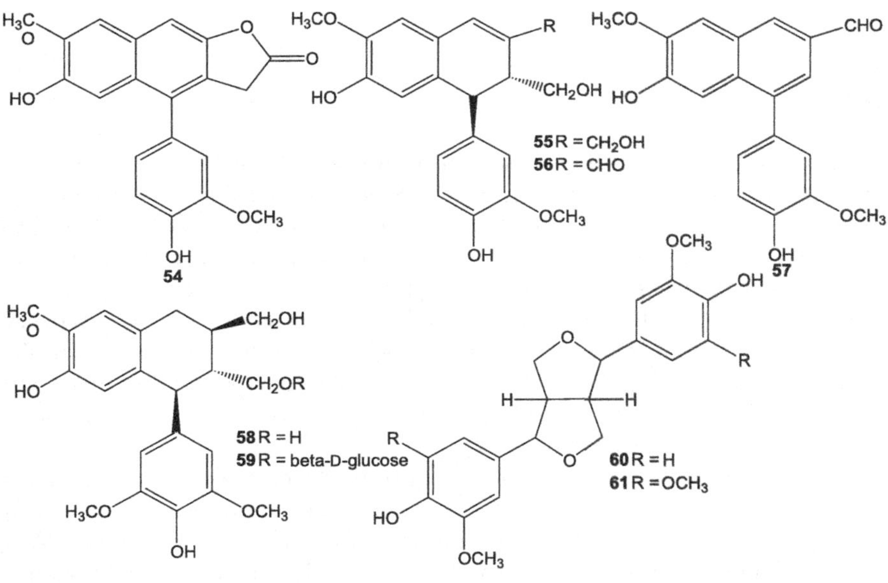

Fig. 10 Structures of the lignans (**54–61**) reported by Haq and coworkers [52]

The tyrosinase inhibitory potency of these compounds was also evaluated by the authors and attempts were made to justify their structure–activity relationships [52]. Their inhibitory potential is shown in Table 3. Compound **58** was found to be the most potent (IC_{50} = 3.21 μM), while other compounds demonstrated moderate to potent inhibitions [52]. It was reported that the substitution of functional group(s) at the C-2 and C-3 positions and the presence of the – CH_2OH group play a vital role in the potency of these compounds. Compound **58** can act as a potential lead molecule to develop new drugs for the treatment of hyperpigmentation associated with the high production of melanocytes [52].

Table 3 The structure–activity relationships of the lignans reported by Haq and coworkers against the enzyme tyrosinase [52]

Lignans	IC_{50} (in μM)
54	10.06
55	6.72
56	7.81
57	9.76
58	3.21
59	Not active
60	15.13
61	5.61

4 Conclusion

Most of the above mentioned research results suggest that there is extensive ongoing research work on mushroom tyrosinase. This is due to the vast range of clinical and industrial applications and also to the easy availability of the enzyme. To accomplish these targets, diverse molecules from both natural and synthetic sources have been investigated in the last few decades. Perceptibly, much more endeavor is still desirable in this direction for the discovery of better and potent inhibitors. Overall, much more research work on mushroom tyrosinase is required to find the role of this enzyme in other unknown fields, which will be accommodating in designing or improving enzymatic activities for various applications [9].

Besides being used in the treatment of some dermatological disorders associated with melanin hyperpigmentation, tyrosinase inhibitors have found an important role in the cosmetics industry for their skin whitening effect and depigmentation after sunburn [5, 37, 41, 53–70]. However, more tangible research work with human tyrosinase is essential from a clinical point of view. An additional important clinical application of mushroom tyrosinase includes its role in the treatment of vitiligo, as the enzyme acts as the marker of this disease [65, 71, 72]. A number of studies have been conducted on animal models, but still more research has to be done to cure vitiligo in human beings [9].

References

1. Gadd GW (1980) Melanin production and differentiation in batch cultures of the polymorphic fungus. FEMS Microbiol Lett 9:237–240
2. Bell AA, Wheeler MH (1986) Biosynthesis and functions of fungal melanins. Annu Rev Phytopathol 24:411–451

3. Zimmerman WC et al. (1995) Melanin and perithecial development in *Ophiostoma piliferum*. Mycologia 87:857–863
4. Zarivi O et al. (2003) Truffle thio-flavours reversibly inhibit truffle tyrosinase. FEMS Microbiol Lett 220(1):81–88
5. Ohguchi K et al. (2003) Effects of hydroxystilbene derivatives on tyrosinase activity. Biochem Biophys Res Commun 307(4):861–863
6. Xu Y et al. (1997) Tyrosinase mRNA is expressed in human substantia nigra. Mol Brain Res 45(1):159–162
7. Chen QX, Kubo I (2002) Kinetics of mushroom tyrosinase inhibition by quercetin. J Agric Food Chem 50(14):4108–4112
8. Rescigno A et al. (2002) Tyrosinase inhibition: general and applied aspects. J Enzyme Inhib Med Chem 17(4):207–218
9. Seo SY, Sharma VK, Sharma N (2003) Mushroom tyrosinase: recent prospects. J Agric Food Chem 51(10):2837–2853
10. Martinez MV, Whitaker JR (1995) The biochemistry and control of enzymatic browning. Trends Food Sci Technol 6:195–200
11. Yaropolov AI et al. (1995) Flow-injection analysis of phenols at a graphite electrode modified with co-immobilized laccase and tyrosinase. Anal Chim Acta 308:137–144
12. Mayer AM (1987) Polyphenol oxidases in plants: recent progress. Phytochemistry 26:11–20
13. van Gelder CWG, Flurkey WH, Wichers HJ (1997) Sequence and structural features of plant and fungal tyrosinases. Phytochemistry 45:1309–1323
14. Sanchez-Ferrer A et al. (1995) Tyrosinase: a comprehensive review of its mechanism. Biochim Biophys Acta 1247:1–11
15. Raper HS (1928) The anaerobic oxidases. Physiol Rev 8:245–282
16. Mason HS (1948) The chemistry of melanin. III. Mechanism of the oxidation of trihydroxyphenylalanine by tyrosinase. J Biol Chem 172:83–99
17. Lerner AB et al. (1949) Mammalian tyrosinases preparation and properties. J Biol Chem 179:185–195
18. Kobayashi T et al. (1995) Modulation of melanogenic protein expression during the switch from eu- to pheomelanogenesis. J Cell Sci 108(Pt6):2301–2309
19. Olivares C et al. (2001) The 5,6-dihydroxyindole-2-carboxylic acid (DHICA) oxidase activity of human tyrosinase. Biochem J 354(Pt1):131–139
20. Liangli YU (2003) Inhibitory effects of (S)- and I-6-hydroxy-2,5,7,8-tetramethylchroman-2-carboxylic acids on tyrosinase activity. J Agric Food Chem 51:2344–2347
21. Lee SE et al. (2000) Inhibition effects of cinnamomum cassia bark-derived materials on mushroom tyrosinase. Food Sci Biotechnol 9:330–333
22. Sugumaran M (1988) Molecular mechanism for cuticular sclerotization. Adv Insect Physiol 21:179–231
23. Barrett FM (1984) Wound-healing phenoloxidase in larval cuticle of *Calpodes ethlius* (Lepidoptera: Hesperiidae). Can J Zool 62:834–838
24. Pawelek JM, Korner AM (1982) The biosynthesis of mammalian melanin. J Am Chem Soc 70:136–145
25. Mosher DB, Pathak MA, Fitzpatric TB (1983) Update: dermatology in general medicine. McGraw Hill, New York, pp 205–225
26. Maeda K, Fukuda M (1991) In vitro effectiveness of several whitening cosmetic components in human melanocytes. J Soc Cosmet Chem 42:361–368
27. Palumbo A et al. (1991) Mechanism of inhibition of melanogenesis by hydroquinone. Biochim Biophys Acta 1073(1):85–90

28. Ohyama Y, Mishima Y (1990) Melanogenesis inhibitory effects of kojic acid and its action mechanism. Fragrance J 6:53
29. Maeda K, Fukuda M (1991) In vitro effectiveness of several whitening cosmetic components in human melanocytes. J Soc Cosmet Chem 42:361–368
30. Arung ET, Shimizu K, Kondo R (2006) Inhibitory effect of artocarpanone from Artocarpus heterophyllus on melanin biosynthesis. Biol Pharm Bull 29(9):1966–1969
31. Boissy RE, Visscher M, DeLong MA (2005) DeoxyArbutin: a novel reversible tyrosinase inhibitor with effective in vivo skin lightening potency. Exp Dermatol 14(8):601–608
32. Choi H et al. (2005) Inhibition of skin pigmentation by an extract of *Lepidium apetalum* and its possible implication in IL-6 mediated signaling. Pigment Cell Res 18(6):439–446
33. Farooqui JZ et al. (1995) Isolation of a unique melanogenic inhibitor from human skin xenografts: initial in vitro and in vivo characterization. J Invest Dermatol 104(5):739–743
34. Funasaka Y et al. (1999) The depigmenting effect of alpha-tocopheryl ferulate on human melanoma cells. Br J Dermatol 141(1):20–29
35. Hamed SH et al. (2006) Comparative efficacy and safety of deoxyarbutin, a new tyrosinase-inhibiting agent. J Cosmet Sci 57(4):291–308
36. Ichihashi M et al. (1999) The inhibitory effect of DL-alpha-tocopheryl ferulate in lecithin on melanogenesis. Anticancer Res 19(5A):3769–3774
37. Imokawa G et al. (1986) Differential analysis of experimental hypermelanosis induced by UVB, PUVA, and allergic contact dermatitis using a brownish guinea pig model. Arch Dermatol Res 278(5):352–362
38. Imokawa G et al. (1997) The role of endothelin-1 in epidermal hyperpigmentation and signaling mechanisms of mitogenesis and melanogenesis. Pigment Cell Res 10(4):218–228
39. Lee JY, Kang WH (2003) Effect of cyclosporin A on melanogenesis in cultured human melanocytes. Pigment Cell Res 16(5):504–508
40. Okombi S et al. (2006) Discovery of benzylidenebenzofuran-3(2*H*)-one (aurones) as inhibitors of tyrosinase derived from human melanocytes. J Med Chem 49(1):329–333
41. Schallreuter KU, Wood JW (1990) A possible mechanism of action for azelaic acid in the human epidermis. Arch Dermatol Res 282(3):168–171
42. Usuki A et al. (2003) The inhibitory effect of glycolic acid and lactic acid on melanin synthesis in melanoma cells. Exp Dermatol 12(Suppl 2):43–50
43. Saboury AA, Alijanianzadeh M, Mansoori-Torshizi H (2007) The role of alkyl chain length in the inhibitory effect of *n*-alkyl xanthates on mushroom tyrosinase activities. Acta Biochim Pol 54(1):183–191
44. Ha SK et al. (2005) Inhibition of tyrosinase activity by *N*,*N*-unsubstituted selenourea derivatives. Biol Pharm Bull 28(5):838–840
45. Ahn SJ et al. (2006) Regulation of melanin synthesis by selenium-containing carbohydrates. Chem Pharm Bull (Tokyo) 54(3):281–286
46. Koketsu M et al. (2002) Inhibitory effects of 1,3-selenazol-4-one derivatives on mushroom tyrosinase. Chem Pharm Bull (Tokyo) 50(12):1594–1596
47. Khan MT et al. (2005) Structure–activity relationships of tyrosinase inhibitory combinatorial library of 2,5-disubstituted-1,3,4-oxadiazole analogues. Bioorg Med Chem 13(10):3385–3395
48. Shaheen F et al. (2005) Alkaloids of Aconitum laeve and their anti-inflammatory antioxidant and tyrosinase inhibition activities. Phytochemistry 66(8):935–940

49. Sultankhodzhaev MN et al. (2005) Tyrosinase inhibition studies of diterpenoid alkaloids and their derivatives: structure–activity relationships. Nat Prod Res 19(5):517–522
50. Casanola-Martin GM et al. (2006) New tyrosinase inhibitors selected by atomic linear indices-based classification models. Bioorg Med Chem Lett 16(2):324–330
51. Ahmad VU et al. (2004) Tyrosinase inhibitors from *Rhododendron collettianum* and their structure–activity relationship (SAR) studies. Chem Pharm Bull (Tokyo) 52(12):1458–1461
52. Azhar UH et al. (2006) Tyrosinase inhibitory lignans from the methanol extract of the roots of *Vitex negundo* Linn. and their structure–activity relationship. Phytomedicine 13(4):255–260
53. Briganti S, Camera E, Picardo M (2003) Chemical and instrumental approaches to treat hyperpigmentation. Pigment Cell Res 16(2):101–110
54. Fechner GA et al. (1992) Antiproliferative and depigmenting effects of the histamine (H2) agonist dimaprit and its derivatives on human melanoma cells. Biochem Pharmacol 43(10):2083–2090
55. Imokawa G, Mishima Y (1985) Analysis of tyrosinases as asparagin-linked oligosaccharides by concanavalin A lectin chromatography: appearance of new segment of tyrosinases in melanoma cells following interrupted melanogenesis induced by glycosylation inhibitors. J Invest Dermatol 85(2):165–168
56. Inoue S et al. (1990) Mechanism of growth inhibition of melanoma cells by 4-S-cysteaminylphenol and its analogues. Biochem Pharmacol 39(6):1077–1083
57. Mishima Y (1994) Molecular and biological control of melanogenesis through tyrosinase genes and intrinsic and extrinsic regulatory factors. Pigment Cell Res 7(6):376–387
58. Monji A et al. (2005) Tyrosinase induction and inactivation in normal cultured human melanocytes by endothelin-1. Int J Tissue React 27(2):41–49
59. Nerya O et al. (2004) Chalcones as potent tyrosinase inhibitors: the effect of hydroxyl positions and numbers. Phytochemistry 65(10):1389–1395
60. Ohyama Y, Mishima Y (1993) Isolation and characterization of high molecular weight melanogenic inhibitors naturally occurring in melanoma cells. Pigment Cell Res 6(1):7–12
61. Park YD et al. (2006) TXM13 human melanoma cells: a novel source for the inhibition kinetics of human tyrosinase and for screening whitening agents. Biochem Cell Biol 84(1):112–116
62. Parvez S et al. (2006) Survey and mechanism of skin depigmenting and lightening agents. Phytother Res 20(11):921–934
63. Prince S, Illing N, Kidson SH (2001) SV-40 large T antigen reversibly inhibits expression of tyrosinase, TRP-1, TRP-2 and Mitf, but not Pax-3, in conditionally immortalized mouse melanocytes. Cell Biol Int 25(1):91–102
64. Saha B et al. (2006) Transcriptional activation of tyrosinase gene by human placental sphingolipid. Glycoconj J 23(3–4):259–268
65. Schallreuter KU et al. (1994) Defective tetrahydrobiopterin and catecholamine biosynthesis in the depigmentation disorder vitiligo. Biochim Biophys Acta 1226(2):181–192
66. Song KK et al. (2006) Inhibitory effects of *cis*- and *trans*-isomers of 3,5-dihydroxystilbene on the activity of mushroom tyrosinase. Biochem Biophys Res Commun 342(4):1147–1151
67. Townsend E et al. (1992) Reversible depigmentation of human melanoma cells by halistanol trisulphate, a novel marine sterol. Melanoma Res 1(5–6):349–357

68. Tsuji-Naito K et al. (2007) Modulating effects of a novel skin-lightening agent, alpha-lipoic acid derivative, on melanin production by the formation of DOPA conjugate products. Bioorg Med Chem 15(5):1967–1975
69. Vogel FS et al. (1977) γ-L-Glutaminyl-4-hydroxybenzene, an inducer of cryptobiosis in *Agaricus bisporus* and a source of specific metabolic inhibitors for melanogenic cells. Cancer Res 37(4):1133–1136
70. Yang F, Boissy RE (1999) Effects of 4-tertiary butylphenol on the tyrosinase activity in human melanocytes. Pigment Cell Res 12(4):237–245
71. Chakraborty DP et al. (1978) Interrelationship of tryptophan pyrrolase with tyrosinase in melanogenesis of *Bufo melanostictus*. Clin Chim Acta 82(1–2):55–59
72. Schallreuter KU, Wood JM, Berger J (1991) Low catalase levels in the epidermis of patients with vitiligo. J Invest Dermatol 97(6):1081–1085

Top Heterocycl Chem (2007) 9: 139–178
DOI 10.1007/7081_2007_079

Published online: 4 July 2007

Xanthones in *Hypericum*: Synthesis and Biological Activities

Ozlem Demirkiran

Department of Chemistry, Faculty of Science and Arts, Trakya University, Gullapoglu Campus, 22030 Edirne, Turkey
ozlemdemirkiran@yahoo.com

Abstract There has been an increasing interest in the genus *Hypericum*, because it is a source of a variety of compounds with different biological activities. Xanthones are one of these compounds within *Hypericum* species. Recently, growing attention has been given to these heterocyclic compounds containing oxygen because of their many interesting pharmacological and biological properties, such as monoamine oxidase inhibition and antioxidant, antifungal, cytotoxic, and hepatoprotective activities.

Keywords Xanthones · *Hypericum* · Pharmacological properties · Biosynthesis · Synthesis

Abbreviations

AA	Arachidonic acid
ABTS	2,2′-Azinobis(3-ethylbenzothiozoline-6-sulfonate)
AcOH	Acetic acid
APTT	Active partial thromboplastin time reagent
CC	Column chromatography
CL	Chemiluminescence
CoMFA	Comparative molecular field analysis
COX	Cyclooxygenase
CPC	Centrifugal partition chromatography
DMSO	Dimethyl sulfoxide
ED_{50}	Effective dose
h	Hour
HPLC	High-pressure liquid chromatography
IC_{50}	Inhibitor concentration
MAO	Monoamine oxidase
MCPBA	*meta*-Chloroperoxybenzoic acid
MIC	Minimum inhibition concentration
PG	Prostaglandin
PT	Prothrombin time reagent
QSAR	Quantitative structure–activity relationship
RP	Reverse phase
TEAC	Trolox equivalent antioxidant capacity

1 Introduction

Xanthones are secondary metabolites commonly occurring in a few higher plant families. Their high taxonomic value and their pharmacological properties, such as monoamine oxidase inhibition, in vitro toxicity, and in vivo antitumor activity, have provoked great interest [1].

Their structures are related to those of flavonoids and their chromatographic behavior is also similar. While flavonoids are frequently encountered in nature, xanthones have been found in a small number of families. They occur in eight families, namely in the Gentianaceae, Guttiferae, Polygalaceae, Leguminosae, Lythraceae, Moraceae, Loganiaceae, and Rhamnaceae. Extensive studies on xanthones have been made in the Gentianaceae and Guttiferae families [2]. A number of species of *Hypericum* genus (Guttiferae) contain xanthones which have been reported to possess several biological activities [3].

The aim of this chapter is not only to provide a review of the distribution of xanthones in *Hypericum* genus, but also to cover some more specific aspects such as biosynthesis, chemotaxonomic significance, biological activities, and synthesis.

2 Chemistry and Classification of Xanthones

2.1 Chemistry

Chemically xanthones (9*H*-xanthen-9-ones) are heterocyclic compounds with the dibenzo-γ-pyrone framework (**1**, Fig. 1). The xanthone nucleus is numbered according to a biosynthetic convention with carbons 1–4 being assigned to acetate-derived ring A and carbons 5–8 to the shikimate-derived ring B. The other carbons are indicated as 4a, 4b, 8a, 8b, 9, and 9a for structure elucidation purposes [4].

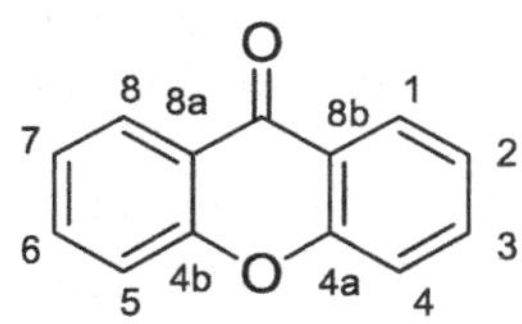

Fig. 1 Dibenzo-γ-pyrone (**1**)

2.2 Classification of Xanthones

Xanthones isolated to date can be classified into five major groups [5]:

1. Simple oxygenated xanthones
2. Xanthone glycosides
3. Prenylated and related xanthones
4. Xanthonolignoids
5. Miscellaneous

2.2.1 Simple Oxygenated Xanthones

They can be further subdivided into six groups depending on the degree of oxygenation pattern of the basic skeleton. Genus *Hypericum* contains all

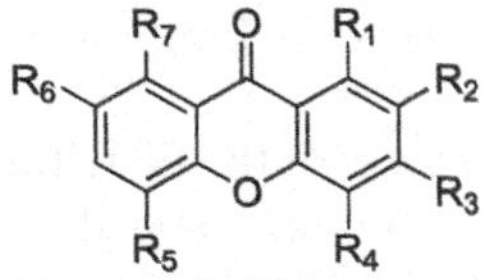

$R_1=R_3=R_4=R_5=R_6=R_7=$ H, $R_2=$OH; 2-hydroxyxanthone (2)
$R_1=R_3=R_4=R_6=R_7=$H, $R_2=$OMe, $R_5=$OH; 5-hydroxy-2-methoxyxanthone (3)
$R_1=R_2=$OMe, $R_3=R_5=R_6=R_7=$H, $R_4=$OH; 4-Hydroxy-1,2-dimethoxyxanthone(4)
$R_1=R_3=$OMe, $R_2=R_4=R_5=$H, $R_6=R_7=$OH; 6,7-dihydroxy-1,3-dimethoxyxanthone(5)
$R_1=R_6=R_7=$OMe, $R_4=R_5=$H, $R_2=R_3=$OH; 2,3-Dihydroxy-1,6,7-trimethoxyxanthone(6)
$R_1=$OH, $R_2=R_3=R_4=R_5=R_6=$OMe;1-hydroxy-2,3,4,5,7-pentamethoxyxanthone(7)

Fig. 2 Examples of oxygenated xanthones

classes of oxygenated xanthones except for the hexa-oxygenated (examples of each class are given in Fig. 2):

(a) Mono-oxygenated xanthones: They are unusual and only a small number of mono-oxygenated xanthones have been isolated from natural sources up to now. 2-Hydroxyxanthone (**2**) can be given as an example of this class [6, 7].

(b) Di-oxygenated xanthones: They are more common derivatives of xanthones. 5-Hydroxy-2-methoxyxanthone (**3**) was isolated from *Hypericum roeperanum* [8].

(c) Tri-oxygenated xanthones: They can be more frequently encountered in nature. 4-Hydroxy-1,2-dimethoxyxanthone (**4**) was isolated from *H. geminiflorum* [9].

(d) Tetra-oxygenated xanthones: They are more numerous than tri-oxygenated xanthones. To date many tetra-oxygenated xanthones have been isolated. 6,7-Dihydroxy-1,3-dimethoxyxanthone (**5**) isolated from *H. geminiflorum* [9] can be shown as an example of tetra-oxygenated xanthones.

(e) Penta-oxygenated xanthones: Only a small number of this class was found in nature. 2,3-Dihydroxy-1,6,7-trimethoxyxanthone (**6**) was isolated from *H. geminiflorum* [10].

(f) Hexa-oxygenated xanthones (**7**) [11]: They have the highest degree of oxygenation observed so far and only a few compounds were identified. No hexa-oxygenated xanthones have been isolated from *Hypericum* species.

2.2.2 Xanthone Glycosides

They might be divided into *O*-glycosides and *C*-glycosides according to the nature of the glycosidic linkage:

(a) *O*-Glycoside xanthones: Although most *O*-glycoside xanthones have the sugar moiety attached to position 1 of the xanthone nucleus, which is difficult to explain considering the vicinity of the carbonyl function, since this might create a strain, it is also possible to observe the glycosyl moiety at any position of the xanthone nucleus. They are easily hydrolyzed in enzymatic or acid environment [13]. Only four *O*-glycoside xanthones have been isolated from *Hypericum* species. One of them is patuloside A (**8**) isolated from *H. patulum* [12].

(b) *C*-Glycoside xanthones: They are more resistant to hydrolysis compared with *O*-glycoside xanthones, but their occurrence is very much limited. Mangiferin (**9**) [5] was the first glycoside xanthone isolated in 1908 from *Mangifera indica* (Anacardiaceae) and has also been isolated from *H. montbretii* Spach. for the first time from genus *Hypericum* [14].

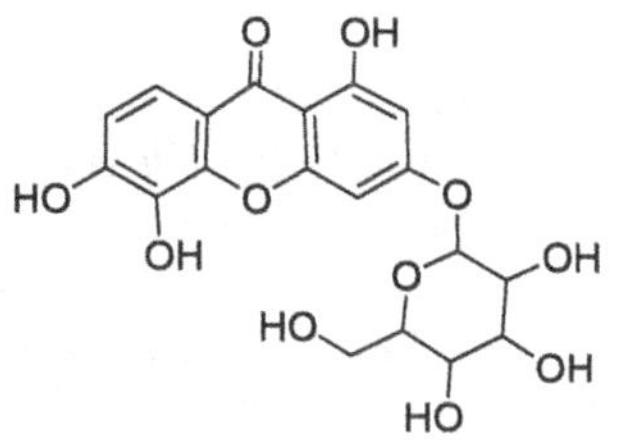

Scheme 1 3-*O*-*β*-D-Glucopyranosyl-1,5,6-trihydroxyxanthone; patuloside A (**8**)

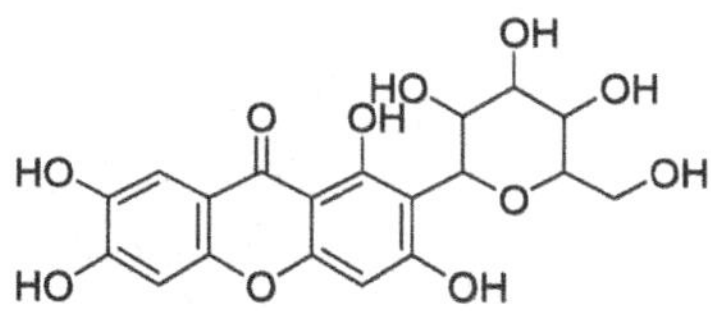

Scheme 2 2-*β*-D-Glucopyranosyl-1,3,6,7-tetrahydroxyxanthone; mangiferin (**9**)

2.2.3 Prenylated Xanthones

The family Guttiferae appears to produce a large number of xanthones with isopentenyl and geranyl substituents. Prenylated xanthones **10** [15] and the corresponding pyranoxanthones **11** [16] have been reported to occur in *H. japonicum* and *H. brasilianse*, respectively.

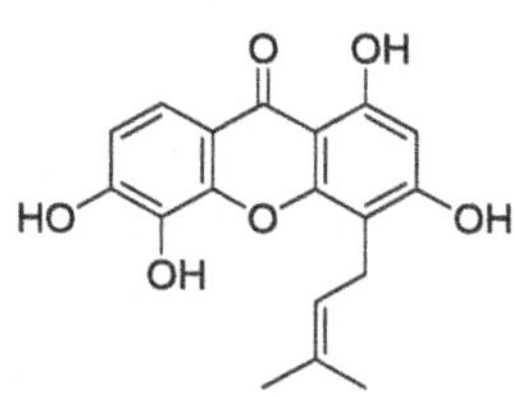

Scheme 3 1,3,5,6-Tetrahydroxy-4-(3-methyl-2-butenyl)-xanthone; ugaxanthone (**10**)

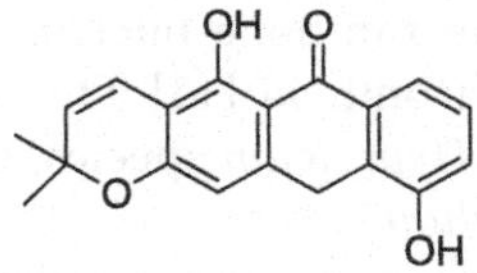

Scheme 4 6-Deoxyjacareubin (11)

2.2.4
Xanthonolignoids

They are a relatively rare group of natural products and principally occur in some genera of the Guttiferae family: *Kielmeyera*, *Caraipa* [17], *Psorospermum* [18], and *Hypericum* [19]. These compounds are very close in the skeletal patterns formed from the association of the xanthone nucleus and the lignoid pattern (coniferyl alcohol or syringenin). The most representative ones are cadensin D (**12**) [15, 19] and kielcorin [15, 19].

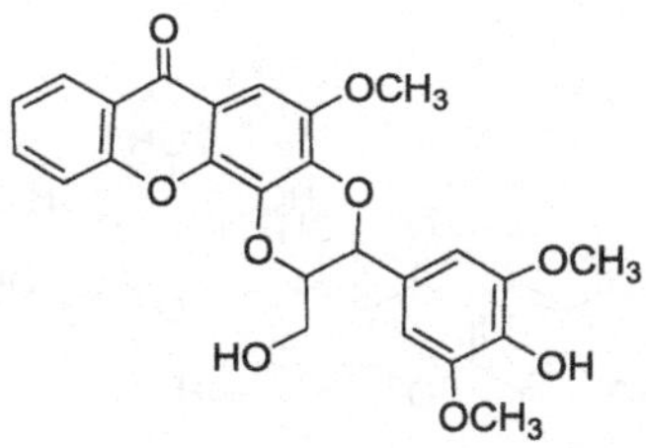

Scheme 5 Cadensin D (12)

2.2.5
Miscellaneous

Besides these groups, some xanthones with unusual substitutions have been isolated from different plant sources including lichens, which could not be classified in the usual manner. These compounds have been grouped as miscellaneous. We can show a chloride compound, 4-chloro-3,8-dihydroxy-

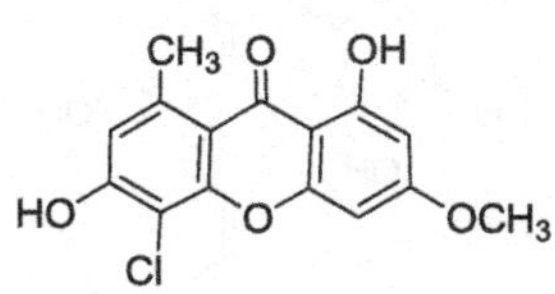

Scheme 6 4-Chloro-3,8-dihydroxy-6-methoxy-1-methylxanthone (13)

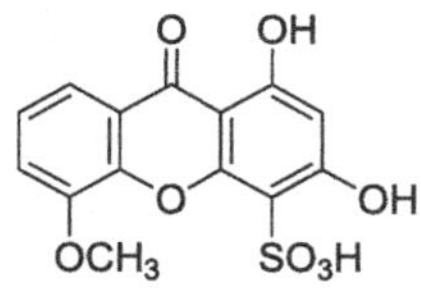

Scheme 7 1,3-Dihydroxy-5-methoxyxanthone-4-sulfonic acid (**14**)

6-methoxy-1-methylxanthone (**13**) from *H. ascyron* [20] and a sulfonated xanthone (**14**) from *H. sampsonii* [21] as examples of this group.

3 Biosynthesis and Synthesis of Xanthones

3.1 Biosynthesis

The biosynthetic pathways to xanthones have been discussed for 40 years. Biosynthetically, the xanthones of higher plants are formed from shikimate and acetate origins [22]. Thus phenylalanine, which is formed from shikimate by losing two carbon atoms from the side chain, is oxidized to form *m*-hydroxybenzoic acid. This combines with three units of acetate (probably via malonate) to produce the intermediate. Suitable folding and ring closure gives a substituted benzophenone, which generates the central ring of the xanthone moiety by an oxidative phenol coupling reaction [23]. Atkinson et al. [24] and Gupta and Lewis [25] performed some experiments on *Gentiana lutea* and obtained an important proof for this pathway. When plants were fed ^{14}C-labeled phenylalanine, the label was observed only in the ring B (Fig. 3). Conversely, feeding of ^{14}C-labeled acetate incorporated the label into ring A.

The other mechanisms for the intramolecular reaction of benzophenone involve quinone addition [26], dehydration between hydroxyl groups on acetate- and shikimate-derived rings (2,2′-dihydroxybenzophenone) [27], or spirodienone formation and subsequent rearrangement to form the xanthone [22, 28].

With the pioneering isolation of maclurin, 2,4,6,3′,4′-pentahydroxybenzophenone, together with 1,3,5,6- and 1,3,6,7-tetrahydroxyxanthone from the heartwood of *Symphonia globulifera* (Guttiferae), the biogenetic role of maclurin as precursor in xanthone biosynthesis was discussed by Locksey et al. [29]. Then, the following studies on the biosynthesis of xanthones in Guttiferae were reported by Bennett and Lee [30]. Cinnamic acid, benzoic acid, *m*-hydroxybenzoic acid, malonic acid, and 4′-deoxymaclurin as the intermediate benzophenone were found to be efficient precursors. This appeared to be also true for xanthone formation in cultured cells of *H. patulum*.

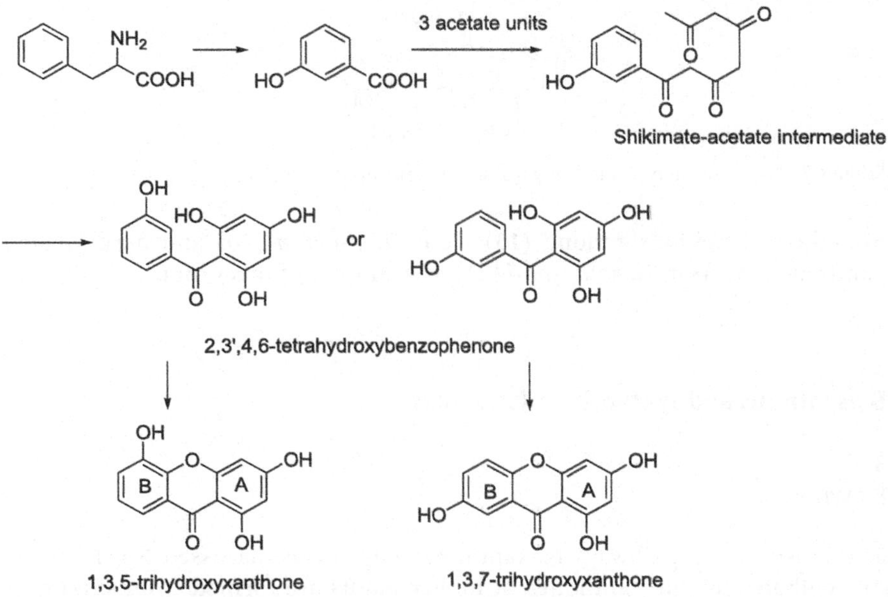

Fig. 3 Biosynthetic pathways leading to the parent xanthones 1,3,5- and 1,3,7-trihydroxyxanthone

It has been suggested that the xanthones isolated from *H. patulum* cultures could be biosynthesized from 4′-deoxymaclurin or maclurin [31–33].

Further investigation of xanthone biosynthesis was carried out by Beerhues with the detection of an enzyme named benzophenone synthase from cultured cells of *Centaurium erythraea* [34]. The formation of 2,3′,4,6-tetrahydroxybenzophenone, which is a central step in xanthone biosynthesis, was shown in cell-free extracts from cultured cells of *C. erythraea* (Fig. 4).

As part of their continuing study the same research group have reported the detection and partial characterization of another enzyme named benzophenone 3′-hydroxylase, leading to the formation of 2,3′,4,6-tetrahydroxybenzophenone in cultured *H. androsaemum* cells as well as benzophenone synthase. In contrast to the enzyme from *C. erythraea*, benzophenone synthase from *H. androseamum* acts more efficiently on benzoyl CoA than 3′-hydroxybenzoyl CoA which is supplied by 3-hydroxybenzoate:CoA ligase [35]. In *C. erythraea*, 2,3′,4,6-tetrahydoxybenzophenone is converted to 1,3,5-trihydroxyxanthone by xanthone synthase; however, in *H. androsaemum*, it is cyclized to 1,3,7-trihydoxyxanthone. Since these two isomers are precursors of the majority of higher plant xanthones [36], 2,3′,4,6-

Fig. 4 Postulated reaction mechanism of xanthones in cell cultures of *C. erythraea* and *H. androseamum* ▶

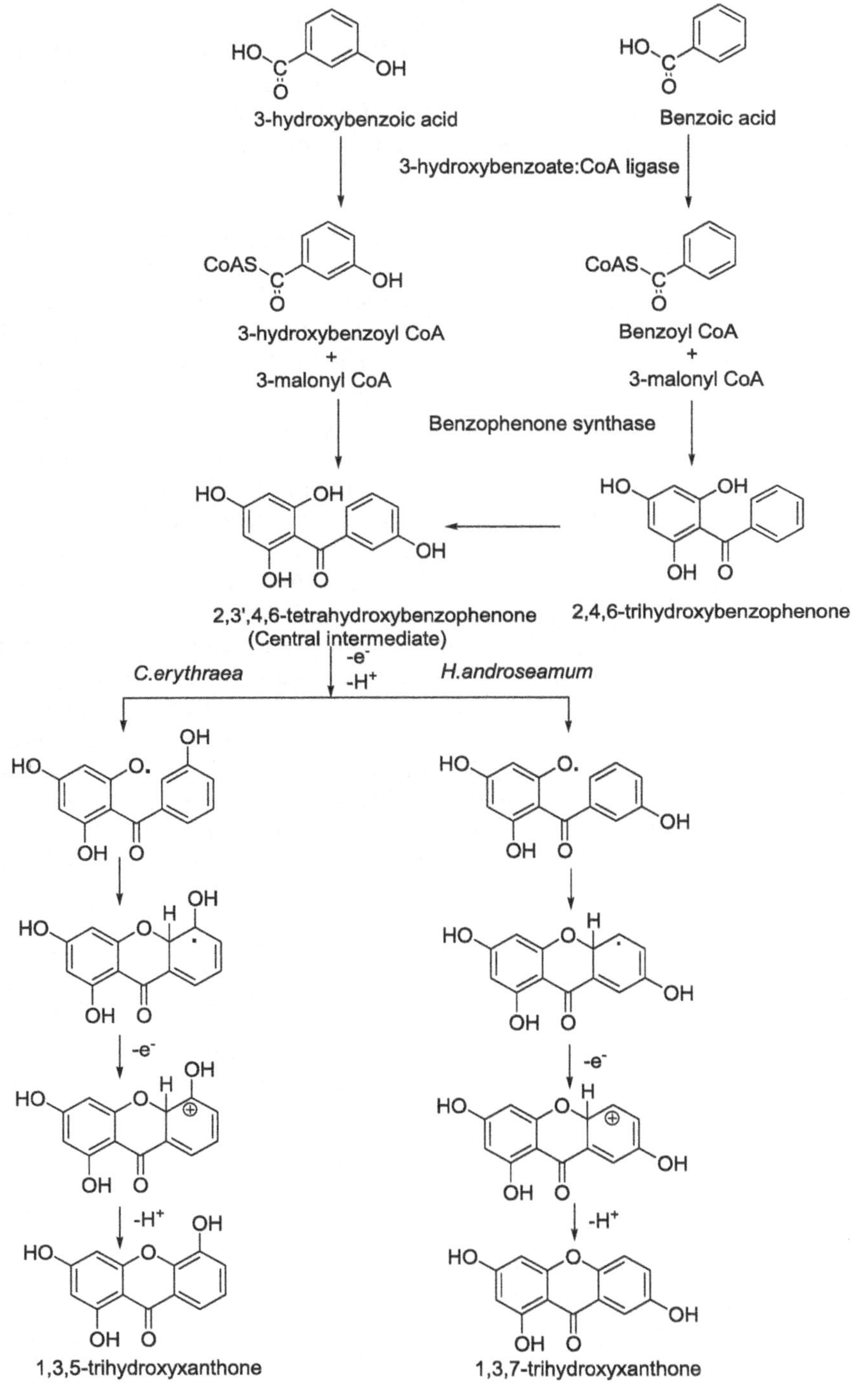

3-hydroxybenzoic acid
Benzoic acid
3-hydroxybenzoate:CoA ligase
3-hydroxybenzoyl CoA
+
3-malonyl CoA
Benzoyl CoA
+
3-malonyl CoA
Benzophenone synthase
2,3',4,6-tetrahydroxybenzophenone
(Central intermediate)
2,4,6-trihydroxybenzophenone
-e⁻
-H⁺
C.erythraea
H.androseamum
1,3,5-trihydroxyxanthone
1,3,7-trihydroxyxanthone

tetrahydoxybenzophenone (THBP) represents a central intermediate in xanthone biosynthesis [37]. The formation of these isomers was explained by the regioselectivity of xanthone synthases, which are cytochrome P_{450} enzymes catalyzing the coupling of benzophenone in both cell cultures. The reaction mechanism underlying the regioselective intramolecular benzophenone cyclizations has been previously proposed by Lewis [23] to be an oxidative phenol coupling. This mechanism involves two one-electron oxidation steps (Fig. 4). The first one-electron transfer and deprotonation generate a phenoxy radical whose electrophilic attack at C-2′ or C-6′ leads to the cyclization of benzophenone. These intermediate hydroxy-cyclohexadienyl radicals are transformed by the loss of a further electron and proton to 1,3,5- and 1,3,7-trihydroxyxanthones. The results indicate that THBP is oxidatively coupled via the *ortho* position to the 3′-hydroxy group in *C. erithraea*, but via the *para* position to the 3′-hydroxy group in *H. androsaemum*. These observations indicated that THBP was the preferred substrate of xanthone synthases, and it was suggested that an oxidative phenol coupling mechanism was strongly favored by the presence of the *ortho–para* directing 3′-hydroxy group of THBP which originated from 3-hydroxybenzoic acid [38].

One of the other pathways of xanthone biosynthesis is dehydration between the hydroxyl groups of the acetate and shikimate derived rings (2,2′-dihydroxybenzophenones) via intermediates such as *O*-pyrophosphates [22, 39]. Up to 2001, however, there was no evidence for in vivo formation of polyhydroxyxanthones from benzophenones by a dehydration mechanism. In the herb *H. annulatum*, Kitanov et al. found the co-occurrence of large amounts of hypericophenoside (**15**) together with 1,3,7-trihydroxyxanthone, gentisein (**16**). The benzophenone *O*-glycoside **15** was easily transformed into gentisein (**16**) by acid or enzymatic hydrolysis (Fig. 5). This fact supports the evidence that 2,4,5′,6-tetrahydroxybenzophenone-2′-*O*-glycoside

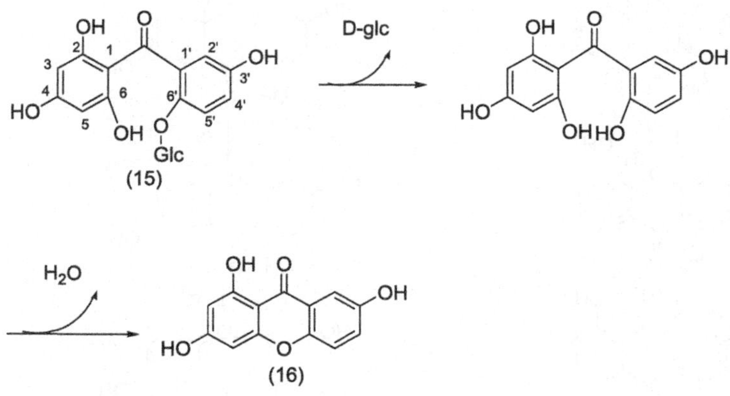

Fig. 5 Transformation of hypericophenoside (**15**) to gentisein (**16**) by a dehydration mechanism

is a precursor of 1,3,7-trihydroxyxanthone. With these results it can be concluded that some xanthones are formed in vivo by dehydration of 2,2′-dihydroxybenzophenones. This is a spontaneous reaction which appears to be regulated by deglucosylation of the precursor, which is a benzophenone with O glycosylation *ortho* to the carbonyl function [40]. The deglucosylation occurs at an earlier stage of the benzophenone biosynthesis before cyclization of two rings [40].

In the case of *C*-glycosylxanthones, it has been suggested that mangiferin is biogenetically related to flavonoids. Indeed, often mangiferin co-occurs with related *C*-glycosyl flavonoids [13]. Fujita and Inoue have reported the biosynthesis of mangiferin in *Anemarrhena asphodeloides* Bunge (Liliaceae) [41]. The results showed that the xanthone nucleus is really formed from a flavonoid-type C_6–C_3 precursor coupled with two malonate units. All the carbon atoms of phenylalanine as well as cinnamic acid and *p*-coumaric acid are incorporated into the xanthone nucleus, but benzoic acid is clearly not on the pathway which is distinct from that of normal xanthones (Fig. 6). Glycosylation seems to occur at the benzophenone stage and is followed by oxidative cyclization. For many years, the glycosylated benzophenone was considered as a hypothetical intermediate. Tanaka et al. found evidence confirming this postulated benzophenone intermediate [42]. 3-*C*-*β*-D-Glucosylmaclurin, which is a benzophenone *C*-glycoside, was isolated together with mangiferin in *Mangifera indica*. Moreover, 2,3,4′,5,6-pentahydroxybenzophenone-4-*C*-glycoside, isolated for the first time from nature, was also recently found together with mangiferin in *Gnidia involucrata* [43]. These results are important evidence for the hypothesis of Fujita and Inoue.

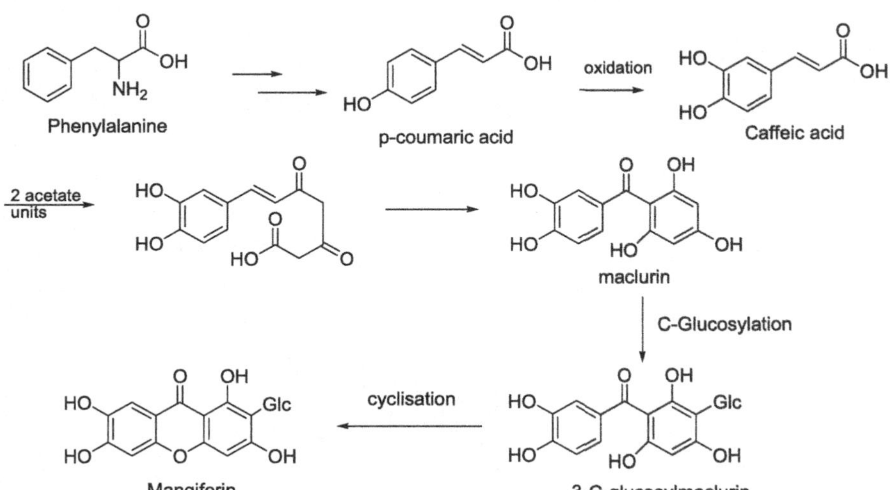

Fig. 6 Biosynthetic pathways leading to mangiferin

3.2 Synthesis of Xanthones

Since xanthones are from natural origins, they have limited type and position of substituents imposed by the biosynthetic pathways. Synthesis of new compounds enables enlargement of the possibilities of having different natures and positions of substituents on the xanthone nucleus. This will allow us to have different structures with a variety of biological activities. Based on these considerations, many xanthone compounds have been synthesized in recent years. In this section, the general methods for synthesizing xanthones and the synthesis of some pharmacologically important xanthones from genus *Hypericum* will be presented.

3.2.1 Standard Methods for the Synthesis of Xanthones

The standard methods for the synthesis of xanthones are via the benzophenone **17** and diaryl ether intermediates **18** (Fig. 7). The intermediate benzophenone derivatives **17** can be obtained by condensation between an *ortho*-oxygenated benzoic acid and an activated phenol, in the presence of phosphorus oxychloride and zinc chloride (a) [44]. This intermediate is also accessible through condensation by the Friedel–Crafts acylation of appropriately substituted benzoyl chlorides with phenolic derivatives (b) [45]. Then the oxidative or dehydrative processes cause the cyclization of 2,2′-di-oxygenated benzophenone to xanthone (c) [46].

The other method can be carried out via a suitable diaryl ether intermediate **18**, which can be obtained from the condensation of a phenol and an

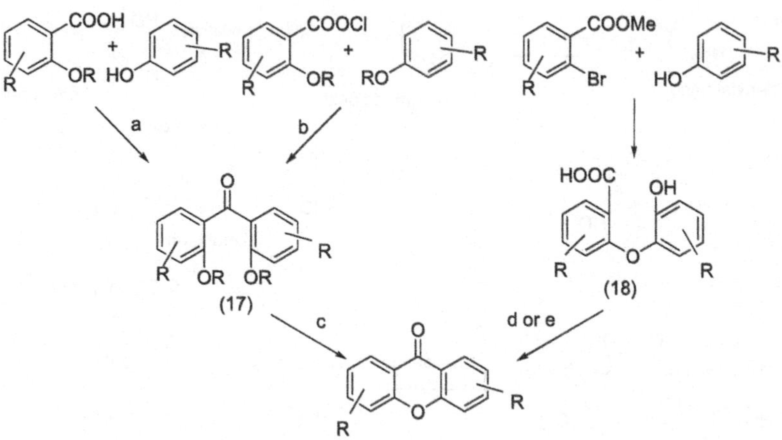

Fig. 7 General methods for the synthesis of xanthones (taken from [50] by permission from Elsevier)

o-chloro or -bromobenzoic acid. Then this biphenyl intermediate converts to the xanthone with ring formation by a one-step reaction with lithium diisopropylamide [47] or by acetyl chloride (e) [48]. Since this method was successfully applied for the first time to the synthesis of euxanthone by Ullmann and Pauchaud [49], it is called Ullmann synthesis.

The general methods for the synthesis of xanthones were shown as Fig. 7 by Pedro et al. [50]. The other well-known methods for synthesizing xanthone derivatives have also been mentioned below.

Asahina–Tanase Method

This is a useful method for the synthesis of some methoxylated xanthones or xanthones with acid-sensitive substituents [51]. Recently Vitale et al. [52] modified the procedure as shown in Fig. 8.

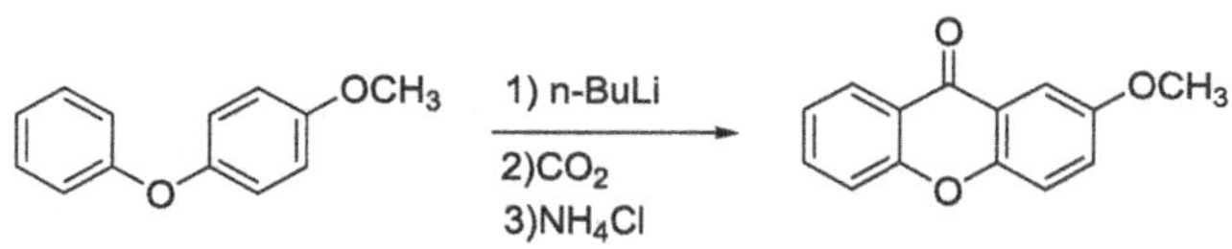

Fig. 8 Asahina–Tanase method [52]

Tanase Method

The Tanase method enables the synthesis of polyhydroxyxanthones. It has been used for the preparation of partially methylated polyhydroxyxanthones with pre-established orientation of some substituents, for example, the synthesis of 1,3-dihydroxyxanthone (Fig. 9) [53].

3.2.2 Synthesis of Prenylated Xanthones

1. O-prenylated xanthones: They have been obtained by O alkylation of a hydroxyxanthone with prenyl bromide in the presence of potassium carbonate [54–56], but very little work has been done in this area.
2. Three methods of C prenylation are known. They are as follows:
   (i) C prenylation with 2-methylbut-3-en-2-ol in the presence of boron trifluoride in ether. Only one case has been reported in the literature with 1,3-dihydroxyxanthone leading to prenylated xanthones **19–21** (Fig. 10) [57].
   (ii) The reaction of 1,3-dihydroxy-5,8-methoxyxanthone with prenyl bromide in the presence of sodium methoxide yields a mixture of O- and C-prenylated xanthones **22–24** (Fig. 11) [58].
   (iii) Another method of prenylation of the aromatic ring starts with an O prenylation followed by a C prenylation by Claisen rearrangement (Fig. 12) [55].

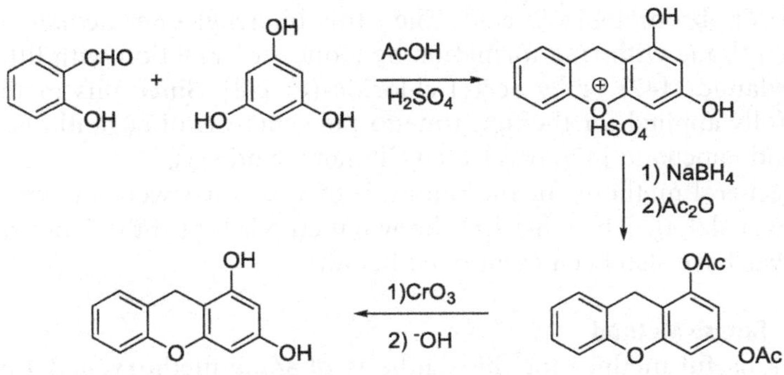

Fig. 9 Synthesis of 1,3-dihydroxyxanthone with the Tanase method [53]

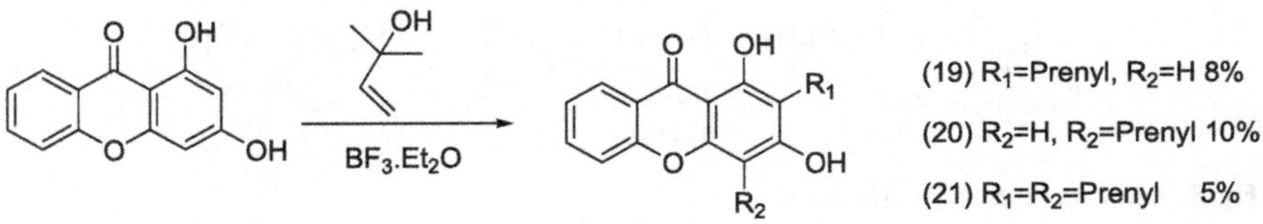

Fig. 10 C prenylation with 2-methylbut-3-en-2-ol [57]

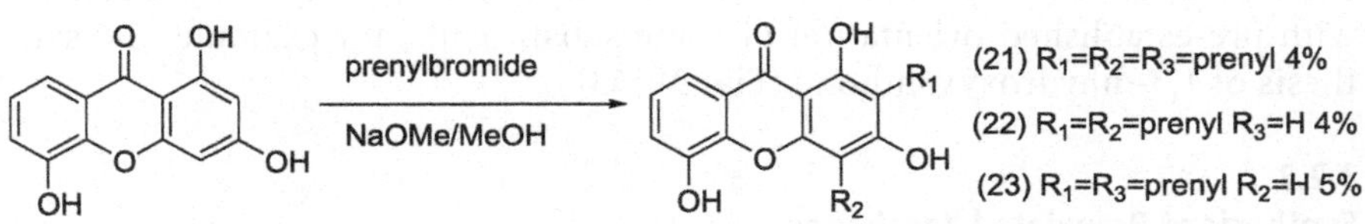

Fig. 11 C prenylation with prenyl bromide in the presence of strong base [58]

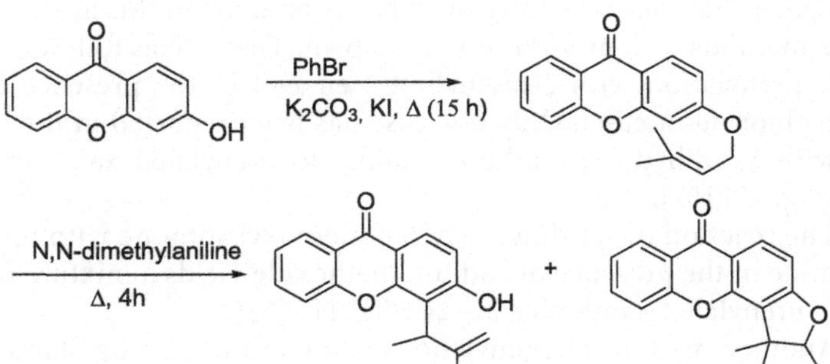

Fig. 12 C prenylation through Claisen rearrangement [55]

3.2.3 Synthesis of Xanthones Isolated from Genus *Hypericum*

2-Hydroxy-5,6,7-trimethoxyxanthone, which was isolated for the first time from *H. ericoides*, was successfully synthesized by Gil et al. [59]. The synthesis was performed from the new benzophenone precursor **25** (Fig. 13). Compound **25** was prepared by Friedel–Crafts acylation of 1,2,3,4-tetramethoxybenzene with 2,5-dibenzyloxybenzoic acid and oxalyl chloride. When benzophenone **25** was heated with Me_4NOH, 5-benzoyloxy-2-hydroxy-2′,3′,4′,5′-tetramethoxybenzophenone (**25**) underwent cyclization to 2-benzyloxy-5,6,7-trimethoxyxanthone (**26**) [60]. Hydrogenolysis of this compound with H_2/Pd-C [61] finally afforded 2-hydroxy-5,6,7-trimethoxyxanthone (**27**) which is identical to natural xanthone [60].

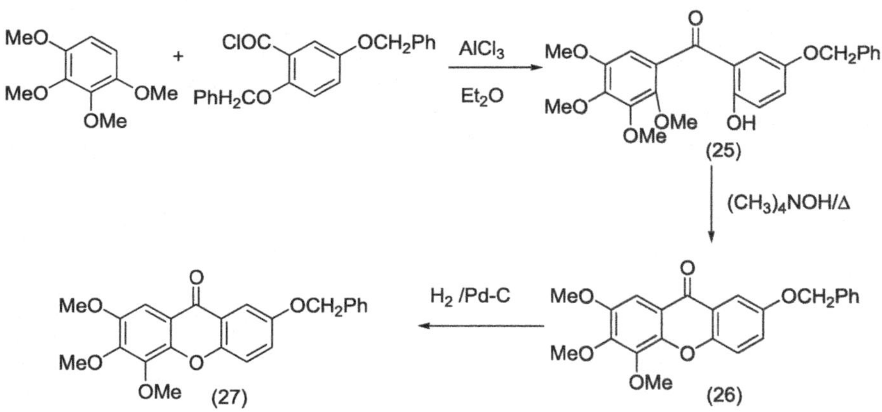

Fig. 13 Synthesis of 2-hydroxy-5,6,7′-trimethoxyxanthone [59]

The xanthonolignoid kielcorin (**35**) has been isolated from several *Hypericum* species [15, 19]. Gottlieb et al. accomplished the synthesis of kielcorin in low yield by oxidative coupling of 3,4-dihydroxy-2-methoxyxanthone and coniferyl alcohol with silver oxide [62]. Then, a facile synthesis of kielcorin (**35**) from readily available materials 3-benzyloxy-4-hydroxy-2-methoxyxanthone (**30**) and ethyl 2-bromo-3-(4-benzyloxy-3-methoxyphenyl)-3-oxopropionate (**31**) was carried out by Tanaka et al. [63]. The starting material **30** was synthesized by benzylation of 4-formyl-3-hydroxy-2-methoxyxanthone (**28**) followed by treatment with *m*-chloroperbenzoic acid in CH_2Cl_2 (the Baeyer–Villager reaction). Compound **30** was then condensed with **31** in acetonitrile in the presence of potassium *tert*-butoxide to give **32** in 74% yield; this product was subjected to catalytic hydrogenation, affording a debenzylation product **33**. Reduction of **33** with lithium borohydride in THF at 0 °C provided an inseparable mixture of alcohols (**34a,b**). The mixture **34a,b** cyclized upon heating in acetic acid in the presence of concentrated hydrochloric acid to furnish kielcorin (30% yield) and *cis*-kielcorin with 3% yield (Fig. 14).

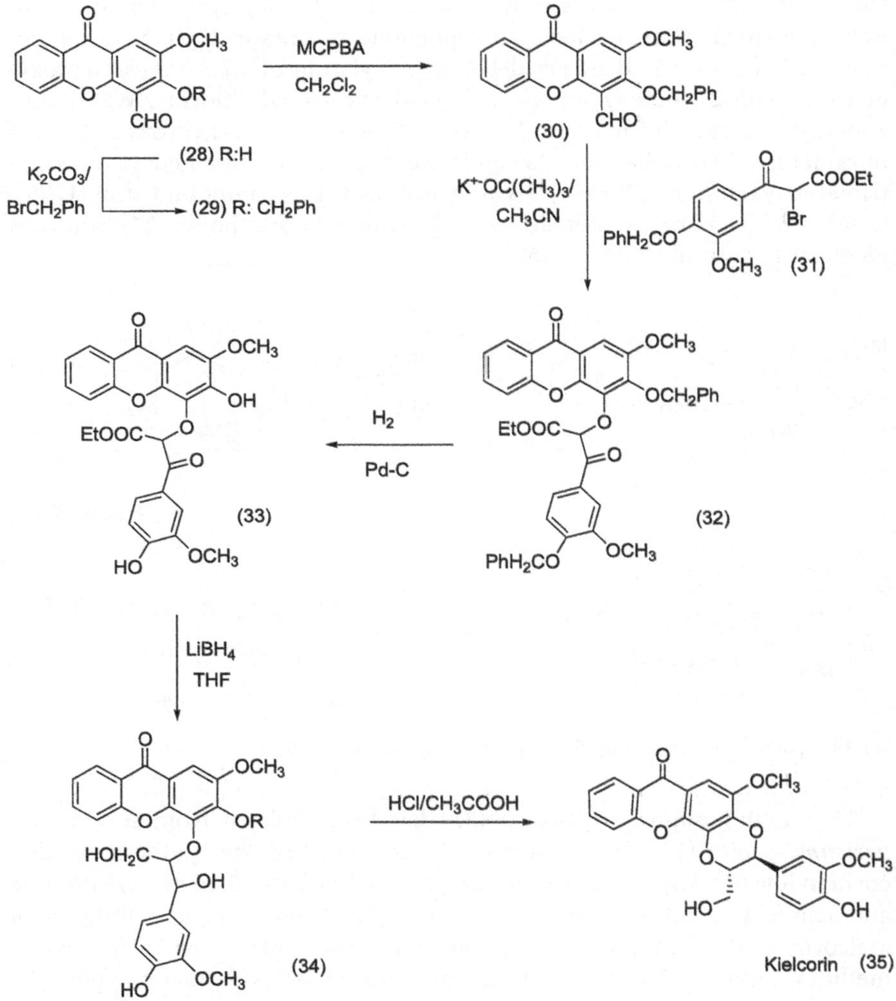

Fig. 14 Synthesis of kielcorin [63]

Recently, a study on the synthesis of novel natural secondary allylic alcohol derivatives was carried out by Helesbeux et al. [64]. The photooxygenation–reduction sequence was applied in the prenylated xanthone series; also 6-deoxyisojacareubin, which was already isolated from *H. japonicum*, was synthesized in this work. 1,3,5-Trihydroxyxanthone (**39**) was synthesized via a polymethoxybenzophenone intermediate, easily obtained by Friedel–Crafts acylation. Thus, 2,3-dimethoxybenzoyl chloride (**36**), prepared in situ from the corresponding acid in the presence of oxalyl chloride, reacted with 1,3,5-trimethoxybenzene (**37**) to give 2-hydroxy-2′,3′,4′,6′-tetramethoxybenzophenone (**38**) with 86% yield (Fig. 15). As already described by Quillian [45], a mono-demethylation occurred on the ring provided by the acid moiety in the position *ortho* to the carbonyl function. Then subsequent base-catalyzed cyclization [65] of **38** led to 1,3,5-trimethoxyxanthone (**39**) with 93% yield along with methanol elimination. Demethylation of **39** was completed in the presence of iodhydric acid and phenol [66, 67] leading to 1,3,5-trihydoxyxanthone (**40**) with 95% yield. In the presence of an aqueous potassium hydroxide solution, **40** reacted with 4-bromo-2-methyl-2-butene to give prenylated derivatives **41–43**, already known as natural prod-

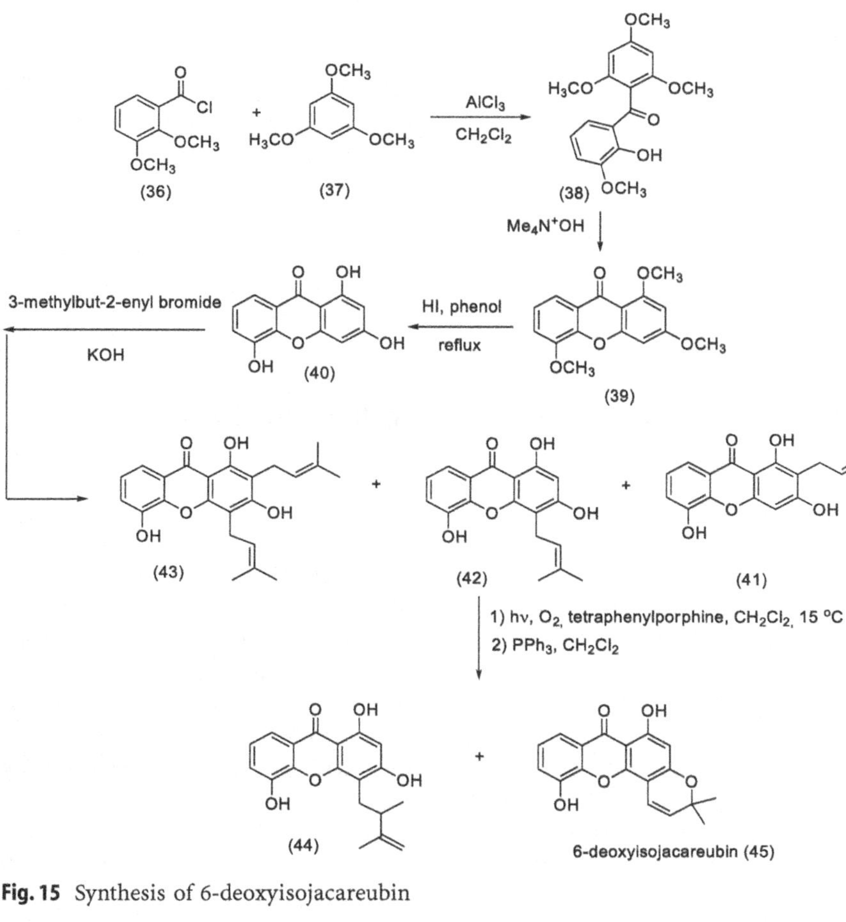

Fig. 15 Synthesis of 6-deoxyisojacareubin

ucts (Fig. 15). When the photooxygenation–reduction sequence conditions were applied to C-4-monoprenylated xanthone **42**, two products were obtained (Fig. 15): secondary allylic alcohol **44**, and as major product with 17% yield the pyranoxanthone **45** already known as a natural compound named 6-deoxyisojacareubin [15].

The C-2-monoprenylated xanthone **41** gave caledol (**46**), which is a secondary allylic alcohol, as oxidation product (Fig. 16) and similar experimental conditions led to the xanthone **47** and dicaledol (**48**) from (C-2, C-4)-diprenylated xanthone **43** (Fig. 17).

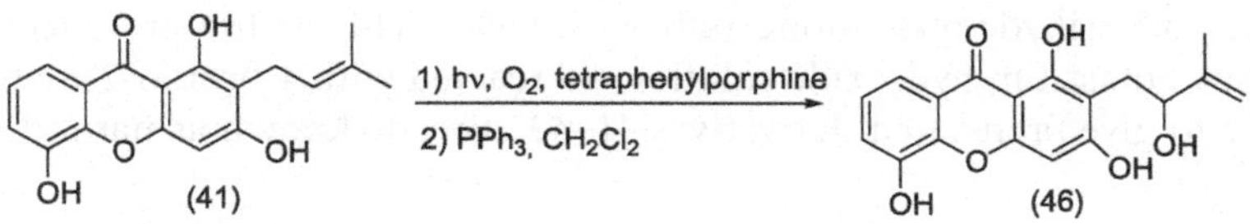

Fig. 16 Synthesis of caledol (**46**) by photooxygenation reaction

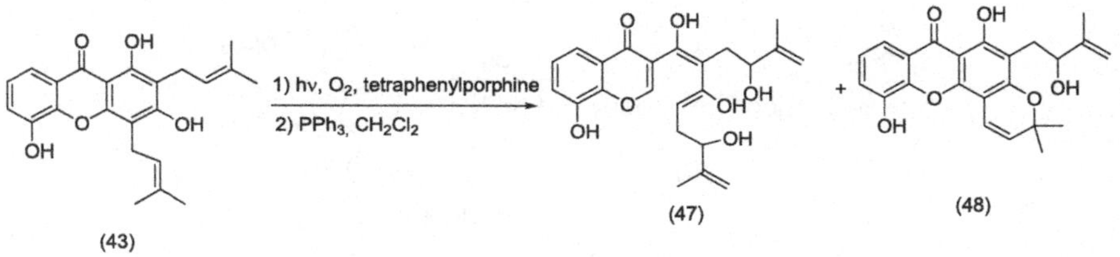

Fig. 17 Photooxygenation reaction of **43**

4
Isolation and Structures of Xanthones from *Hypericum* Species

4.1
Isolation Methods

Xanthones can be found in all parts of the plant. Extraction of xanthones from *Hypericum* species is usually carried out on dried plant material. The classical method using increasingly polar solvents has been used effectively. Lipophilic xanthones were generally separated by silica gel column chromatography (CC). For the polar xanthones, such as glycosidic and those containing hydroxyl groups, polyamide column chromatography, gel filtration chromatography on Sephadex LH-20, reversed-phase chromatography by preparative HPLC on an RP-18 column, and centrifugal partition chromatography (CPC) have been successfully employed.

Table 1 shows the part of the plant, the extracts containing xanthonoids in *Hypericum* species which have been worked, and the analytical methods used.

Table 1 Analytical methods for the isolation of xanthones from *Hypericum* species

Plant source	Plant part used	Extraction	Analysis	Refs.
H. androseamum	Roots	$CHCl_3$: MeOH	Si gel CC	[68]
H. annulatum	Aerial parts	EtOAc	Polyamide, Sephadex LH-20	[40]
H. ascyron	Whole plant	$CHCl_3$	Si gel, sephadex LH-20, ODS	[20]
H. balearicum	Twigs and leaves	Acetone	Si gel	[69]
H. beanii	Aerial parts	CH_2Cl_2	Sephadex LH-20 CC	[70]
H. brasiliense	Root and stem	CH_2Cl_2	Sephadex LH-20 CC	[16]
H. canariensis	Aerial parts	$CHCl_3$	Si gel CC	[19, 71]
H. chinense	Leaves	Hexane	Si gel CC, Toyopearl HW-20, HPLC	[72]
H. chinense	Stems	EtOAc	Si gel CC, Toyopearl HW-20, HPLC	[72]
H. erectum	Aerial parts	$CHCl_3$	Si gel CC	[73]
H. ericoides	Aerial parts	Ether	Si gel CC	[60]
H. geminiflorum	Aerial parts	MeOH	Si gel CC	[9]
H. henryi	Aerial parts	$CHCl_3$	Polyamide, Sephadex LH-20 CC	[15]
H. hookerianum	Woody stems	$CHCl_3$	Si gel column	[74]
H. inodorum	Not defined	$CHCl_3$	Si gel column	[75]
H. japonicum	Whole plant	$CHCl_3$	Si gel CC, Sephadex LH-20 CC	[76]
H. japonicum	Aerial parts	EtOAc	Si gel, polyamide, Sephadex LH-20 CC	[15]
H. japonicum	Aerial parts	CH_2Cl_2	Si gel CC, polyamide CC	[15]
H. maculatum	Not defined	CH_2Cl_2	Si gel CC	[77]
H. mysorence	Leaves and twigs	Ether	Si gel CC	[78]
H. mysorence	Timber	$CHCl_3$	Si gel CC	[79]
H. patulum	Cell cultures	EtOAc	Flash CC on Si gel	[80]
H. patulum	Cell cultures	CH_2Cl_2	Flash CC on Si gel	[80]
H. perforatum	Callus	Hexane	Si gel CC	[81]
H. reflexum	Aerial parts	$CHCl_3$	Si gel CC	[82]
H. roeperanum	Roots	CH_2Cl_2	Sephadex LH-20, HPLC on RP-18	[8]
H. sampsonii	Whole plant	EtOH	Sephadex LH-20	[21]
H. sampsonii	Whole plant	EtOAc	Si gel CC	[83]
H. scabrum	Aerial parts	EtOAc	Si gel CC, Sephadex LH-20, Toyopearl HW-20, Si gel HPLC	[84]
H. styhelioides	Leaves	BuOH	Sephadex LH-20, RP-HPLC on C18	[85]
H. subalatum	Whole plant	EtOH	Charcoal CC, Si gel CC	[86]
H. umbellatum	Aerial parts	EtOAc	Sephadex LH-20	[87]

4.2 Isolated Xanthones from *Hypericum* Species

Table 2 presents the structures of xanthones isolated and the *Hypericum* species from which they were obtained.

4.3 Structural Elucidation of Xanthones

^{1}H and ^{13}C NMR spectroscopies are the most useful methods in the structure elucidation of xanthones. ^{1}H NMR spectroscopy has been used for the determination of the substituents on each ring and it also gives information about the oxidation pattern. The observation of the signal at δ12–13 in the spectra shows chelated OH with hydroxyl substitution at position 1 or 8. When these positions are unsubstituted, aromatic protons appear between δ7.70 and 8.05 [101].

Application of the nuclear Overhauser effect (NOE) to the aromatic system may be used to determine the positions of substituent groups. The recently developed two-dimensional techniques, which can be found in numerous applications in the xanthone field, are very useful for the determination of these compounds. ^{13}C NMR analysis has played a major role in the rapid structure elucidation of xanthones. It gives information about the substitution pattern of aglycon and also about the positions of the glycosidic chains on aglycon, as well as the configuration and conformation of interglycosidic linkages [2].

The ^{13}C NMR spectra of a great number of naturally occurring xanthones have been reported and all chemical shifts assigned [102–104]. The carbonyl carbon shift is very important for identifying the oxygenation pattern of xanthone derivatives. The general observation is a decrease in electron density at the carbonyl carbon due to chelation:

1. Double chelation (1,8-di-OH): carbonyl carbon δ184
2. Monochelation (1- or 8-OH): carbonyl carbon δ178–181
3. No chelation: carbonyl carbon δ174–175

^{13}C NMR spectroscopy has now become a routine method for the structure elucidation of new xanthones. Hambloch and Frahm introduced a computer program called SEOX 1, which rapidly identifies unknown xanthones with the help of additivity rules that represents a remarkable facility in structure elucidation [105].

The UV spectrum varies in a characteristic manner depending on the oxygenation pattern. It is basically useful for locating free hydroxyl groups on the xanthone nucleus. Particularly, a free hydroxyl group at position 3 or position 6 can be easily detected by addition of NaOAc, which results in a bathochromic shift of the 300–345-nm band. A strong base such as NaOMe is capable of deprotonating all phenolic hydroxyl groups except those at-

Table 2 Structures of xanthones isolated from *Hypericum* species

	Compound	Plant source	Refs.
1	Bijaponicaxanthone 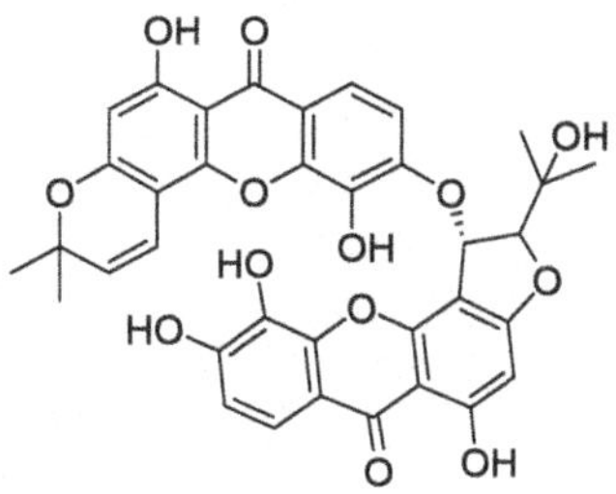	*H. japonicum*	[15]
2	Jacarelhypherol A: diasteromer of bijaponicaxanthone	*H. japonicum*	[88]
3	Jacarelhyperol B: 6-deoxy 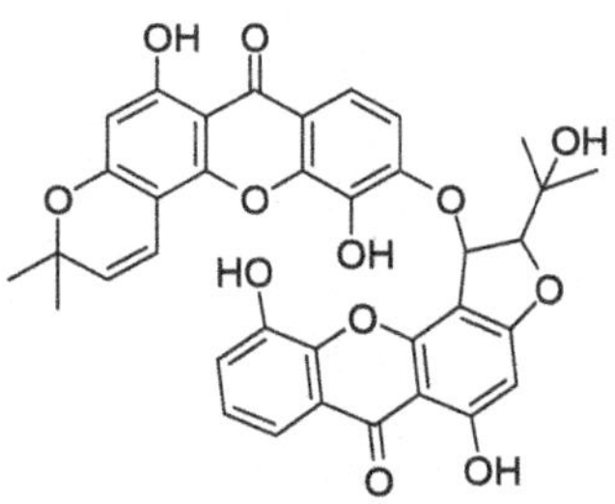	*H. japonicum*	[88]
4	Bijaponicaxanthone C 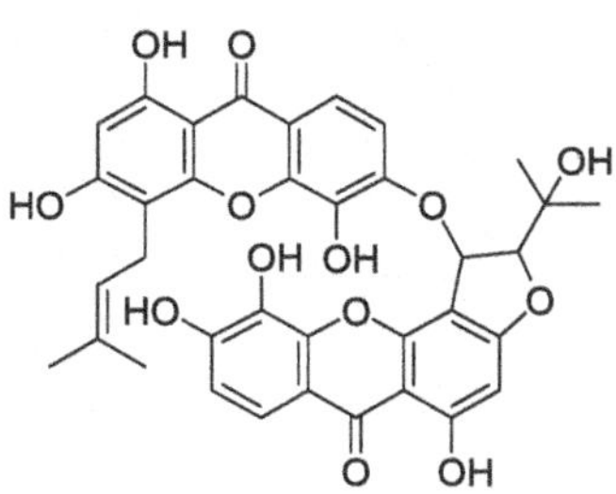	*H. japonicum*	[89]
5	Cadensin D	*H. canariensis* and *H. mysorense*	[19, 90]

Table 2 (continued)

	Compound	Plant source	Refs.
6	Calycinoxanthone D; 1,3,5,6-tetrahydroxy-4-lavandulylxanthone	*H. roeperanum*	[8]
	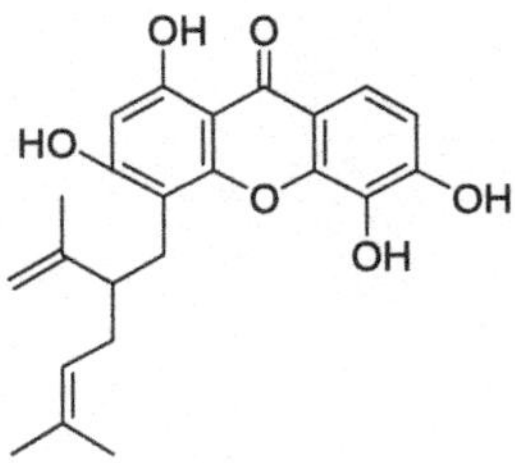		
7	4-Chloro-3,8-dihydroxy-6-methoxy-1-methylxanthone	*H. ascyron*	[20]
8	Deprenylrheediaxanthone B: 1,2-dihydro-5,9,10-trihydroxy-1,1,2-trimethyl-6*H*-furo[2,3-*c*]xanthen-6-one	*H. japonicum* and *H. roeperanum*	[8, 15]
	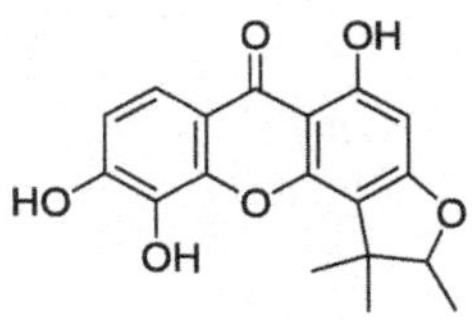		
9	5-*O*-Methyldeprenylrheediaxanthone B	*H. roeperanum*	[8]
10	6-Deoxyisojacareubin: 6,11-dihydroxy-3,3-dimethyl-3*H*,7*H*-pyrano[2,3-*c*]xanthen-7-one	*H. japonicum*	[15]
	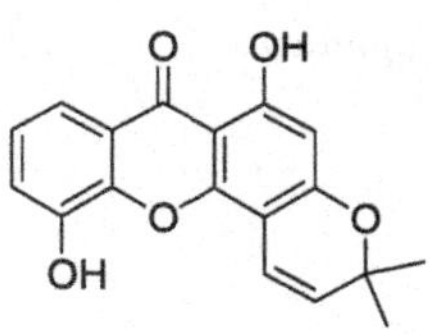		
11	1,2-Dimethoxyxanthone	*H. mysorense*	[79]
	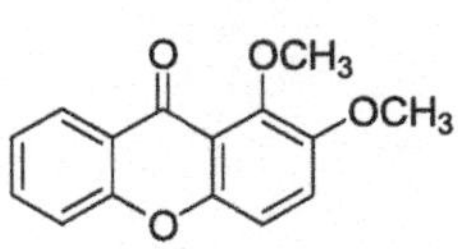		
12	1-Hydroxy-7-methoxyxanthone	*H. mysorense*	[79]

Table 2 (continued)

	Compound	Plant source	Refs.
13	2-Hydroxy-3-methoxyxanthone	*H. mysorense*	[79]
14	2,3-Methylenedioxyxanthone	*H. mysorense*	[91]
	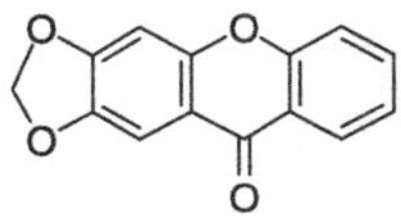		
15	2,5-Dihydroxyxanthone	*H. canariensis*	[71]
16	5-Hydroxy-2-methoxyxanthone	*H. inodorum*, *H. roeperanum*	[8, 75]
17	2-Hydroxy-5-methoxyxanthone	*H. canariensis*, *H. androsaemum*	[68, 71]
18	5-(1,1-Dimethyl-2-propenyl)-3,6,8-trihydroxy-1,1-bis(3-methyl-2-butenyl)-1*H*-xanthene-2,9-dione	*H. erectum*	[73]
	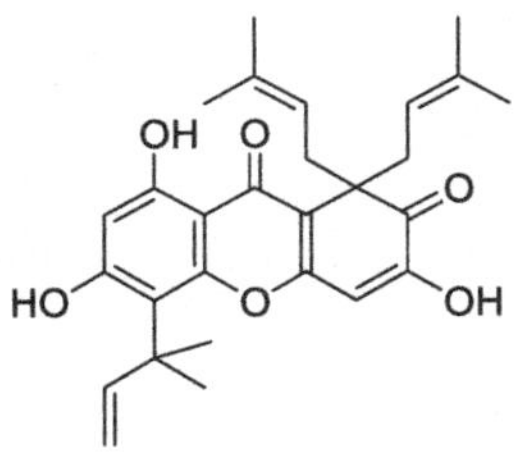		
19	Garcinone B	*H. patulum*	[32]
	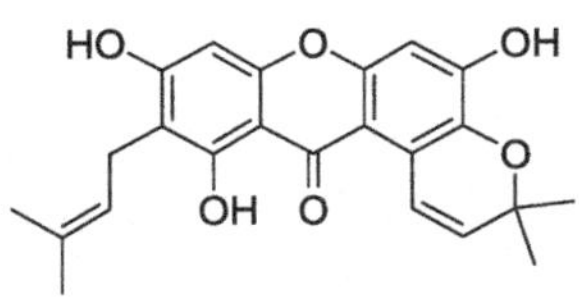		
20	Gemixanthone A	*H. geminiflorum*	[9]
	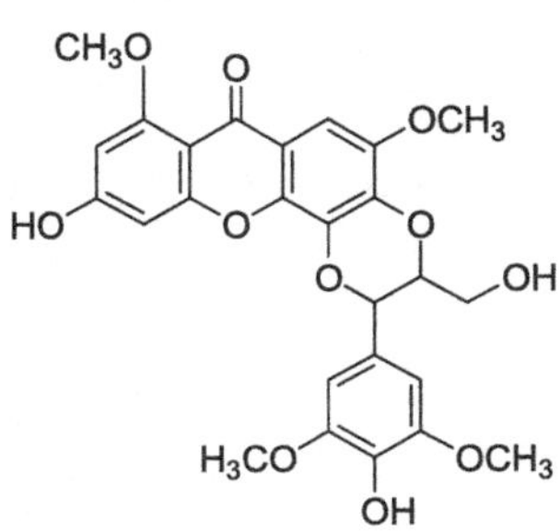		

Table 2 (continued)

	Compound	Plant source	Refs.
21	2-Hydroxyxanthone	*Hypericum* spp.	[71, 79]
22	2-Methoxyxanthone	*Hypericum* spp.	[20]
23	Hypericanarin 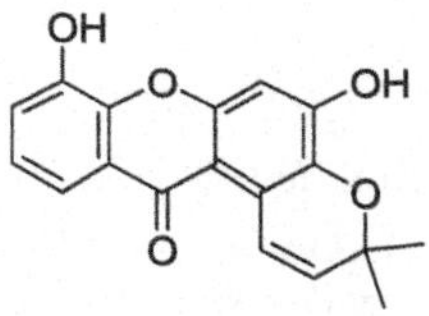	*H. canariensis*	[19]
24	Hypericanarin B 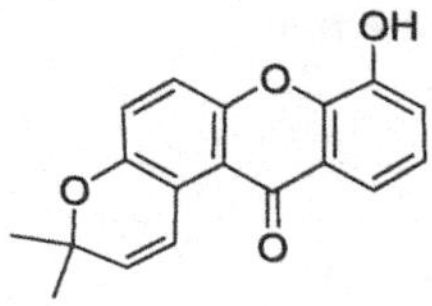	*H. canariensis*	[92]
25	Hyperxanthone: 5,9-dihydroxy-3,3-dimethylpyrano-[3,2-*a*]xanthen-12(3*H*)-one 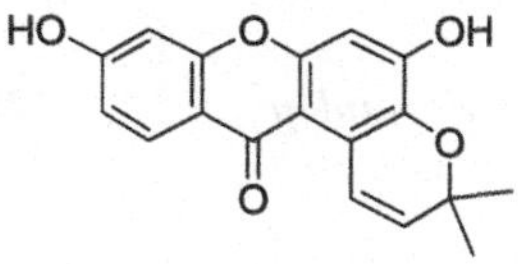	*H. sampsonii*	[93]
26	Hyperireflexin 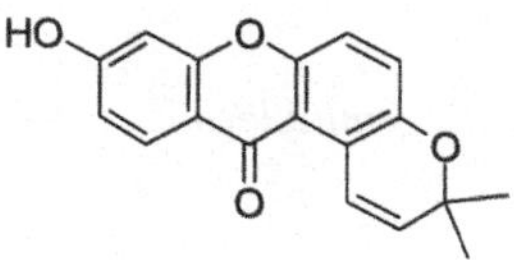	*H. reflexum*	[82]
27	Hyperxanthone B 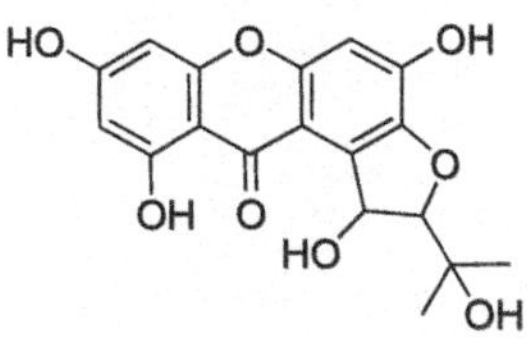	*H. scabrum*	[84]

Table 2 (continued)

	Compound	Plant source	Refs.
28	Hyperxanthone A (1′-deoxyhyperxanthone)	*H. scabrum*	[84]
	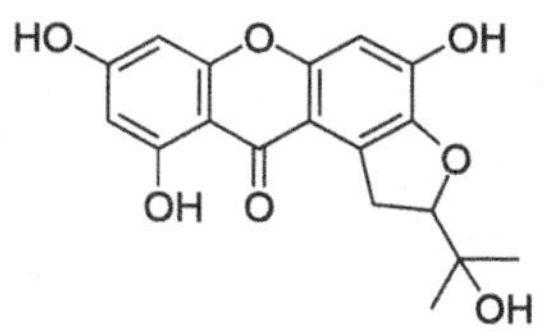		
29	Jacarelhyperol A	*H. japonicum*	[88]
	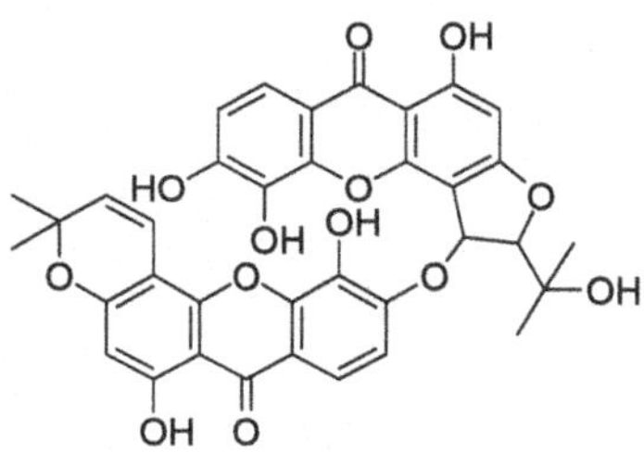		
30	Jacarelhyperol B: 6-deoxyjacarelhyperol	*H. japonicum*	[88]
31	Kielcorin (±) form	*Hypericum* spp.	[3, 15, 19]
32	Methoxykielcorin	*H. reflexum*	[82]
	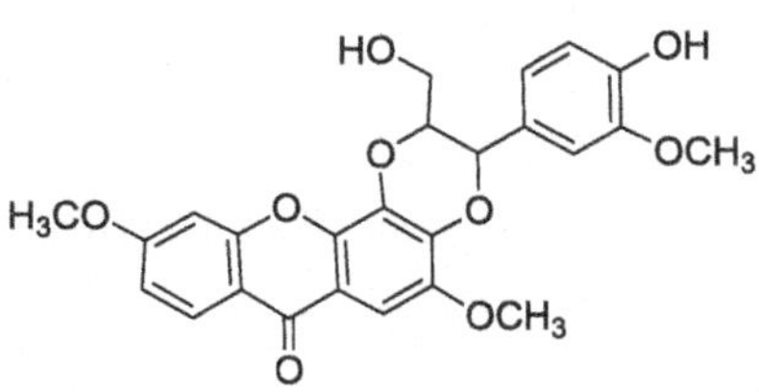		
33	Maculatoxanthone	*H. maculatum*	[77]
	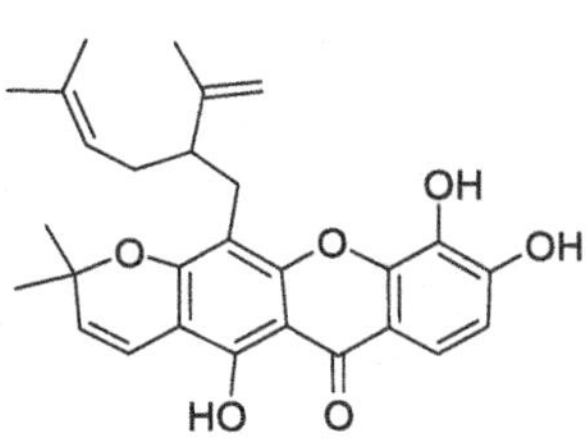		

Table 2 (continued)

	Compound	Plant source	Refs.
34	Paxanthone B 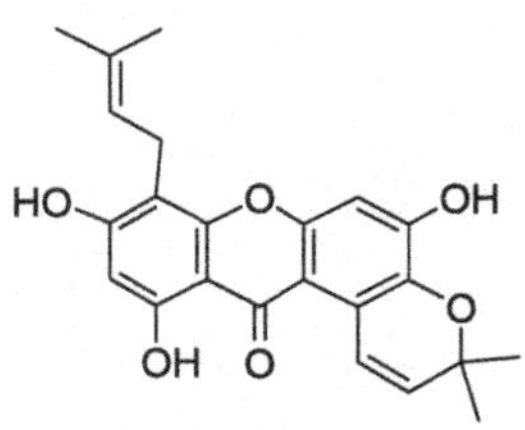	*H. patulum*	[32]
35	Paxanthonin 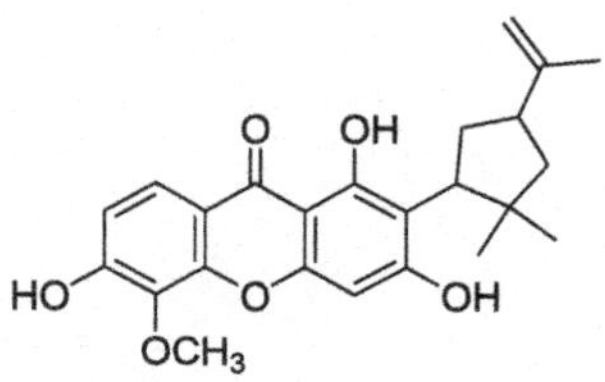	*H. patulum*	[33, 94]
36	5-*O*-Demethylpaxanthonin	*H. patulum*, *H. roeperanum*, and *H. styphelioides*	[8, 85, 94]
37	5-*O*-Demethyl-6-deoxypaxanthonin	*H. styphelioides*	[85]
38	2,3-Dihydroxy-1,6,7-trimethoxyxanthone 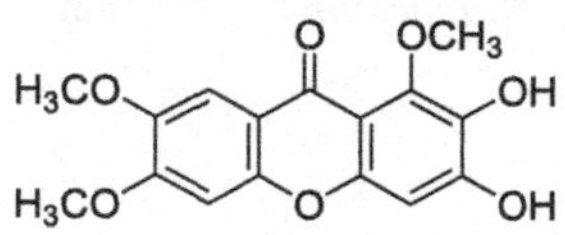	*H. geminiflorum*	[10]
39	3,6-Dihydroxy-1,5,7-trimethoxyxanthone	*H. geminiflorum*	[10]
40	Roeperanone (+),(*E*) form 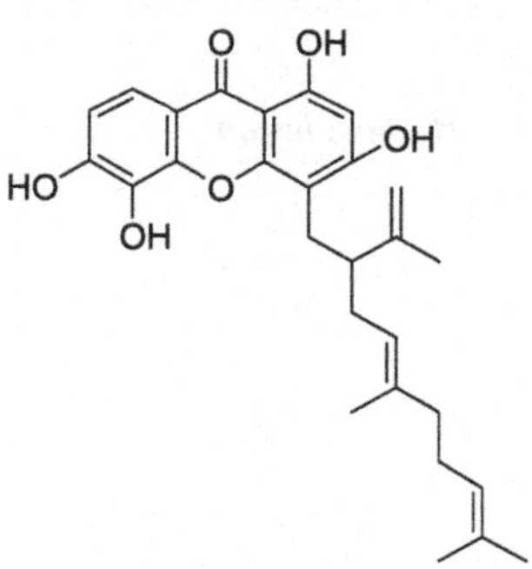	*H. roeperanum*	[8]

Table 2 (continued)

	Compound	Plant source	Refs.
41	Subalatin	*H. subalatum*	[86]
	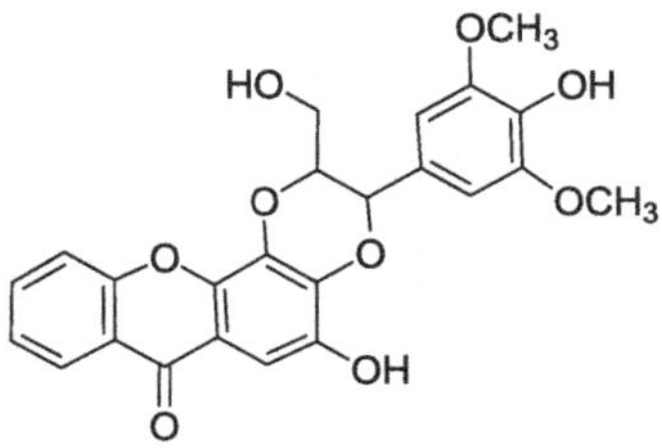		
42	1,4,6-Trihydroxy-3-methoxy-2-prenylxanthone	*H. sampsonii*	[83]
	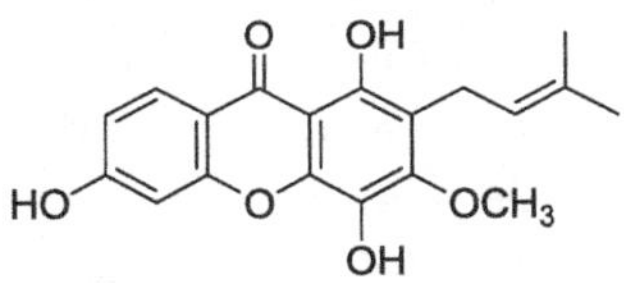		
43	Morusignin D; 1,3,6-trihydroxy-5-methoxy-2-prenylxanthone	*H. patulum*	[33]
	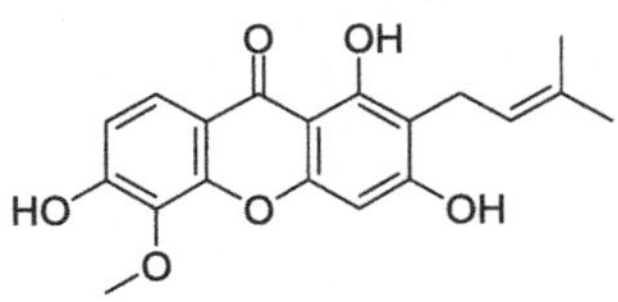		
44	1,3,5,6-Tetrahydroxy-4-prenylxanthone; ugaxanthone	*H. japonicum*	[15]
45	2,3,6,8-Tetrahydroxy-1-prenylxanthone	*H. androsaemum* and *H. patulum*	[32, 68]
	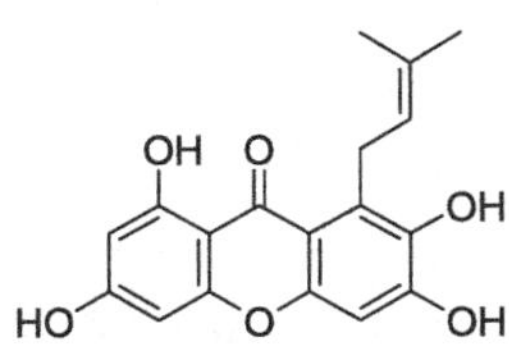		
46	Hyperxanthone C; 2,3,6,8-tetrahydroxy-1-(2-hydroxy-3-methyl-3-butenyl)xanthone	*H. scabrum*	[84]
47	Patuloside A; 1,5,6-trihydroxyxanthone 3-*O*-*β*-D-glucopyranoside	*H. patulum*	[12]

Table 2 (continued)

	Compound	Plant source	Refs.
48	Patuloside B; 1,5,6-trihydroxyxanthone 3-*O*-[α-L-rhamnopyranosyl-(1 → 2)-α-D-glucopyranoside] 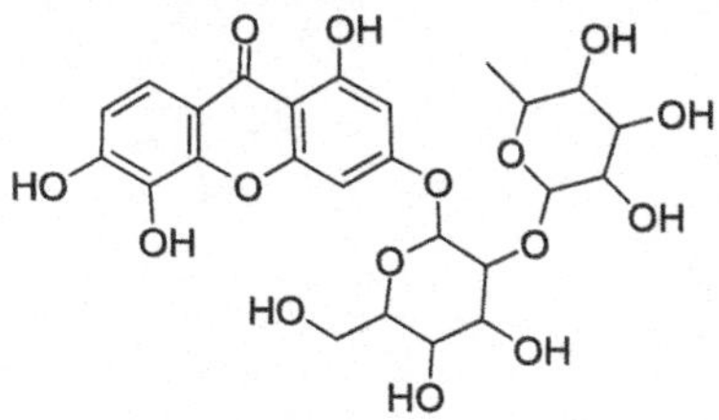	*H. patulum*	[12]
49	6,7-Dihydroxy-1,3-dimethoxyxanthone	*H. geminiflorum*	[9]
50	3,6-Dihydroxy-1,7-dimethoxyxanthone	*H. ascyron*	[20]
51	1,6-Dihydroxy-5,7-dimethoxyxanthone	*H. subalatum*	[95]
52	1-Hydroxy-5,6,7-trimethoxyxanthone	*H. perforatum* ssp. *Perforatum*	[81]
53	2,4-Dihydroxy-3,6-dimethoxyxanthone	*H. reflexum*	[82]
54	4-Hydroxy-2,3,6-trimethoxyxanthone	*H. reflexum*	[82]
55	2,7-Dihydroxy-3,4-dimethoxyxanthone	*H. subalatum*	[95]
56	7-Hydroxy-2,3,4-trimethoxyxanthone	*H. ericoides*	[60]
57	Padiaxanthone 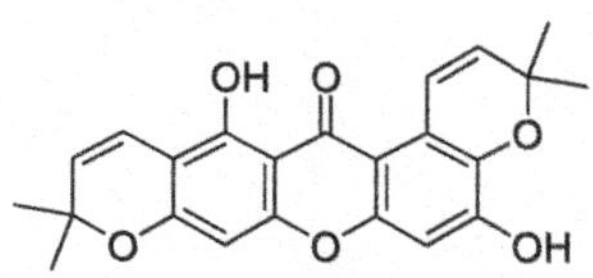	*H. patulum*	[94]
58	Patulone; 3,6,8-trihydroxy-1,1-bis(3-methyl-2-butenyl)-1*H*-xanthene-2,9-dione 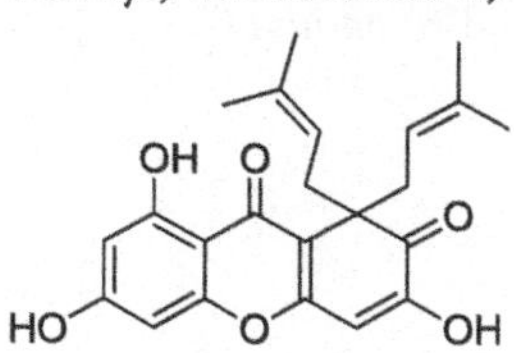	*H. patulum*	[96]
59	Toxyloxanthone B; 5,9,11-trihydroxy-3,3-dimethyl-pyrano [3,2-*a*]xanthen-12(3*H*)-one 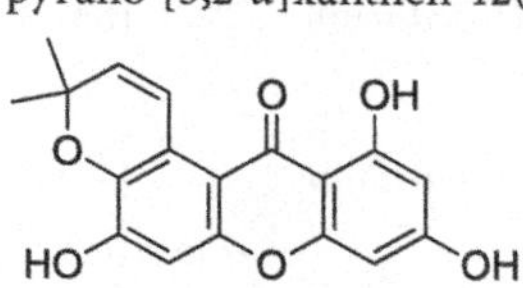	*H. androsaemum*, *H. patulum*	[68]

Table 2 (continued)

	Compound	Plant source	Refs.
60	Paxanthone 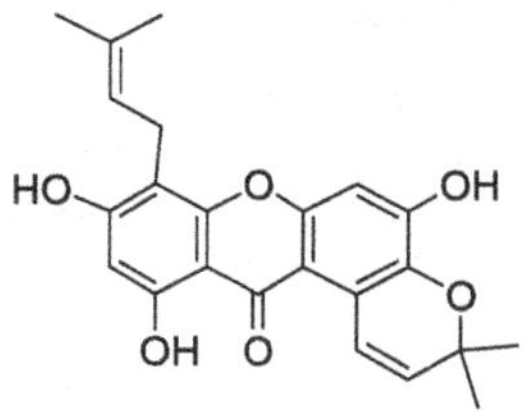	*H. patulum*	[31]
61	3-*O*-Methylpaxanthone	*H. perforatum* ssp. *Perforatum*	[81]
62	Hyperxanthone E 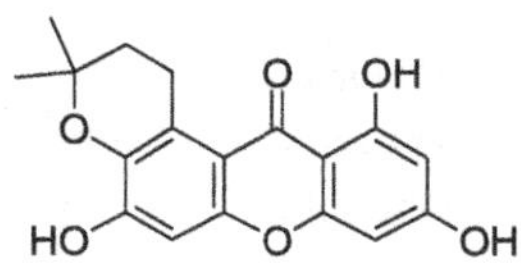	*H. scabrum*	[84]
63	Isojacareubin 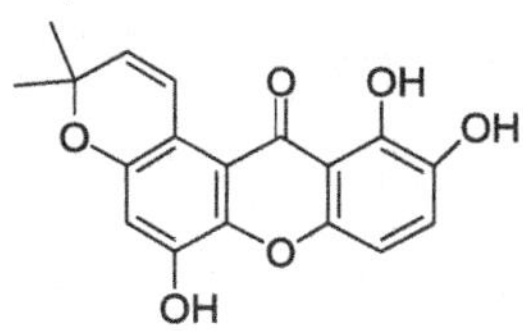	*H. japonicum* and *H. roeperanum*	[8, 15]
64	5-*O*-Methylisojacareubin 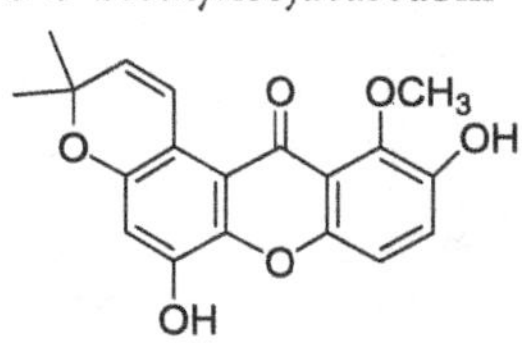	*H. roeperanum*	[8]
65	2,6,8-Trihydroxy-1-(2-hydroxy-3-methyl-3-butenyl)xanthone: hyperxanthone D 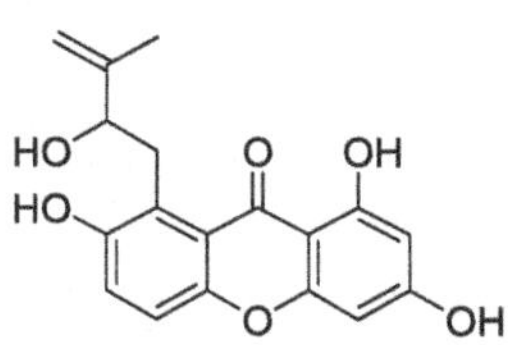	*H. scabrum*	[84]
66	4-Hydroxy-1,2-dimethoxyxanthone	*H. geminiflorum*	[9]

Table 2 (continued)

	Compound	Plant source	Refs.
67	1,2,5-Trihydroxyxanthone	*H. balearicum*	[69]
68	1,5-Dihydroxy-2-methoxyxanthone	*H. roeperanum*	[8]
69	1,3,7-Trihydroxyxanthone: gentisein	*Hypericum* spp.	[3]
70	Xanthohypericoside: 1,7-dihydroxyxanthone 3-*O*-β-D-glucopyranoside 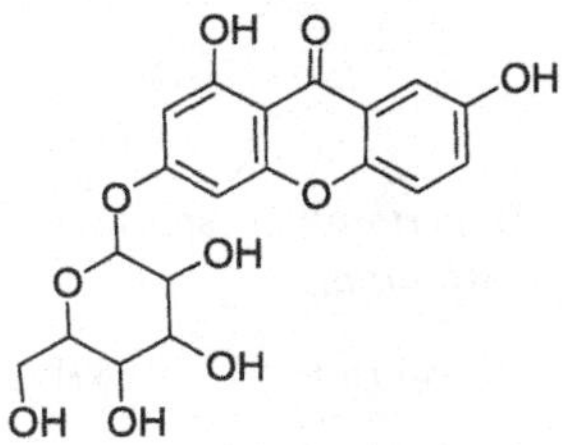	*H. annulatum*	[97]
71	1,5-Dihydroxyxanthone 6-*O*-β-D-glucopyranoside	*H. japonicum*	[15]
72	1-Hydroxy-6,7-dimethoxyxanthone	*H. mysorense*	[79]
73	2-Hydroxy-3,4-dimethoxyxanthone	*H. sampsonii*	[93]
74	3-Hydroxy-2,5-dimethoxyxanthone	*H. androsaemum*	[68]
75	3,6-Dihydroxy-2-methyoxyxanthone	*H. reflexum*	[83]
76	1,3,5-Trihydroxyxanthone-4-sulfonic acid; 5-*O*-β-D-glucopyranoside 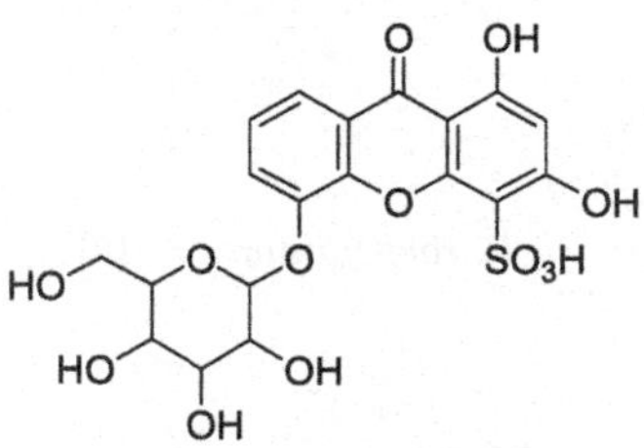	*H. sampsonii*	[21]
77	1,3-Dihydroxy-5-methoxyxanthone-4-sulfonic acid	*H. sampsonii*	[21]
78	Jacarelhyperol D 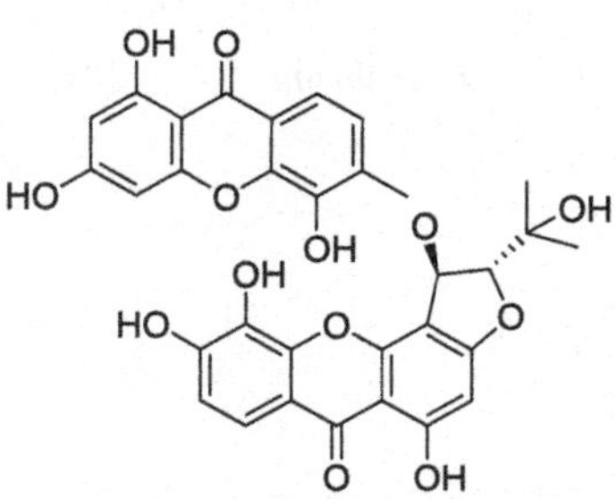	*H. japonicum*	[98]
79	4,6-Dihydroxy-2,3-dimethoxyxanthone	*H. chinense*	[99]

Table 2 (continued)

	Compound	Plant source	Refs.
80	2,6-Dihydroxy-3,4-dimethoxyxanthone	*H. chinense*	[99]
81	6-Hydroxy-2,3,4-trimethoxyxanthone	*H. chinense*	[99]
82	3,6-Dihydroxy-1,2-dimethoxyxanthone	*H. chinense*	[99]
83	4,7-Dihydroxy-2,3-dimethoxyxanthone	*H. chinense*	[99]
84	3,7-Dihydroxy-2,4-dimethoxyxanthone	*H. chinense*	[99]
85	1,6-Dihydroxyisojacareubin-5-*O*-*β*-D-glucoside 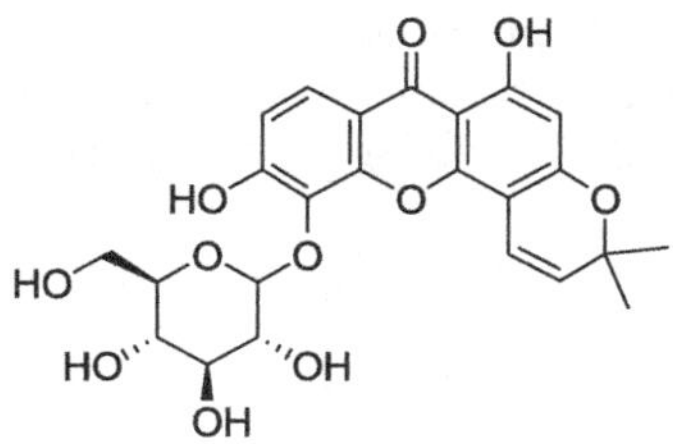	*H. japonicum*	[100]
86	3,6,7-Trihydroxy-1-methoxyxanthone	*H. japonicum*	[100]

tached at positions 1 and 8. These hydroxyls can be detected by the complex formed on addition of $AlCl_3$ which is stable to HCl. *Ortho*-dihydroxyl groups similarly may give this complex, but can be distinguished from the former by the instability of the complexes in HCl [2].

Mass spectrometry (MS) has not been applied extensively to the study of naturally occurring xanthones, but the mass spectral data provide valuable information about the structure elucidation of xanthones. As well as electron impact MS, which is a routine technique for the structure elucidation of xanthones, recently developed soft ionization techniques, such as desorption–chemical ionization MS (D/CI-MS) and fast atom bombardment MS (FAB-MS), are of great interest for the analysis of glycosides. Molecular ion peaks can be observed without derivatization. Tandem MS/MS can be extensively employed in directly characterizing constituents of complex mixtures. Recently, xanthone profiles of *H. perforatum* cell cultures were identified by HPLC-MS/MS analysis [106].

5
Biological Activities of Xanthones Isolated from *Hypericum* Species

The study of xanthones in *Hypericum* species is interesting not only for the chemotaxonomic investigation but also from the pharmacological point of view. Several biological properties, such as strong and selective inhibition of MAO-A, in vitro toxicity, in vivo antitumor activity, as well as antibacterial and antifungal activities, have been attributed to xanthones [2]. Also, the xanthones isolated from genus *Hypericum* have been found to possess various biological activities. These pharmacological properties are explained below.

5.1 MAO Inhibitor Activity

Monoamine oxidase (MAO), which exists as two isoenzymes, MAO-A and MAO-B, plays a key role in the regulation of some physiological amines [107] and acts by desamination of some neurotransmitters, such as catecholamine, serotonin, or tyramine, as shown in Fig. 18 [108]. Inhibitors of MAO are used as antidepressive drugs [107].

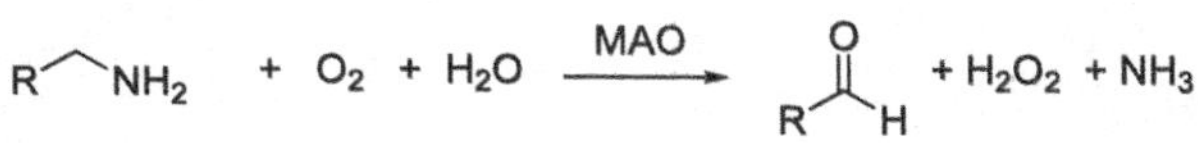

Fig. 18 Mechanism of desamination of monoamine by MAO

The antidepressive activity of *H. perforatum* has been demonstrated in vivo [109]. This activity was first explained by the presence of hypericin, which has been shown to inhibit MAO-A and B in vitro [110]. But recent investigations indicate that certain xanthones and flavonoid aglycones which have potent MAO inhibitory properties may have a more important contribution to play in the antidepressive activity of *H. perforatum* [111]. As evidence confirming these investigations, the xanthones of *H. brasiliense* showed differing degrees of inhibition of MAO-A and B. Moreover, they seemed to act in a reversible and time-independent manner. The tetracyclic xanthone, 6-deoxyjacareubin, was more potent versus MAO-A and MAO-B than the tricyclic xanthone 5-hydroxy-2-methoxyxanthone. 1,5-Dihydroxyxanthone emerged as the most potent and selective inhibitor, with an IC_{50} value of 0.73 μM for MAO-A and a selectivity index of 0.01. Thus, the internally H-bonded hydroxyl group might be operative in the mechanism of inhibition. These activities may have relevance in the search for new drugs suitable for the treatment of depression [16].

Recently, 59 xanthones (= 9*H*-xanthen-9-ones) of natural or synthetic origin were investigated for their inhibitory activity toward MAO-A and MAO-B. The majority of the compounds demonstrated reversible, time-independent activities, with selectivity toward MAO-A. The most active inhibitor (1,5-dihydroxy-3-methoxyxanthone) had an IC_{50} of 40 nM. 3D-QSAR studies revealed the importance of an OH substituent in position 1 or 5 instead of a MeO substituent and the contrary is true for position 3, where MeO substituents lead to more active compounds than OH substituents. The CoMFA/GOLPE procedure provided information about the importance of an electron-rich zone between positions 4 and 5, and the unfavorable effect of an electron-rich zone around position 7. The ALMOND procedure showed the importance of the distance between two H-bond acceptor groups in modulating activity. These promising MAO inhibitory activities should be confirmed by in vivo experiments for the design of new antidepressant drugs [112].

5.2
Antifungal Activity

All the isolated compounds, including xanthones from *H. brasiliense*, are antifungal against *Cladosporium cucumerinum*. The minimum quantities of γ-pyrone (hyperbrasilone), 5-hydroxy-1-methoxyxanthone, 6-deoxyjacareubin, and 1,5-dihydroxyxanthone required to inhibit growth of the fungus in the bioautographic assay on TLC plates were 3, 3, 3, and 0.25 μg, respectively. Propiconazole, a triazole antifungal agrochemical, was active at 0.1 μg in the same bioassay [16].

The isolated compounds from *H. roeperanum* were tested for their antifungal activity against *Candida albicans* and *Cladosporium cucumerinum* in TLC bioautographic assays [113, 114]. The minimum amount of xanthones 2-deprenylreediaxanthone B, 5-*O*-methyl-2-deprenylrheediaxanthone B, calcinoxanthone D, roeperanone, and 5-*O*-demethylpaxanthonin required to inhibit the growth of *C. albicans* on TLC plates was 1 μg, whereas xanthone 5-*O*-methylisojacareubin showed antifungal activity at 5 μg. The reference compounds amphotericin B and miconazole were active at 1 and 0.001 μg, respectively. It must be noted that the crude dichloromethane extract did not show activity against *C. albicans* at the usual test level of 100 μg owing to the low concentration of the xanthones in *H. roeperanum*. None of the xanthones from *H. roeperanum* inhibited growth of *C. cucumerinum* at 10 μg [8].

5.3
Cytotoxic Activity

According to the first report of sulfonated xanthonoids which were isolated from *H. sampsonii*, xanthones 1,3-dihydroxy-5-methoxyxanthone-4-sulfonate and 1,3-dihydroxy-5-*O*-β-D-glycopyranosylxanthone-4-sulfonate exhibited significant cytotoxicity against the P388 cancer cell line. They were evaluated for cytotoxicity proportion against the P388 cancer cell line and were found to be moderately active (ED_{50} of 3.46 and 15.69 $\mu mol\,L^{-1}$, respectively). By contrast, VP-16 (positive control) had an ED_{50} of 0.064 $\mu mol\,L^{-1}$ [21].

5.4 Coagulant Activity

The compounds from *H. japonicum* were tested for their coagulant activity in in vitro systems. They were found to exert an interesting coagulant activity. Compound 1,5-dihydroxyxanthone-6-*O*-*β*-D-glucoside showed activity by promoting coagulation of PT (prothrombin time reagent) and isojacareubin showed anticoagulation of APTT (active partial thromboplastin time reagent) [15].

5.5 Antioxidant Activity

The xanthone compounds 1,3,5-trihydroxy-2-(2′,2′-dimethyl-4′-isopropenyl) cyclopentanylxanthone, 5-*O*-demethyl-6-deoxypaxanthonin, and 5-*O*-demethylpaxanthonin, as well as 3,5-dihydroxybenzophenone-4-*β*-D-glucoside and 3-geranyl-1-(3-methylbutanoyl)phloroglucinol isolated from the leaves of *Hypericum styphelioides*, were evaluated for their antioxidative properties in Trolox equivalent antioxidant capacity (TEAC) and chemiluminescence (CL) assays.

The free-radical-scavenging activity of compounds was evaluated in the TEAC and CL assays. The first measures the relative ability of antioxidant substances to scavenge the radical cation 2,2′-azinobis(3-ethylbenzothiozoline-6-sulfonate) ($ABTS^{\cdot+}$) as compared to a standard amount of the synthetic antioxidant Trolox (6-hydroxy-2,5,7,8-tetramethylchroman-2-carboxylic acid). The CL assay measures the inhibition of iodophenol-enhanced chemiluminescence by a horseradish peroxidase/perborate/luminol system. Trolox was used as the reference antioxidant. The results showed that xanthones exhibited free-radical-scavenging activity at potency levels comparable to those of reference antioxidant compounds quercetin and rutin, while the benzophenone and phloroglucinol type compounds had more moderate activities [87].

5.6 Antimicrobial Activity

Isojacareubin, which was found for the first time in *H. japonicum*, exhibited antimicrobial activity together with a new isopentenylated flavonol, salothranol. The antimicrobial test was carried out by the agar-well method using *Staphylococus aureus*. The compounds were examined as a 50% DMSO solution. The activity was expressed by the inhibitory diameter, which was measured after incubation for 18 h at 37 °C. The minimum inhibitory concentrations (MICs) of isojacareubin and salothranol were 125 μg/mL [77].

5.7
Anti-inflammatory Activity

The five xanthone compounds, demethylpaxantonin, patulone, garcinone B, tripteroside, and 1,3,5,6-tetrahydroxyxanthone, purified from a callus tissue culture of *H. patulum* were evaluated for their anti-inflammatory activity.

It is very important in anti-inflammatory drug development to search for natural leading compounds which exert the following pharmacological actions: (1) prevention of release of prostaglandins (PGs), major chemical mediators in the regulation of inflammation, by direct inhibition of the enzymes responsible for arachidonic acid (AA) and PG biosynthesis, including phospholipase A_2 and cyclooxygenase-2 (COX-2); and (2) suppression of transcription control of genes encoding enzymes responsible for PG biosynthesis or inflammatory cytokines. Two compounds were found with such pharmacological actions: garcinone B, which inhibits both A23187-induced PGE_2 release and LPS-induced NF-κB-dependent transcription, and patulone, which prevents COX-1 activity and A23187-induced PGE_2 release, raising the possibility of its anti-inflammatory action. These results suggested that garcinone B could become a neuropharmacological tool to elucidate intracellular signaling pathways involved in inflammation [115].

5.8
Use of Some *Hypericum* Species Containing Xanthones

It has been observed that a growing number of *Hypericum* species containing xanthones exhibit various biological properties and are used as chemotherapeutic agents in indigenous medicine for the treatment of many diseases. Typical examples are given below:

- *Hypericum* was widely used in the folk medicine of a number of European countries as a shooting agent, an antiphlogistic in the treatment of inflammation of the bronchi and the urogenital tract, a hemorrhoid treatment, and a healing agent in the treatment of traumas, burns and scalds, ulcers of various kinds, and other local and general illnesses [116].
- In the Canary Islands, various species of *Hypericum* genus have been used in folkloric medicine as a vermifuge, diuretic, and wound healer, as well as a sedative, antihysteric, and antidepressant agent [117].
- *H. perforatum* Linn. is used as an antiseptic or anthelmintic in Sri Lanka and India [118].
- *H. japonicum* Thunb. has been used in Chinese herbal medicine for the treatment of some bacterial diseases, infectious hepatitis, and tumors [119].
- Stems, leaves, and flowers of *H. ericoides* are used in Valentian folk medicine [74].

- *H. ascyron* is used in the treatment of numerous disorders, such as abscesses, boils, headache, nausea, and stomach ache in Chinese herbal medicine [119].
- *H. patulum* has been used in Chinese herbal medicine for the treatment of hepatitis, bacterial diseases, and nasal hemorrhage [119].
- *H. roeperanum* is employed, alone or in association with various plants, to cure female sterility [120].
- *H. sampsonii* is a herbal medicine used in the treatment of blood statis, to relieve swelling, and as an antitumor herb in Taiwan [121].
- *H. scabrum* is one of the most popular medicinal herbs in Uzbekistan and is used in the treatment of bladder, intestinal, and heart diseases, rheumatism, and cystitis [122, 123].
- *H. styphelioides* has been employed in traditional Cuban herbal medicine as a depurative, diaphoretic, diuretic, and tonic against blennorrhea, cold, cough, and dysmenorrhea and for the treatment of arthritis, rheumatism, hepatitis, herpes, and syphilis [124].
- Today this medicinal plant is used for these traditional purposes, but it is also largely used for the treatment of depression [118].

6 Conclusion

The recent widespread interest in the antidepressant activity of *H. perforatum* has encouraged the investigation of secondary metabolites from other *Hypericum* species. Since species of this genus occur widely in the temperate regions of the world, they have been used as traditional plants in various parts of the world. They produce several types of secondary metabolites, including flavonoids, biflavonoids, xanthones, anthraquinones, prenylated phloroglucinols, and benzophenones. Among these compounds xanthones show outstanding biological activities, such as monoamine oxidase inhibition and cytotoxic and antitumor properties, although they are relatively rare in nature in comparison with other phenolic compounds. For this reason these compounds have provoked great interest.

In this chapter, as well as the methods currently used for the isolation, separation, and structure elucidation of xanthones, their biosynthesis, synthesis, and importance as therapeutic agents was also discussed. The use of recently developed chromatographic techniques will provide characterization of these compounds and lead to the discovery of new xanthones. There are still many xanthones waiting to be discovered and evaluated by researchers for their many more biological activities.

References

1. Bennett GJ, Lee HH (1989) Phytochemistry 28:967
2. Hostettmann K, Hostettmann M (1989) Xanthones. In: Dey PM, Harborne JB (eds) Methods in Plant Biochemistry, vol 1. Plant Phenolics. Academic, London
3. Peres V, Nagem TJ (1997) Phytochemistry 44:191
4. Pedro M, Cerqueria F, Sausa ME, Nascimento MSJ, Pinto M (2002) Bioorg Med Chem 10:3725
5. Mandal S, Das PC, Joshi PC (1992) J Indian Chem Soc 69:611
6. Gottlieb OR, Stefani GM (1970) Phytochemistry 9:453
7. Gottlieb OR, Mesquita AAL, Oliveira GG, Melo MT (1970) Phytochemistry 9:2537
8. Rath G, Potterat O, Mavi S, Hostettmann K (1996) Phytochemistry 43:513
9. Chung MI, Weng JR, Lai MH, Yen MH, Lin CN (1999) J Nat Prod 62:1033
10. Chung MI (2002) Planta Med 68:25
11. Rodriguez S, Wolfender JL, Odontuya G, Purev O, Hostettman K (1995) Phytochemistry 40:1265
12. Ishiguro K et al. (1999) J Nat Prod 62:906
13. Hostettmann K, Wagner H (1977) Phytochemistry 16:821
14. Demirkiran O (2005) PhD thesis, Trakya University, Turkey
15. Wu QL, Wang SP, Du LJ, Yang JS, Xiao PG (1998) Phytochemistry 49:1395
16. Rocha L, Marston A, Kaplan AC, Stoeckli-Evans H, Thull U, Testa B, Hostettmann K (1994) Phytochemistry 36:1381
17. Castelao JF, Gottlieb OR, De Lima RA, Mesquita AAL (1977) Phytochemistry 16:735
18. Abou-Shoer M, Habib AA, Chang CJ, Cassady JM (1989) Phytochemistry 28:2483
19. Cardona ML, Fernandez MI, Pedro JR, Seoane E, Vidal R (1986) J Nat Prod 49:95
20. Hu LH, Yip SC, Sim KY (1999) Phytochemistry 52:1371
21. Hong D, Yin F, Hu LH, Lu P (2004) Phytochemistry 65:2595
22. Carpenter I, Locksey H, Scheinman F (1969) Phytochemistry 8:2013
23. Lewis JR (1963) Proc Chem Soc, p 373
24. Atkinson JE, Gupta P, Lewis JR (1968) Chem Commun, p 1386
25. Gupta P, Lewis JR (1971) J Chem Soc C, p 629
26. Ellis RC, Whalley WB, Ball K (1967) Chem Commun, p 803
27. Markham KR (1965) Tetrahedron 21:1449
28. Gottlieb OR (1968) Phytochemistry 7:411
29. Locksey HD, Moor I, Scheinmann F (1967) Tetrahedron 23:2229
30. Bennett GJ, Lee HH (1988) J Chem Soc Chem Commun, p 619
31. Ishiguro K, Fukumoto H, Nakajima M, Isoi K (1993) Phytochemistry 33:839
32. Ishiguro K, Fukumoto H, Nakajima M, Isoi K (1995) Phytochemistry 38:867
33. Ishiguro K, Fukumoto H, Nakajima M, Isoi K (1995) Phytochemistry 39:903
34. Beerhues L (1996) FEBS Lett 383:264
35. Schimidt W, Beerhues L (1997) FEBS Lett 420:143
36. Bennett GJ, Lee HH (1989) Phytochemistry 28:967
37. Schmidt W, Beerhues L (1997) FEBS Lett 420:143
38. Peters S, Schmidt W, Beerhues L (1998) Planta 204:64
39. Sultanbawa MUS (1980) Tetrahedron 36:1465
40. Kitanov GM, Nedialkov PT (2001) Phytochemistry 57:1237
41. Fujita M, Inoue T (1980) Chem Pharm Bull 28:2476
42. Tanaka T, Sueyasu T, Nonala G-I, Nishioka I (1984) Chem Bull 32:2676

43. Ferrari J, Terraux C, Sahpaz S, Msonthi JD, Wolfender JL, Hostettmann K (2000) Phytochemistry 54:883
44. Grover PK, Shah GD, Shah RCJ (1955) J Chem Soc, p 3982
45. Quillinan AJ, Scheinmann FJ (1973) J Chem Soc Perkin Trans I, p 1329
46. Jackson WT, Robert JB, Froelich LL, Gapinski DM, Mallett BE, Sawyer JS (1993) J Med Chem 36:1726
47. Familoni OB, Ionica I, Bower JF, Snieckus V (1997) Synlett, p 1081
48. Hassal CH, Lewis JR (1961) J Chem Soc 2:2312
49. Ullmann F, Pauchaud L (1906) Ann 350:108
50. Pedro M, Cerqueria F, Sousa ME, Nascimento MSJ, Pinto M (2002) Bioorg Med Chem 10:3725
51. Granoth I, Pownall HJ (1975) J Org Chem 40:2088
52. Vitale AA, Romanelli GP, Autinio JC, Pomilio AB (1994) J Chem Res S, p 82
53. Pillai RKM, Naiksatam P, Johnson F, Rajagopalan R, Watts PC, Cricchio R, Borras S (1986) J Org Chem 51:717
54. Burling ED, Jefferson A, Scheinmann F (1965) Tetrahedron 21:2653
55. Patel GN, Trivedi KN (1988) J Indian Chem Soc 65:192
56. Noungoue TD, Dijoux-Franca M-G, Mariotte A-M, Tsamo E, Daskiewiez JB, Bayet C, Barron D, Conseil G, Dipietro A (2000) Bioorg Med Chem 10:1343
57. Anand SM, Jain AC (1973) Indian J Chem 11:1237
58. Anand SM, Jain AC (1974) Aust J Chem 27:1515
59. Gil S, Sanz V, Tortajada A (1987) J Nat Prod 50:30 1
60. Cardona ML, Seoane E (1982) J Nat Prod 45:134
61. Quillinan AJ, Scheinmann FJ (1972) J Chem Soc Perkin Trans I, p 1382
62. Pinto MMM, Mesquita AAL, Gottlieb OR (1987) Phytochemistry 26:2045
63. Tanaka H, Ishihara M, Ichino K, Ohiwa N, Ito K (1988) Chem Pharm Bull 37:1916
64. Helesbeux JJ, Duval O, Dartigualengue C, Séraphin D, Oger JM, Richomme P (2004) Tetrahedron 60:2293
65. Crombie L, Games DE, McCormick A (1966) Tetrahedron Lett, p 151
66. Lin CN, Liou SS, Ko FN, Teng CM (1992) J Pharm Sci 81:1109
67. Liou SS, Shieh WL, Cheng TH, Wong SJ, Lin JN (1993) J Pharm Pharmacol 45:791
68. Neilsen H, Arends P (1979) J Nat Prod 42:301
69. Wollenweber E, Dorr M, Roitman JN (1994) Z Naturforsch C 49:393
70. Shiu WKP, Gibbons S (2006) Phytochemistry 67:2568
71. Cardona ML, Pedro JR, Seoane E, Vidal R (1985) J Nat Prod 48:467
72. Tanaka N, Takaishi Y (2006) Phytochemistry 67:2146
73. An TY, Hu LH, Chen ZL (2002) Chin Chem Lett 13:623
74. Wilairat R, Manosroi J, Manosroi A, Kijjoa A, Nascimento MSJ, Pinto M, Silva AMS, Eaton G, Herz W (2005) Planta Med 71:680
75. Cardona L, Fernandez I, Pedro JR (1992) Heterocycles 34:479
76. Ishiguro K, Nagata S, Fukumoto I, Yamaki M, Isoi K, Oyama Y (1993) Phytochemistry 32:1583
77. Arends P (1969) Tetrahedron Lett 55:4893
78. Kikuchi T, Kadota S, Matsuda S, Tanaka K, Namba T (1985) Chem Pharm Bull 33:557
79. Gunatilaka AAL, Jasmin De Silva AMY, Sotheeswaran S (1982) Phytochemistry 21:1751
80. Ishiguro K, Oku H, Isoi K (1999) Medicinal and aromatic plants In: Bajaj YS (ed) Biotechnology in agriculture and forestry, vol 43. Springer, Berlin, pp 199–212
81. Ferrari F, Pasqua G, Monacelli B, Cimino P, Botta B (2005) Nat Prod Res 19:171

82. Cardona L, Fernandez I, Pedro J, Serrano A (1990) Phytochemistry 29:3003
83. Don MJ, Huang YJ, Huang RL, Lin YL (2004) Chem Pharm Bull 52:866
84. Tanaka N et al. (2004) J Nat Prod 67:1870
85. Turro GD, Rubio CO, Gonzalez SP (2004) J Nat Prod 67:869
86. Chen MT (1988) Heterocycles 27:2589
87. Nedialkov PT, Kitanov GM, Dimitrova Z, Girreser U (2007) Biochem Syst Ecol 35: 118
88. Ishiguro K, Nagata SH, Oku HH, Yamaki MH (2002) Planta Med 68:258
89. Fu P, Zhang WD, Li TZ, Liu RH, Li HL, Zhang W, Chen HS (2005) Chin Chem Lett 16:771
90. Vishwakarma RA, Popli SP, Kapil RS (1986) Indian J Chem Sect B 25:1155
91. Balachandran S, Vishwakarma RA (1988) Indian J Chem Sect B 27:385
92. Cardona ML, Fernandez IP Jr, Serrano A (1989) Heterocycles 29:2297
93. Chen M-T, Chen C-M (1985) Heterocycles 23:2543
94. Ishiguro K, Fukumoto H, Suitani A, Nakajima H, Isoi K (1996) Phytochemistry 42:435
95. Chen M-T, Kuoh Y-P, Wang C-H, Chen C-M, Kuoh C-S (1989) J Chin Chem Soc (Taipei) 36:165
96. Ishiguro K, Nagareya N, Suitani A, Fukumoto H (1997) Phytochemistry 44:1065
97. Kitanov GM, Nedialkov PT (2000) Pharmazie 55:397
98. Zhang WD, Fu P, Liu RH, Li TZ, Li HL, Zang W, Chen HS (2007) Fitoterapia 78:74
99. Tanaka N, Takaisi Y (2007) Chem Pharm Bull 55:19
100. Fu P, Zhang WD, Liu RH, Li TZ, Shen YH, Li HL, Zang W, Chen HS (2006) Nat Prod Res 20:1237
101. Barraclough D, Locksey HD, Schienmann F, Taveira-Magalhaes M, Gottlieb OR (1970) J Chem Soc B, p 603
102. Westerman PW, Gunasekera SP, Sultanbawa MUS, Kazlauskas R (1977) Org Magn Reson 9:631
103. Frahm AW, Chaudhuri RK (1979) Tetrahedron 35:2035
104. Castelao JF Jr, Gottlieb OR, Lima RA, Mesquita AAL, Gottlieb HE, Wenkert E (1977) Phytochemistry 16:735
105. Hambloch H, Frahm AW (1980) Tetrahedron 36:3273
106. Conceiçao LFR, Ferreres F, Tavares RM, Dias ACP (2006) Phytochemistry 67:149
107. Strolin-Benedetti M, Dostert PL (1992) In: Testa B (ed) Advances in drug research, vol 23. Academic Press, London, pp 65–125
108. Fowler CJ, Ross SB (1984) Med Res Rev 4:323
109. Okpanyi SN, Weischer ML (1987) Arzneimittelforschung 37:10–13
110. Suzuki O, Katsumata Y, Oya M, Bladt S, Wagner H (1984) Planta Med 50:272
111. Sparenberg B (1993) PhD thesis, University of Marburg, Germany
112. Gnerre C, Thull U, Gaillard P, Carrupt PA, Testa B, Fernandes E, Silva F, Pinto M, Wolfender JL, Hostettmnn K, Cruciani G (2001) Helv Chim Acta 84:552
113. Rahalison L, Hamburger M, Hostettmann K (1991) Phytochem Anal 2:199
114. Homans AL, Fuchs A (1970) J Chromatogr 51:327
115. Yamakuni T, Aoki K, Nakatani K, Kondo N, Oku H, Ishiguro K, Ohizumi Y (2006) Neurosci Lett 394:206
116. Bombardelli E, Morazonni P (1995) Fitoterapia 66:43
117. Rabanal RM, Arias A, Prado M, Perez H, Sanchez-Mateo CC (2002) J Ethnopharmacol 81:287
118. Chistik TA (1957) Farmakol Toksikol 20:76

119. Chang Su (1977) Dictionary of Chinese crude drugs. Shanghai Scientific Technological, Shanghai
120. Iwu MM (1993) Handbook of African medicinal plants. CRC, Boca Raton, Florida, p 355
121. Chiu NY, Chang KH (1986) The illustrated medicinal plants of Taiwan, II. SMC, Taipei, p 126
122. Bandyukova VA, Khalmatov Kh (1966) Khim Prir Soedin 3:214
123. Khodjimatov KK, Aprasidi GS, Khodjimatov OK (1995) Dikorastushie celebniye rasteniya srednei azii, p 122
124. Roig JT (1988) Plantas medicinales aromaticas o venenosas de Cuba. Editorial Cientifico-Tecnica, La Habana, Cuba, pp 716–717

Top Heterocycl Chem (2007) 9: 179–264
DOI 10.1007/7081_2007_081

Published online: 8 September 2007

Chemistry of Biologically Active Isothiazoles

Francesca Clerici · Maria Luisa Gelmi (✉) · Sara Pellegrino · Donato Pocar

Istituto di Chimica Organica "A. Marchesini", Facoltà di Farmacia, Università di Milano, Via Venezian 21, 20133 Milano, Italy
marialuisa.gelmi@unimi.it

To our dear Prof Pocar for his retirement

Abstract The isothiazole ring as well as the corresponding benzo- and heterocondensed rings are present in many chemically interesting compounds. The isothiazole ring can be a substituent of a bioactive scaffold or the pharmacophore of bioactive molecules. New compounds have been designed, synthesised and tested towards different biological targets and, in many cases, they display interesting pharmaceutical activities. Different SAR studies are reported starting from already known isothiazole derivatives or from new compounds characterised by a particular substitution pattern aiming to improve the biological activity. Agrochemical applications are also reported.

Keywords Benzisothiazoles · Biologically active compounds · Isothiazoles · Pyridoisothiazoles · Saccharins · Sultams

1
Introduction

The importance of isothiazoles and of compounds containing the isothiazole nucleus appears to have grown over the years. New synthetic approaches and unprecedented reactions have recently been reported and numerous technical and pharmaceutical applications have been discovered [1–6]. This review is based on the fact that the isothiazole ring is present in many chemically interesting compounds that display biological activity. The isothiazole ring can be a substituent of a bioactive scaffold or the pharmacophore of bioactive molecules. A great number of SAR studies have been carried out starting from already prepared isothiazole derivatives or from new compounds characterised by a particular substitution pattern aiming to ameliorate the biological activity. Among the benzisothiazoles the best-known derivative is the noncaloric sweetener saccharin [7]. Due to the large amount of literature on this compound and its biological activity, it is considered in this review only for particular and original applications.

Regarding other isothiazole derivatives, different activities have been claimed such as antimicrobial, antibacterial, antifungal, antiviral, antiproliferative and anti-inflammatory activities. They have also been tested as inhibitors of proteases, for the treatment of anxiety and depression, for their action on the 5-HT receptor, and as inhibitors of aldoso reductase. Patented compounds for agrochemical applications are of old interest. This review is concerned with recent results, mainly limited to the last decade. Previous references have been taken into account only if they are of particular relevance or necessary for a better understanding of the text. Furthermore, patents have been included only if they add important information to the existing scientific literature. Articles exclusively dedicated to biological investigations without chemical interest have not been considered.

The review is divided into four main chapters concerning isothiazoles, sultams, benzisothiazoles and benzisothiazolones of biological interest. In these last two cases, heterocondensed compounds are included. In each chapter, the corresponding dioxides as well as partially hydrogenated rings are considered. For each class of compounds, the following is reviewed: i) procedures by direct synthesis of the above rings pointed out as synthetic *methods A–V*, and ii) the reactivity insofar as related to biologically interesting transformations of functional groups present on the isothiazole ring or related to the functionalisation of the isothiazole nucleus with a proper pharmacophore.

A chapter on biological applications of selected compounds has also been inserted in which the main activities are depicted and in which, for each class of bioactive compounds, the lead compound has been selected. A short account of the main biological mechanisms in which the target compounds are involved has also been reported when of strong relevance.

2
Isothiazoles

Isothiazoles constitute a relatively novel class of heterocyclic compounds, the first preparation being reported in the mid 1950s [8]. The rapid progress in their chemistry and intense studies on the synthesis and chemical conversion of their derivatives are due primarily to the extraordinarily broad range of useful properties manifested by various representatives of this class of compounds. Data on the chemistry of isothiazoles are documented in several monographs and reviews disclosing both synthetic and reacting aspects [1–6]. Below only few general synthetic methods based on the construction of the ring are reported if useful for the preparation of compounds characterised by a claimed activity. Analogously, the reactivity of the isothiazole system is related to the preparation of active or potentially active compounds.

Many 3-heterosubstituted isothiazoles characterised by interesting biological applications have been prepared, in which the heteroatom can be oxygen, sulphur, nitrogen and a halogen. In particular, depending on the substitution at the nitrogen atom, the oxygen substituted compounds can exist as enoles, when nitrogen is unsubstantiated, or as isothiazol-3(2H)ones when nitrogen is substituted.

2.1
Synthesis of Carbon Linked Isothiazoles

The well-known synthesis from 3-halogeno-α,β-unsaturated aldehydes **1** and ammonium thiocyanate, via intermediate **2**, is very useful for the preparation of substituted isothiazoles **3** and for fused analogues (Scheme 1, *Method A*) [1, 2].

A number of modifications of this method has been described. For example, the alkinyl carbonyl compounds **4** and $Na_2S_2O_3$ afforded the dithion-

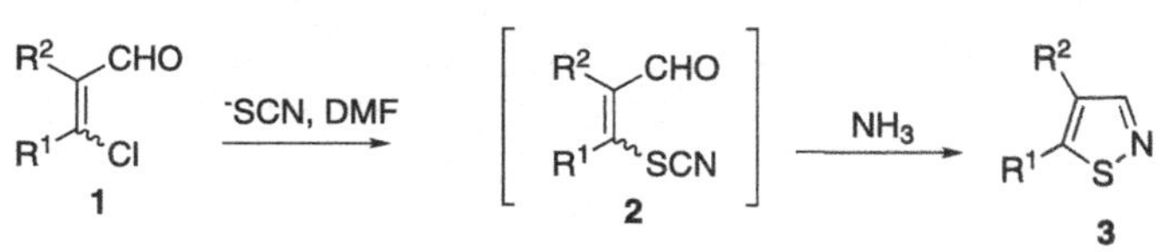

Scheme 1 Synthesis of isothiazoles: *Method A*

ites **5** then cyclised in liquid NH_3 to the corresponding isothiazoles **6** [1,2] (Scheme 2, *Method B*).

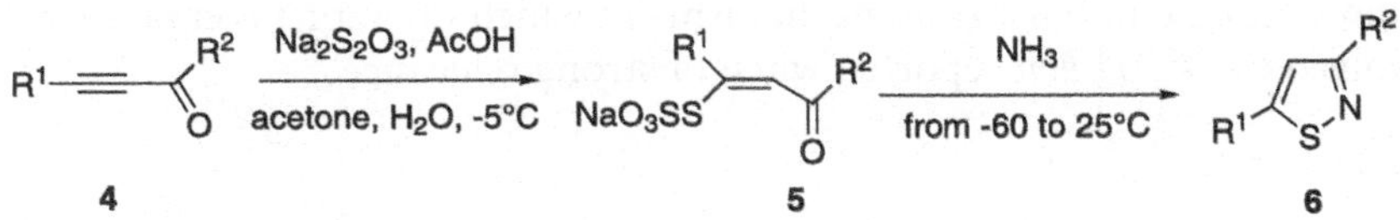

Scheme 2 Synthesis of isothiazoles: *Method B*

Several antifungal 4-arylisothiazoles **7** (25–54% yields) were prepared taking advantage of the classical procedure described in *Method A* [9] from **1** ($R^1 = H$, $R^2 = Ar$) and ammonium thiocyanate in DMF. 4-(2-Methoxyphenyl)isothiazole (**9**) was also prepared from the corresponding 2-nitro derivative **7** (Ar = 2-NO_2Ph) in 26% overall yield via intermediate **8** (Scheme 3) (see Sect. 4.1, analogue compounds of brassilexin.

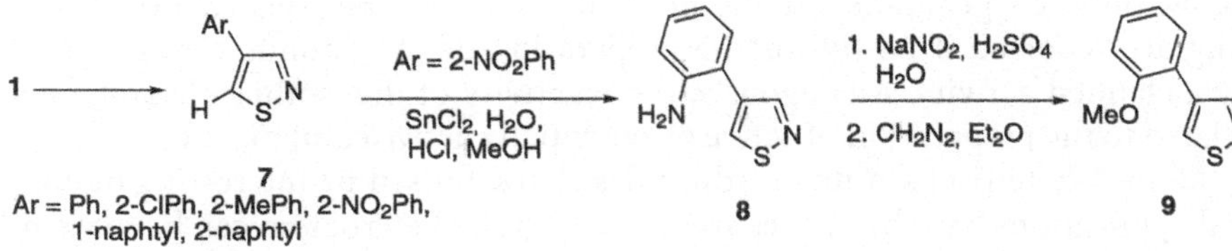

Scheme 3 Synthesis of 4-arylisothiazoles

The 5-substituted isothiazole **12** is a substrate for viral HSV-1 thymidine kinase [10]. The isothiazole ring was built according to *Method B* starting from the alkinyl aldehyde **11** prepared from 2′-desoxy-5-iodouridine **10** and propiolaldehyde diethyl acetal in presence of ($PdCl_2(Ph_3P)_2$), CuI and TEA at 50 °C. Acetal intermediate (91%) was formed, which was then deprotected with aqueous AcOH (80%) giving aldehyde **11** (51%). Its reaction with $Na_2S_2O_3$ in acetone/H_2O/AcOH afforded a *cis/trans* mixture of a thiosulfate (1 : 1, not isolated) then treated with liq. NH_3 at − 70 °C to give **12** (13%) (Scheme 4).

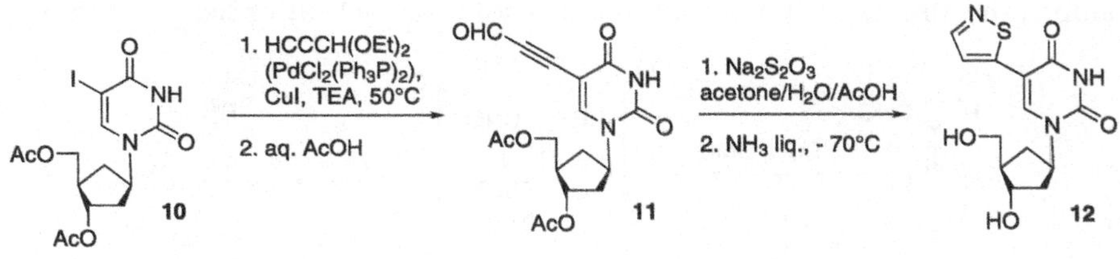

Scheme 4 Synthesis of 2′-deoxy(5-isothiazol-5-yl)uridine

The synthesis of isothiazoles **13a,b** (R = H, 30% yield, R = Me, 38% yield) was performed according to the same *Method B* from the propioaldehyde **4** ($R^1 = R^2 = H$) or 1-butin-3-one **4** ($R^1 = H$, $R^2 = Me$). Compounds **13a,b** were then transformed into the regioisomeric phosphonium salts **14a–b**, which are the key intermediates for the preparation of several antimicrobic agents based on a carbapenem structure [11]. Compounds **13a** was transformed into the phosphonium salt **14a** through formylation (*n*-BuLi, DMF, 73%), reduction with $NaBH_4$ (83%), substitution of the hydroxy group with bromine (PPh_3, CBr_4, 80%) and reaction with PPh_3 (63%). On the other hand, **13b** was first brominated (NBS, AIBN, CCl_4, 45%) and then transformed into the phosphonium salt **14b** (63%) as reported in Scheme 5. The preparation of 1-β-methylcarbapenems **17a,b**, bearing isothiazoloethenyl moieties al C-5 position of the pyrrolidine ring, was achieved using as key reaction the Wittig reaction from salts **14** and aldehyde **15** operating with NaHMDS as the base in THF at – 78 °C. Alkenes **16a,b** were isolated in 76 and 73% yield, respectively. The *E*-isomers were obtained exclusively. Mesylates **16** were then used for the preparation of carbapenem derivatives **17**.

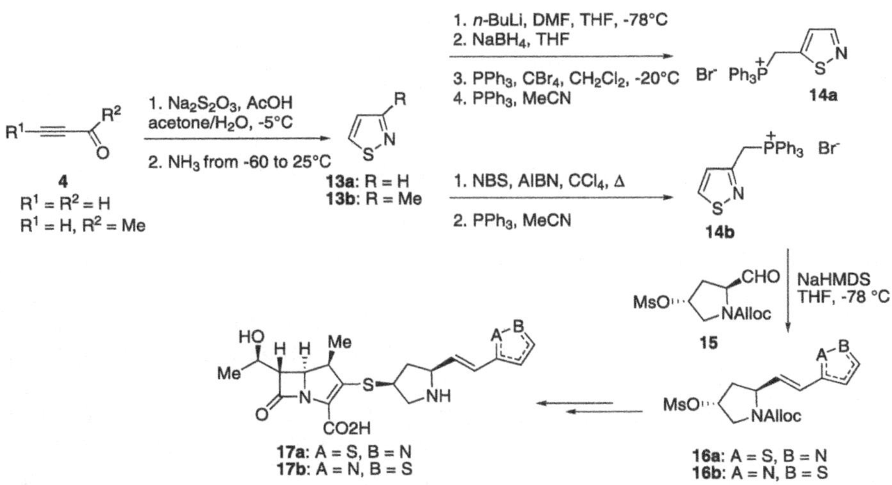

Scheme 5 Synthesis of 1-β-methylcarbapenems with an isothiazoloethenyl side chain

2.2 Synthesis of 3-Heterosubstituted Isothiazoles

The main method for the formation of the 3-heterosubstituted isothiazole ring is, even today, a ring closing reaction based on the formation of an S – N bond [1, 2].

The well-known synthesis based on the oxidative cyclisation of dithiopropionamides **20**, obtained from acids **18** via dichlorides **19**, is considered an efficient and simple procedure affording good yields of isothiazol-3(2H)ones **21** (Scheme 6, *Method C*). Different reagents were tested aiming to ameliorate the yield and the efficiency of the strategy depending on the different dithiopropionic acids **18** and amines used in the process. This synthetic approach was also used for the preparation of the corresponding benzoderivatives [1, 2, 4].

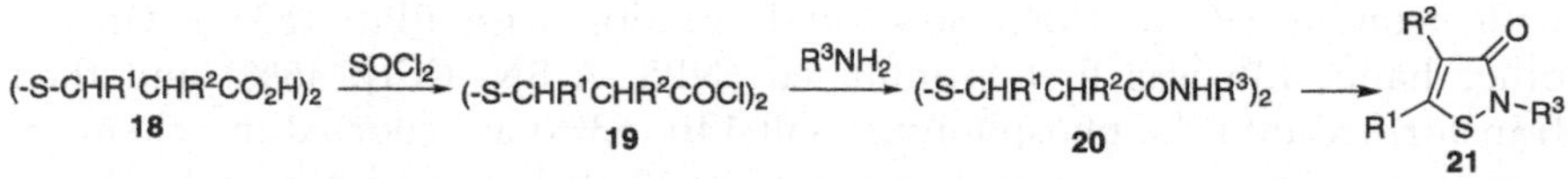

Scheme 6 Synthesis of isothiazol-3(2*H*)ones: *Method C*

The *Methods D* outlined in Scheme 7 are also based on S – N bond formation. Thioenols **23** and enamines **24**, prepared from β-ketoamides **22**, represent the starting compounds in a number of procedures based on oxidative cyclisation with different reagents. 3-Hydroxy substituted isothiazoles **25** can be obtained this way [1, 3].

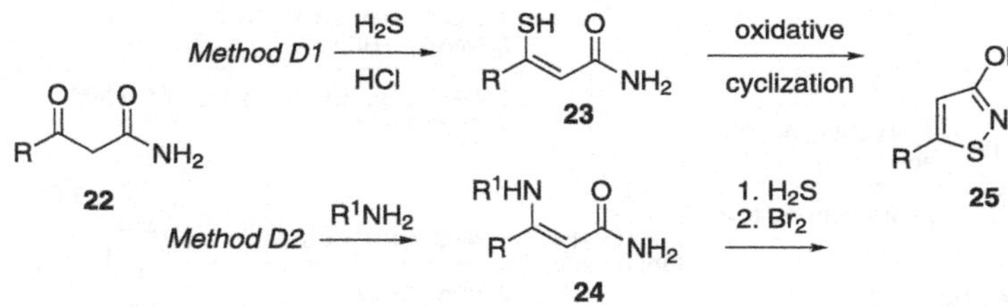

Scheme 7 Synthesis of 3-hydroxy substituted isothiazoles: *Methods D*

A different synthetic pathway, which is useful for the preparation of 4-cyano-isothiazoles **28–31** substituted at C-3 with different heteroatoms (Hal, S, O), is exploited by *Methods E* (Scheme 8). These procedures start from dicyanomalonate **26** and different electrophiles giving the key intermediates **27a–d**, subsequently cyclised to the isothiazole-4-carbonitriles **28–31**. A number of modifications of these procedures are known and are very useful for obtaining different starting materials for the preparation of isothiazole derivatives of biological interest [1, 2, 4].

Several antibacterial compounds containing substituted 3(2H)-isothiazolones as scaffold and, among them, a series of 3(2H)-isothiazolones **32**, substituted at the nitrogen atom with an aryl moiety functionalised with groups that are different in hydrophobicity, size, steric and electronic parameters, were prepared (Scheme 9). They were synthesised adopting *Method C* starting from dithiodipropionamides **20** ($R^1 = R^2 = H$, $R^3 = Ar$). By incremental addition to **20** of a dichloromethane solution of sulfuryl chloride, as

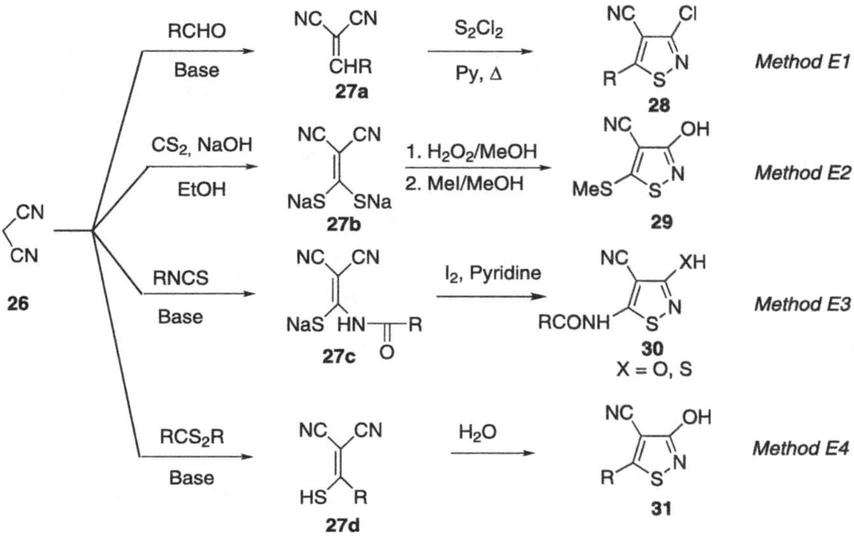

Scheme 8 Synthesis of 4-cyano-3-heterosubstituted isothiazoles: *Methods E*

an oxidizing agent, a mixture of isothiazolones **32a** and **32b**, unsubstituted and chloro substituted in position 5, respectively, was obtained [12]. The authors found that by manipulation of the stoichiometry SO_2Cl_2/amide, a different distribution of the products **32a/b** was observed. The use of an excess of SO_2Cl_2 favours the formation of chloro compounds **32b**, while the 1 : 1 stoichiometry favours the formation of **32a** which were obtained in variable yields (16–70%) depending on the aniline derivative.

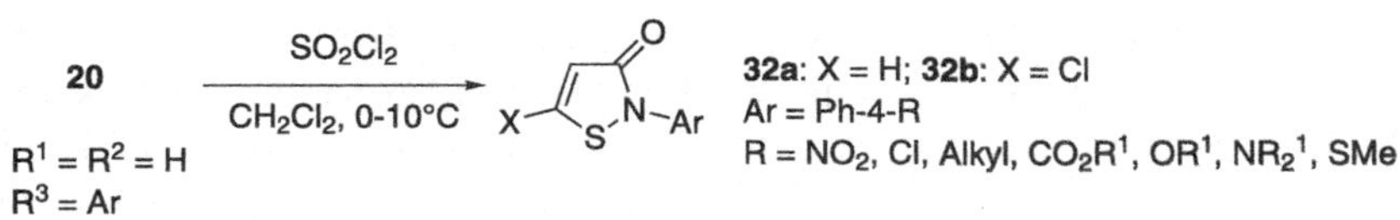

Scheme 9 Synthesis of 2-aryl-isothiazol-3(2*H*)ones

A similar synthetic scheme was adopted by Nadel [13] to prepare isothiazolones **34** and **35** from L-cystine derivatives transformed into the corresponding cystine *bis-N*-(methylamides) **33a–c** ($EtOCO_2Cl$, TEA in CH_2Cl_2 then $MeNH_2/H_2O$, 69–73%) (Scheme 10). As reported in Table 1, several parameters were evaluated in order to find the best reaction conditions to produce the different isothiazolone derivatives **34** (43–63%) or **35** (52–66%). As by-products, the polyhalogenated compounds **36** and **36′** (not separable) were formed in some instances. Finally, the deprotection of the amino group

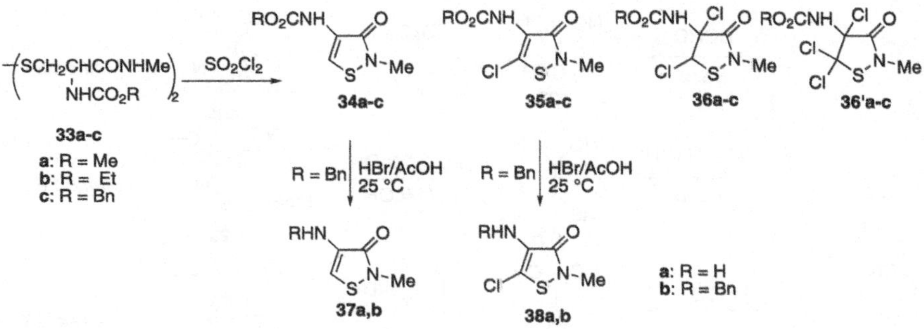

Scheme 10 Synthesis of 4-amino-isothiazol-3(2*H*)ones

Table 1 Summary of ring closing reactions of the amino blocked cystine *bis*(methylamides)

33	Solvent	SO_2Cl_2/45 Molar ratio	Rate of feeding (mL/h)	Temp. (°C)	Product
33a	CH_2Cl_2/C_6H_{12} (1 : 1)	4	9.6	35–40	**34a**
33a	hexane	6	19.2	30	**35a**
33b	$CHCl_3$	3.5	3.4	boiling	**34b**
33b	CCl_4	6.5	20.8	25	**35b**
33c	CH_2Cl_2/C_6H_{12} (1 : 1)	4	6.4	35–40	**34c**
33c	$CHCl_3$	6.5	31.2	50	**35c**
33a	CCl_4	7	20	35–40	**34a** + **35a** + **36/36′a**
33b	CCl_4	7.5	20	20	**34b** + **35b** + **36/36′b**
33c	CCl_4	7.5	25	55	**34c** + **35c** + **36/36′c**

of **34c** and **35c** was performed using a mixture of HBr in AcOH and compounds **37** (78%) and **38** (65%) were formed (Scheme 10).

Recently, a series of analogues of thio-THIP (4,5,6,7-tethrahydroisothiazolo[5,4-*c*]pyridine-3-ol see Sect. 6), 5-(4-piperidyl)isothiazol-3-ol **41** and 5-(1,2,3,6-tetrahydro-pyrid-4-yl)isothiazol-3-ole **43** were developed basing their preparation on the same general strategy as described in *Method D1* from the enolised β-thioxoamide **39**, as shown in Scheme 11. Oxidation of **39** with I_2 under basic conditions afforded **40**, which was deprotected to give **41** by treatment with HBr in AcOH. Compound **42**, obtained from **40** by alkylation of the oxygen atom followed by bromination (EtBr, K_2CO_3, then NBS) and was dehalogenated and deprotected by treatment with 48% HBr to give the target compound **43** [14].

Isothiazol-3-ones **46** were prepared and their activity was evaluated on different serine proteases [15]. The key isothiazole salts **44** were prepared according to *Method A* by reacting isothiocyanate intermediates **2** and several aniline derivatives operating in AcOH in the presence of $HClO_4$. They

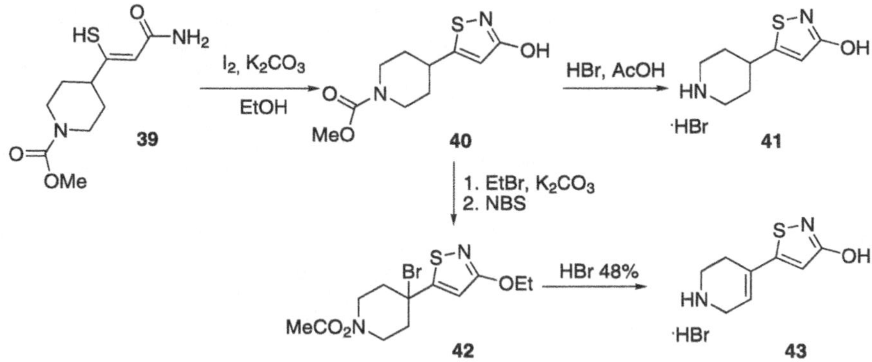

Scheme 11 Synthesis of 5-(4-piperidyl)- and 5-(1,2,3,6-tetrahydro-pyrid-4-yl)isothiazol-3-ols

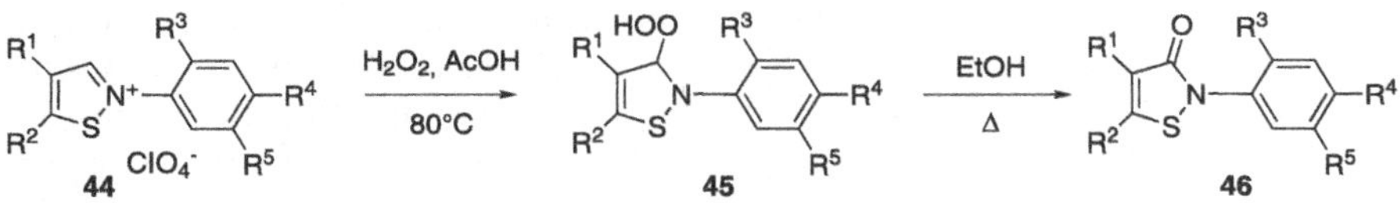

Scheme 12 Synthesis of substituted 2-aryl-isothiazol-3(2*H*)ones

were oxidised to peroxides **45** with H_2O_2 in AcOH and then transformed (EtOH, reflux) into isothiazol-3-ones **46** in moderate to good yields (20–65%) (Scheme 12).

2.3 Reactivity of Isothiazoles

A series of new HIV protease inhibitors were designed and synthesised [16] (Scheme 13). One of the most active compounds is derivative **53** functionalised with the 3,5-dimethylisothiazole-4-methylamide ring. The latter was synthesised from the commercially available ethyl 5-amino-3-methyl-isothiazole-4-carboxylate **47**, which was converted into the 5-methyl derivative **48** (40%) by treatment with first isoamyl nitrite and I_2 and then with Me_4Sn and $PdCl_2(PPh_3)_2$. Its reduction with $LiAlH_4$ followed by treatment of the alcohol intermediate with CBr_4, Ph_3P, NaN_3, afforded the azido intermediate then reduced to amine **49** with Ph_3P (39%). By condensation of amine **49** with **50** (EDC, HOAT, DIEA, 60%), **51** was formed then hydrolyzed at the lactone function (LiOH, H_2O, dioxane), protected at hydroxyl group (TBSOTf, AcOEt, DIEA), condensed with **52** (HBTU, DIEA, HOAT cat.) then deprotected at the oxygen atom with Bu_4NF to give **53**.

An efficient synthesis of DNA binding tetrameric acid derivatives **57** [17], consisting of an isothiazole and three *N*-methylpyrrole carboxamide units,

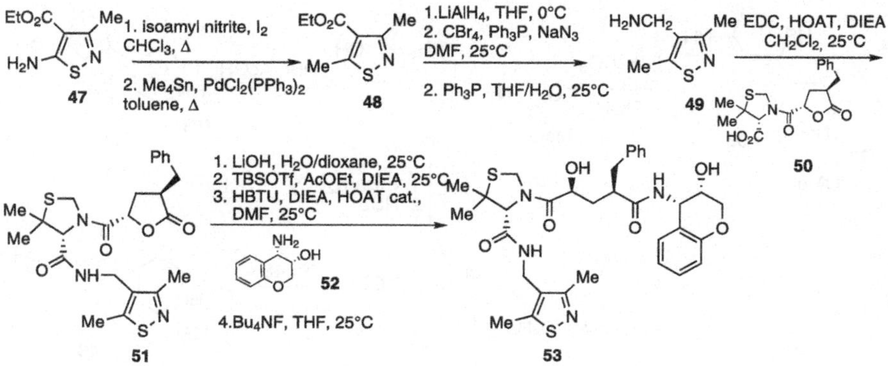

Scheme 13 Synthesis of new HIV protease inhibitors

was performed and various substituents at both termini were introduced (Scheme 14). The coupling of trimeric *N*-methylpyrrole amino acid **54** with isothiazole **55** (HBTU in DMF and *i*-Pr_2EtN, 91%) resulted in the desired tetramer **56**. The two amino groups were then introduced in sequence by performing a "one pot" reaction. First Cl-5 was selectively substituted by the proper amine operating in presence of BOPCl and NMP/*i*-Pr_2EtN. Then the amide function was formed using a large excess of a second amine and **57** was obtained. Benzisothiazole derivative **58** was also prepared [18].

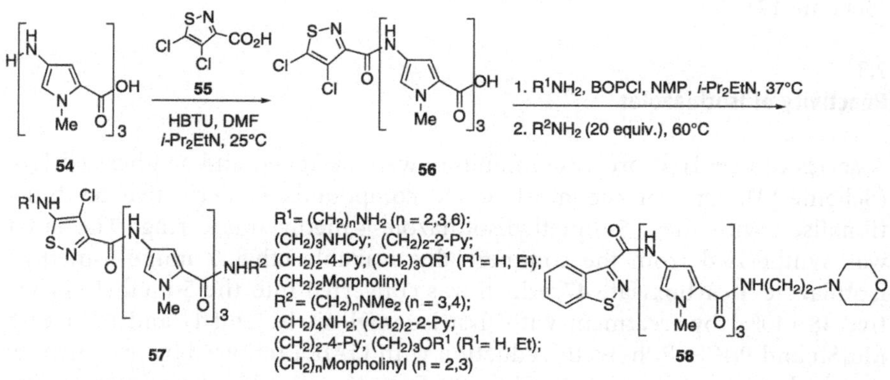

Scheme 14 Synthesis of tetrameric acid containing isothiazole and pyrazole rings

Many isothiazoles containing an amino acid functionality, analogues of isoxazole-based molecules, acting on glutamate receptors have been prepared in the last two decades. The isothiazole ring has mainly been synthesised according to *Method D1*. Afterwards, functionalisation of the ring was performed affording the biologically interesting target compounds, which can be divided into two groups: i) isothiazoles with the characteristic amino acid group at C-4 and ii) isothiazoles with the characteristic amino acid group at C-5. The methodologies for the synthesis of such compounds are described in the following.

i) Isothiazoles with the characteristic amino acid group on C-4 [19]. Thio-AMPA (**61a**) and thio-ATPA (**61b**) were synthesised as outlined in Scheme 15. Compounds **25**, prepared from **23** with I_2 as oxidizing agent, were treated with trioxane, HBr/MeOH affording **59a,b** (27%, 39% respectively). Several reaction conditions (times and temperatures) were tested in attempts to increase the yields without appreciable results. Dimethyl acetamidomalonate reacted easily in basic conditions with **59a,b** affording **60a,b** which were deprotected under acidic conditions providing thio-AMPA (**61a**, 28%) and thio-ATPA (**61b**, 31%). Chiral chromatographic resolution of **61b** was also performed (Scheme 15).

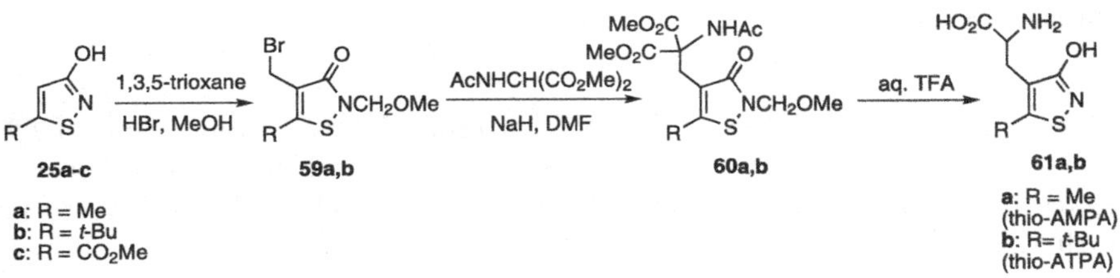

Scheme 15 Synthesis of thio-AMPA and thio-ATPA

From **60a** the isothiazole amino acids **66** and **67a,b** were synthesised as outlined in the Scheme 16 [20]. Treatment of **60a** with $BF_3{\cdot}Et_2O$ in the presence of Ac_2O and subsequent heating under reflux of the intermediate product in methanolic sodium methoxide provided the 3-isothiazolol **62** (60% yield). Alkylation of **62** with ethyl chloroacetate gave a reaction mixture containing two main components **63** (26%) and **64** (31%), which were deprotected using 1M TFA to give **66** (14%) and **67a** (31%), respectively. The synthesis of **65** by alkylation of compound **62** with diethyl 4-toluenesulfonyloxymethylphosphonate under basic conditions was accompanied by extensive decomposition products. Compound **65** was thus obtained in a low yield (15%) without the formation of the isomeric *N*-alkylated product and then deprotected to give the target phosphono-amino acid **67b** (32% yield).

ii) Isothiazoles with the characteristic amino acid group at C-5 [19]. The functionalisation of C-5 with an amino acid function linked to the ring through a methylene bridge was accomplished by elaboration of a preexisting carboxy group. Methylation of **25c** (R = CO_2Me) followed by reduction and substitution of alcohol with $SOCl_2$ provided **68**, which was further converted into the fully protected compound **69**. The deprotection of **69**, which required conc. HBr, was accompanied by a marked decomposition of the

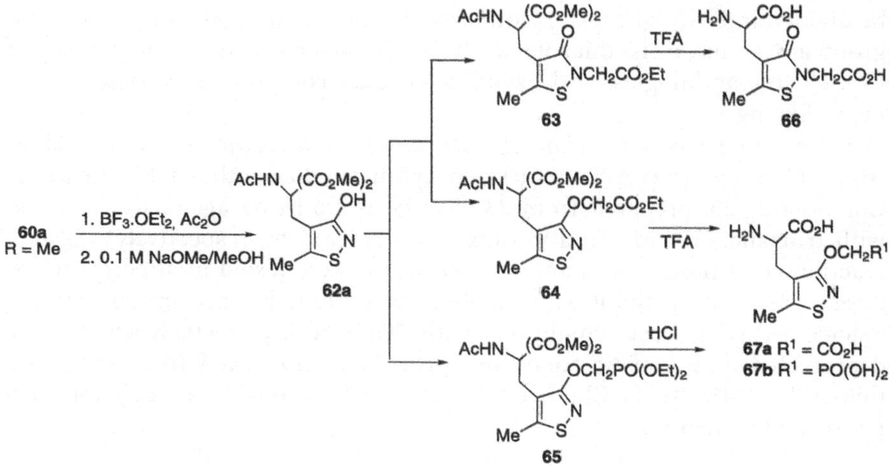

Scheme 16 Synthesis of thio-AMPA and thio-ATPA derivatives

product and provided **70a** as the hydrobromide (25%). Bromination of **69** and deprotection by refluxing in 4 M HCl for a period of less than 4 h, resulted in a pronounced decomposition and in a low yield (20%) of **70b** as the hydrochloride (Scheme 17).

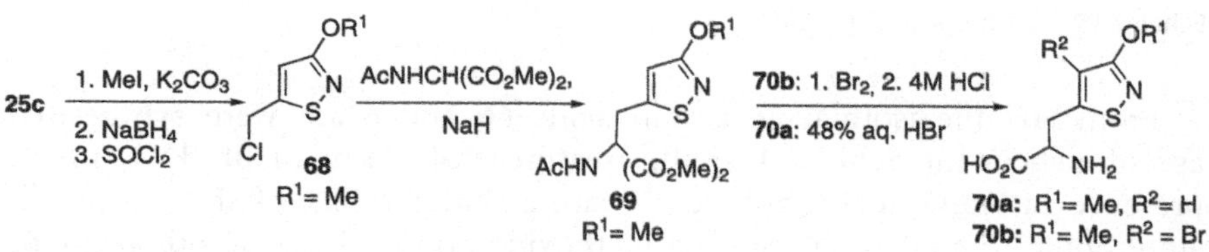

Scheme 17 Preparation of β-isothiazol-5-yl alanines

Differently from the case reported above (compounds **70**), thioibotenic acid derivatives like **75** and **83a–g**, NO the isothiazole analogues of ibotenic acid, were prepared taking advantage of a general methodology enabling the direct introduction of a substituent in the 5-position of 3-(benzyloxy)isothiazoles **71** [21]. The reaction of a series of electrophiles (**E1-6**) with **71** in the presence of LDA (1.1 equiv., 0.1 M in Et_2O) produced compounds **72a–d** (54–68% yields). When an amino acid synthon, i.e. **73**, was used in the reaction as the electrophile, the protected amino acid **74** (36%) was obtained. The ester **74** was hydrolyzed under basic conditions and subsequent treatment with HBr in AcOH removed the Boc and benzyl groups to give compound **75** (18%) (Scheme 18).

The 4-substituted Thio-Ibo derivatives **83a–g** (Scheme 20) were prepared with similar chemistry from the corresponding 4-substituted 3-benzyloxy-isothiazoles **76**, **77** and **78** synthesised as outlined in Scheme 19. The use

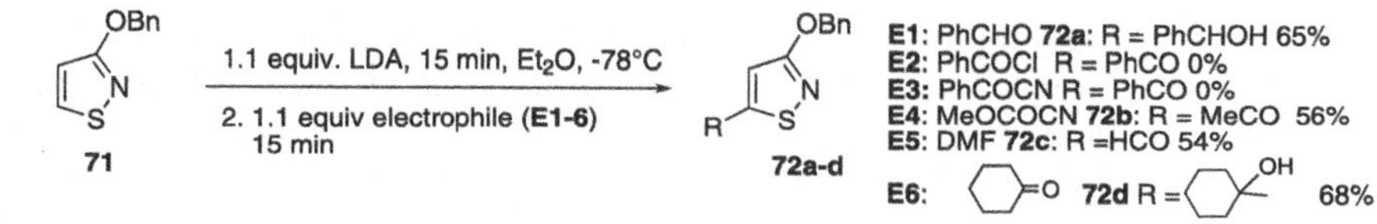

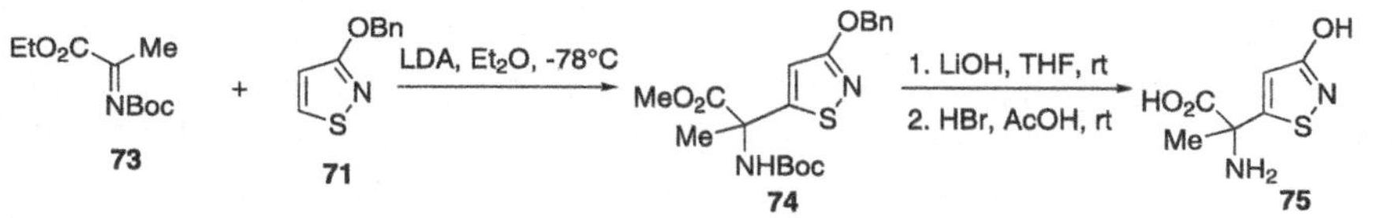

Scheme 18 Preparation of α-isothiazol-5-yl alanine

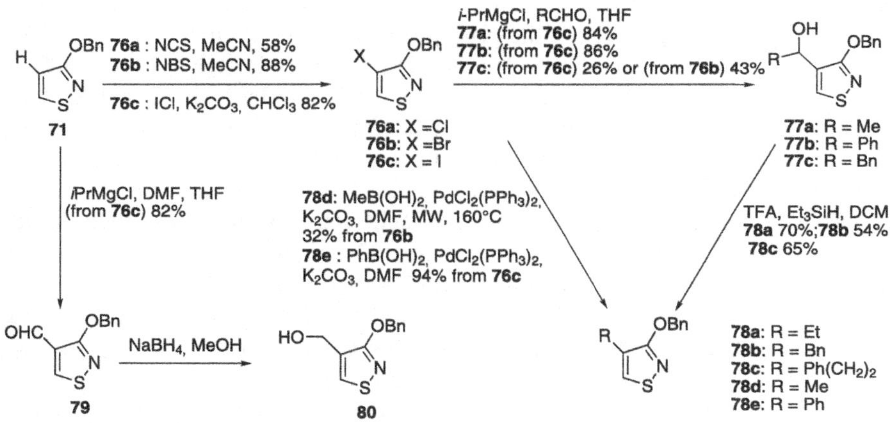

Scheme 19 Synthesis of 4-substituted 3-benzyloxyisothiazoles

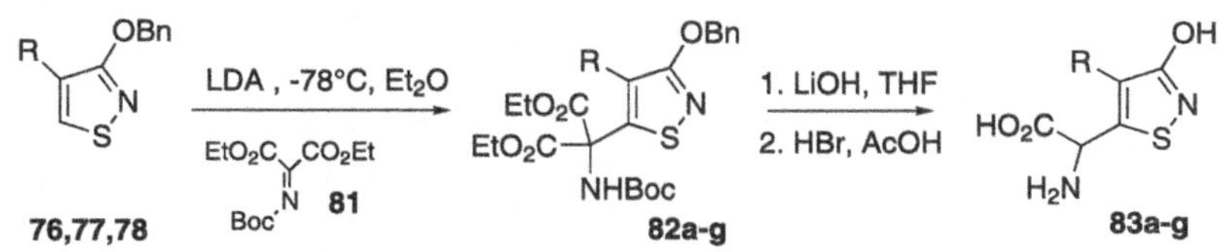

Scheme 20 Synthesis of 4-substituted Thio-Ibo

of 3-*O*-protected-isothiazolols **71** as key reagents to functionalise the C-4 position represent a convenient approach. Compound **71** can be synthesised through cyclisation reaction of 3,3-dithiodipropionamide according to *Method C* followed by *O*-protection. The benzyl protecting group in compound **71** did not withstand acidic halogenation conditions and the syntheses of **76a–c** were accomplished employing non-acidic conditions (ICl or NCS or NBS). The 4-iodo **76c** or 4-bromo **76b** compounds were suitable starting materials for magnesium–halogen exchange followed by Grignard reactions with various aldehydes leading to alcohols **77**. Treatment of **77** with TFA and Et_3SiH gave the respective isothiazoles **78a–c**. However, the final step was not applicable for the synthesis of the methyl compound **78d**; for this reason alcohol **80** was effectively prepared via aldehyde **79** and various reductive methods were tried aiming to the synthesis of **78d** from **80**, but all were unsuccessful. Due to the difficulty of reducing the alcohol functionality in **80**, the possibility of introducing the methyl group by a Suzuki cross-coupling reaction from **76** was investigated. Microwave heating and the use of methyl boronic acid and $PdCl_2(PPh_3)_2$ limited by-product formation sufficiently to give an isolated yield of 32% of **78d**. Standard aryl–aryl Suzuki reaction was applied to give the phenyl compound **78e** in excellent yield.

Starting from **76, 77, 78**, in the presence of LDA (1.1 equiv., 0.1 M in Et_2O), the 2-(*N*-*t*-butoxycarbonylimino) malonic acid diethyl ester (**81**) was added and the desired products **82a–g** were obtained. The hydrolysis of the ester functionality and the deprotection of nitrogen and oxygen atoms gave **83a–g** as the zwitterion upon treatment of the corresponding hydrobromides with propylene oxide.

Isothiazole esters of phosphorothioic acid were prepared for their interest as harmful organism-controlling agents [22]. Numerous examples of these compounds are patented and the syntheses of the isothiazole nucleus were usually done by the way of *Method E4* or slightly modified methodologies. As examples of the above interesting derivatives, the preparation of compounds **84** and **85** are reported in Scheme 21. Compound **31** was treated with anhydrous K_2CO_3 in acetone and dialkyl-chlorothiophosphate was added affording **84**. For the preparation of **85**, compound **31** was firstly transformed into the corresponding carbamoyl derivative (H_2SO_4), which was treated in DMF with NaH and subsequently made to react with dialkyl-chlorothiophosphate.

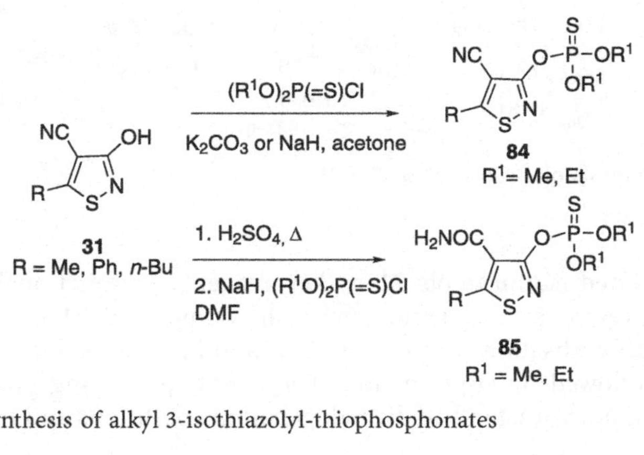

Scheme 21 Synthesis of alkyl 3-isothiazolyl-thiophosphonates

3-Hydroxy-4-carboxyalkylamidino-5-amino-isothiazole derivatives **87** were discovered as potent MEK1 inhibitors (Scheme 21) [23–25]. 5-Amino-isothiazoles **86** were prepared according to *Method E3* from the isothiocyanate, which was treated with cyanoacetamide (KOH in DMF) and then cyclised with bromine. By reaction of the nitrile group with an amine, usually cyclohexylmethylamine, in EtOH at reflux, the corresponding amidines **87** were formed. Following the same synthetic scheme a series of amidine compounds of general formula **88** were also prepared [26].

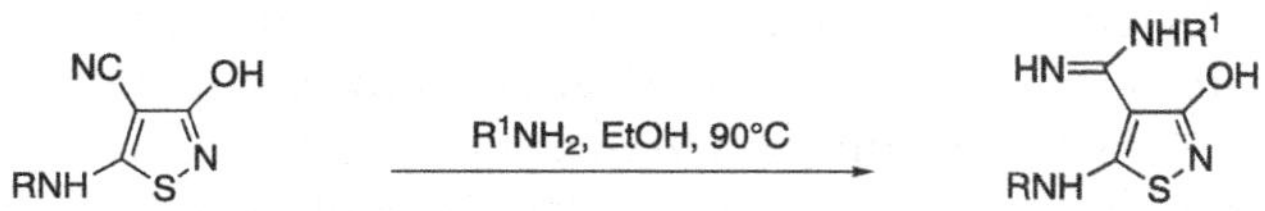

86 **87**

R = *n*-Pr, 4-MeO-Bu, *i*Pr, 3-morpholinyl-$(CH_2)_3$, cyclohexyl-CH_2,2-thienyl-$(CH_2)_2$, Ph-$(CH_2)_2$, 4-HO-Ph-$(CH_2)_2$, MeCH(Ph), 3-Py, pyrazolyl, Ar
Ar = 2- or 3- or 4-Me-Ph; 2- or 3- or 4-MeO-Ph; 2- or 3- or 4-O_2N-Ph, 4-NC-Ph, 2-F_3C-Ph, 4-EtO_2C-Ph, 3-Br-Ph, 4-Cl-Ph, 3,5-Cl_2-Ph, 2,5-Cl_2-Ph, 2,4-Cl_2-Ph, 3-F_3C-4-Cl-Ph

R^1 = Cyclohexyl-CH_2, R = 4-Me-Ph, 2-MeO-Ph, 3- or 4-Cl-Ph; 3- or 4-Br-Ph; 2-F-Ph; 3-O_2N-4-Cl-Ph, 4-O_2N-2-MeO-Ph, 2,4-Me_2-Ph, 3,5-Cl_2-Ph, 2,4-Cl_2-Ph

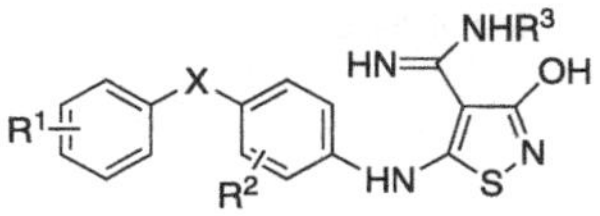

88

X = O, NH, CH_2, CO, S, SO_2, CMe_2, N=N
R^3 = alkylalcohols, alkylamines,alkyldiols, alkyl
R^1 = H, Cl, F, Me
R^2 = Cl, F, CF_3, NMe_2

Scheme 22 Synthesis of 3-hydroxy-4-carboxyalkylamidino-isothiazoles

Starting from **86** (R = Ph), a mixture of 3-methoxy compound **89** and 3-isothiazolone **90** was obtained using NaH and MeI (Scheme 23) [23].

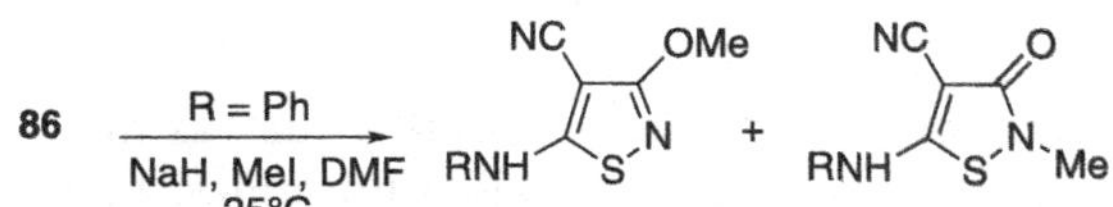

89 **90**

Scheme 23 Methylation of 3-hydroxy-isothiazoles

An alternative and more general approach to functionalisation at C-5 with the nucleophilic group was found from sulfone **91**, which derives from oxidation of **29** (H_2O_2 in Ac_2O/AcOH). By the way of a nucleophilic substitution reaction with the proper nucleophile, compounds **92** were prepared (Scheme 24) [23].

3-Chloroisothiazoles **28** functionalised with a cyano group at C-4 were the starting materials for the preparation of a series of molecules characterised by antiviral activity [27]. Depending on the substituent at C-5, they were prepared in 60–72% yields according to *Method E1*. They were then functionalised at C-3 both with an amino group and with a thio group. The reaction of

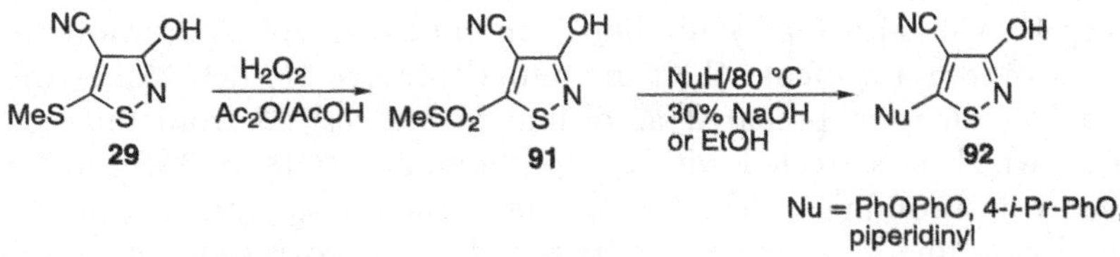

Scheme 24 General procedure for nucleophilic substitution at C-5

chloroisothiazole **28** with secondary amines in EtOH at reflux gave derivatives **93** (70–84%). Instead, the functionalisation at C-3 with a sulphur atom was made by making **28** react first with $Na_2S{\cdot}9H_2O$, affording the intermediate **94**. Intermediate **94** can be also prepared by condensation of O-alkyl thioates **95** with $CH_2(CN)_2$ in EtONa/EtOH and treatment with S_8 at reflux (*Method E*) (Scheme 25).

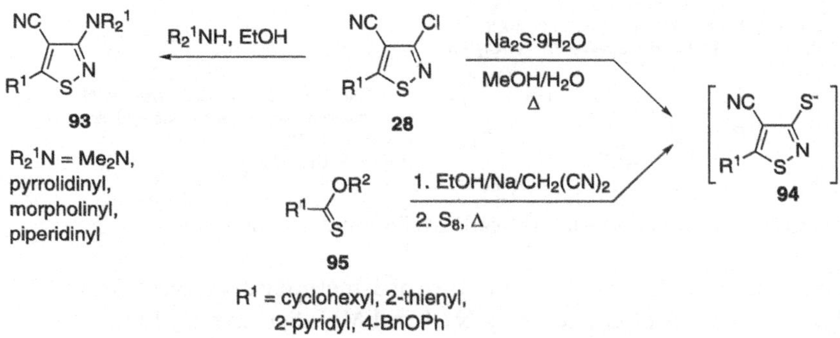

Scheme 25 Preparation of 4-cyano-isothiazole synthons

From **94**, several compounds were generated such as the 3-thioalkyl compounds **96** (40–67%) by reaction with alkyl bromides. Intermediate **94** was also treated with HCl, iodine, 1,3-dibromopropane or with ethyl chlorocarbonate affording the mercaptane **97** (68%), the disulfide **98** (45%, $n = 0$), compound **99** (37%, $n = 3$) and the thiocarbonate **100** (67%), respectively. By a partial hydrolysis of the nitrile group with H_2SO_4 at reflux, carboxamides **101a–b** (92%) were prepared from **96**. The same reagents **96** can be transformed into methyl esters **102** (Scheme 26).

Because compound **96** (R^1 = Me, R = BnOPh) exhibited a broad antipicornavirus spectrum of action, it was selected as a model and a series of new functionalised aryl compounds **104** (66–82%) was synthesised from compound **103**, which was functionalised at the oxygen atom with a proper chain operating in acetone in the presence of K_2CO_3 at reflux (Scheme 27) [27].

Anti-inflammatory compounds **107** and **108** (Scheme 28) were prepared from the known amino acid **105** following three different strategies, the first

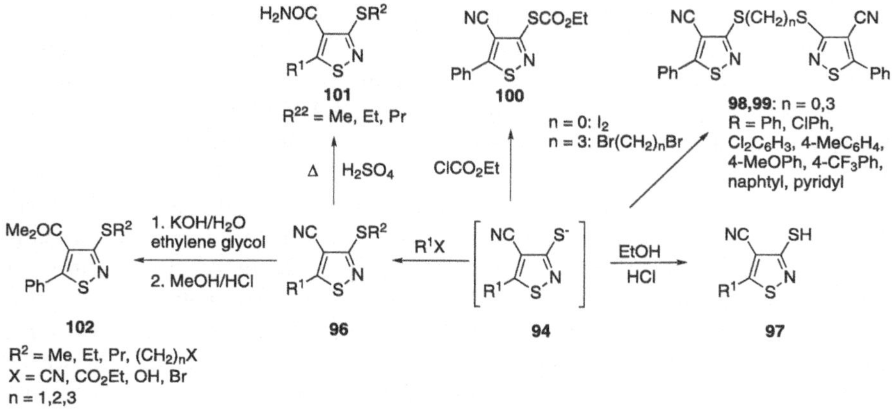

Scheme 26 Chemical applications of synthon **94**

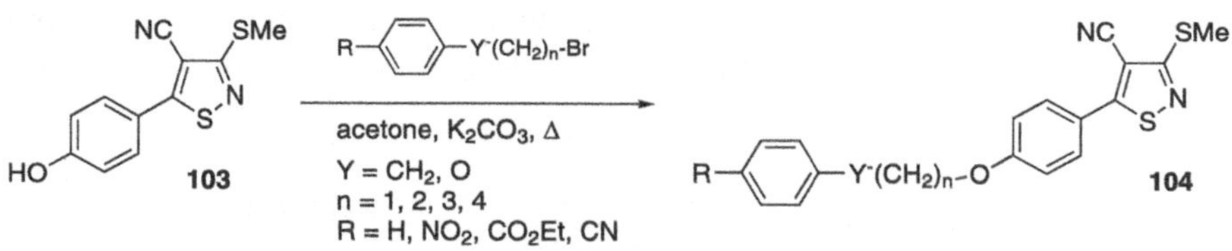

Scheme 27 Synthesis 4-cyano-3-methylthio-5-substituted isothiazoles as potential antipicornavirus agents

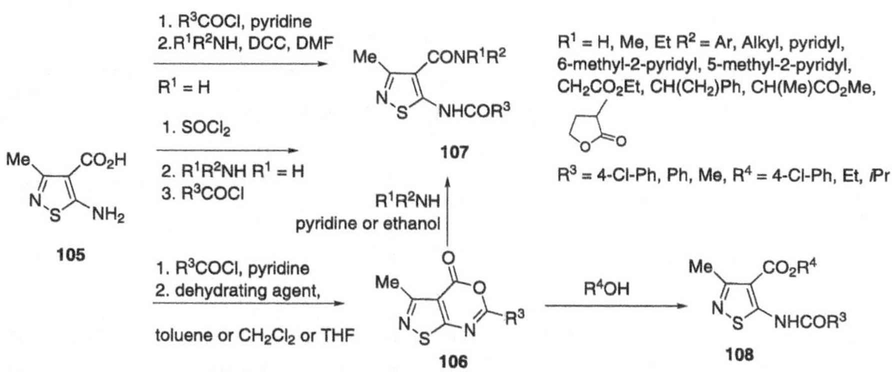

Scheme 28 Synthesis of 5-acylamino-isothiazol-4-carboxylic acid derivatives

one consisting in the preparation of 4-carboxamides via 4-carbonyl chloride, then acylated to nitrogen in position 5. Alternatively, **105** can be firstly reacted with acyl chloride affording 5-carboxyamides then converted into **107** with amines in DMF/DCC. The same 5-carboxyamides can be converted into the semi-anhydride **106** under the influence of dehydrating reagents such as $SOCl_2$, DCC, $POCl_3$, and ethyl chloroformate in solvents such as benzene, toluene, chloroform, methylene chloride and THF. Oxazinone **106** was used to obtain the diamides **107** and the proper ester **108** by reacting with alcohols [28].

Other 5-carboxamidoisothiazoles, which are interesting as insecticides, functionalised with a chloro atom in position 4, were prepared with a standard procedure [29, 30]. Some of them were prepared by reacting **109b** with an acyl chloride in boiling xylene. Alkylation of **110** with several alkyliodides or bromides resulted in the formation of both the *N*-substituted isomeric product **111a** and **b** [31] (Scheme 29). In order to obtain exclusively the structures **111a**, the alkyl group was introduced at the amine stage using reductive amination on **109b** affording **112**, which was subsequently acylated. To enhance biological selectivity, other derivatives were prepared with substituents at the α-methylene position [32]. Compound **110** was converted to the enamineamide **113**, via the Mannich intermediate, followed by substitution of the dimethylamino group with several amines. The same Mannich intermediate was transformed into derivatives **114**.

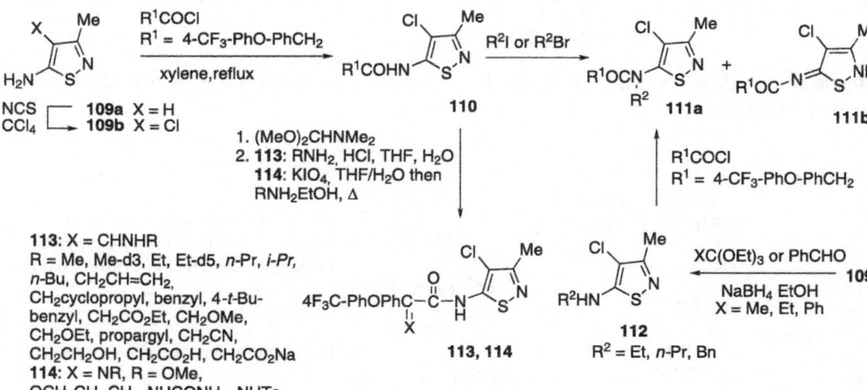

Scheme 29 Synthesis of 5-carboxamido-4-chloroisothiazole

CP 547, 632 was identified as a potent inhibitor of the VEGFR-2 and basic fibroblast growth factor (FGF) kinases [33, 34] (Scheme 30). The synthesis of this compound, and of several other derivatives, starts from compound **115** prepared according to *Method E3*. The hydrolysis of benzyl ether of **115** (H_2SO_4) followed by alkylation with 2,6-difluoro-4-bromobenzylalcohol in THF/DEAD/P(Ph)$_3$ afforded the intermediate **116**, which was reacted with 4-pyrrolidin-1-yl-butylamine affording the target compound **CP 547, 632**.

A similar chemistry was employed for the preparation of a series of isothiazoles **117** and **118** under study as TrkA kinase inhibitors (Scheme 30) [35].

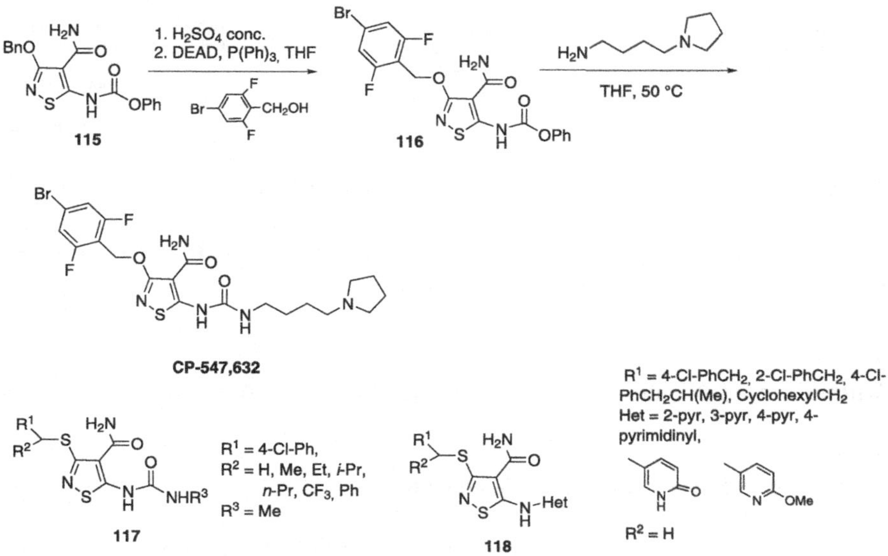

Scheme 30 Synthesis of **CP 547,632** and TrkA kinase inhibitors

Following *Method D2*, **119** (Scheme 31) was obtained, which could be transformed into the corresponding 3-ketoderivative **120** with $POCl_3$ and then with $Na_2Cr_2O_7/H_2SO_4$. Transformation of **120** afforded 4-amino derivatives **121** [37].

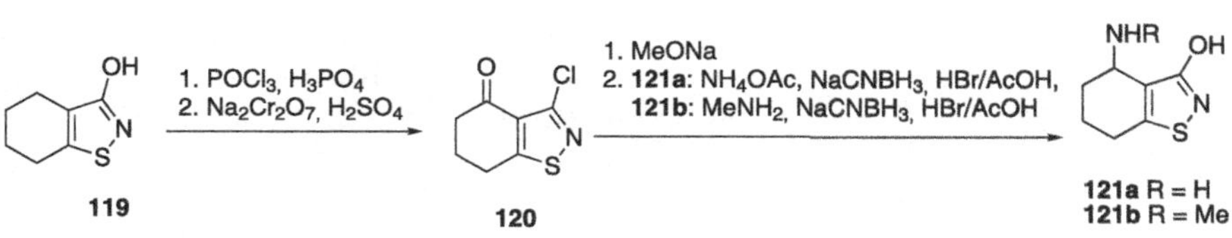

Scheme 31 Synthesis of 4-amino-3-hydroxy-4,5,6,7-tetrahydrobenzo[*d*]isothiazoles

A series of 4-hydroxy-4,5,6,7-tetrahydroisothiazolo[4,5-*c*]pyridines **124** and **125** were synthesised and pharmacologically characterised as conformationally restricted mAChR ligands (Scheme 32) [38]. Their synthesis was based on the corresponding enamines **122** transformed according to *Method D2* into isothiazoles **123**. From **123** with different alkylating reagents and employing proper conditions (RX, K_2CO_3, Bu_4NHSO_4; RX, K_2CO_3; R_2SO_4, Bu_4NHSO_4; NaOH, CH_2N_2) **124** was obtained (R^3 = Boc), which was

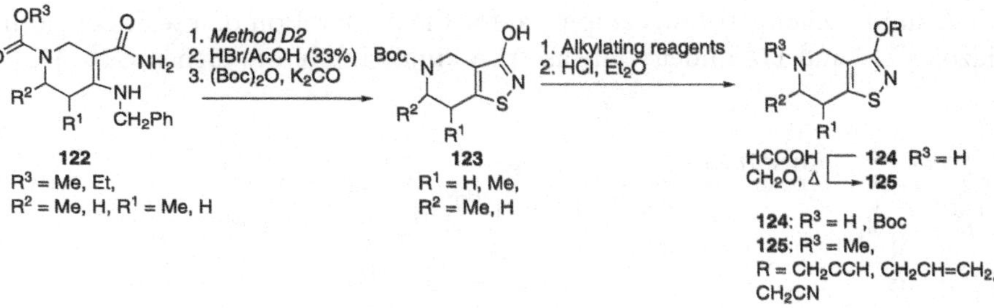

Scheme 32 Synthesis of 3-alkoxy-4,5,6,7-tetrahydroisothiazolo[4,5-*c*]pyridines

deprotected to nitrogen ($R^3 = H$) and then methylated at the nitrogen atom to afford **125** (HCHO/HCOOH at reflux).

2.4 Synthesis of 2,3-Dihydro-, 4,5-Dihydro- and Isothiazole *S*,*S*-Dioxides

2.4.1 Isothiazole and dihydroisothiazole *S*,*S*-dioxides

The main synthetic method to prepare the isothiazole *S*,*S*-dioxide ring and analogous dihydro compounds takes advantage of the intramolecular condensation with different electrophiles of the anion generated in position α to the SO_2 group. A very interesting class of isothiazoles is represented by the 4-amino-2,3-dihydroisothiazole *S*,*S*-dioxide series **127**. These compounds were synthesised from **126** through the well-studied [39] procedure known as CSIC (carbanion-mediated-sulfonamide-intramolecular cyclisation) shown in the Scheme 33 (*Method F*).

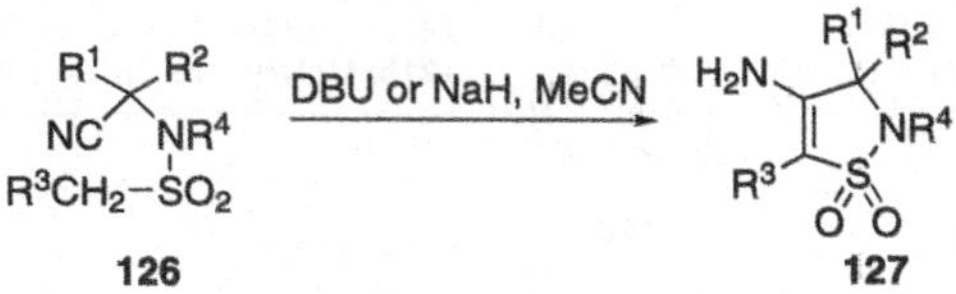

Scheme 33 General synthesis of 4-amino-2,3-dihydroisothiazole *S*,*S*-dioxides: *Method F*

An intramolecular cyclisation is also at the basis of the synthetic strategy outlined in Scheme 34 (*Method G*), which can afford 2,3-dihydroisothiazoles **129** from **128** operating with NaH, DMF (Scheme 34) [40–42].

When the starting compounds are sulfonylamidines **132**, prepared via a cycloaddition reaction of azide **130** and enamine **131**, 3-aminosubstituted-4,5-dihydroisothiazol *S*,*S*-dioxides **133** can be obtained by base catalyzed in-

Scheme 34 General synthesis of 2,3-dihydroisothiazole *S,S*-dioxides: *Method G*

tramolecular cyclisation (*Method H*). Substitution of the hydroxy group with a halogen (compound **134**) followed by dehydrohalogenation, affords the corresponding unsaturated isothiazole *S,S*-dioxides **135** [43] (Scheme 35).

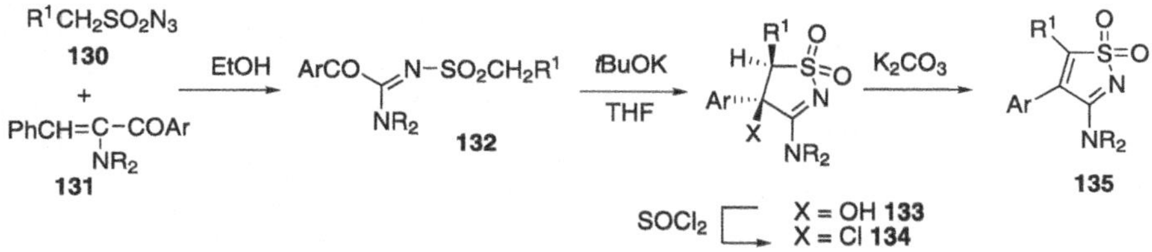

Scheme 35 General synthesis of 3-amino-isothiazole *S,S*-dioxide derivatives: *Method H*

Isothiazole dioxides **135** are the key intermediates for the preparation of a series of compounds for which the inhibitory activity on protein farnesyltransferase from *Tripanosoma brucei* was evaluated (Scheme 36) [44, 45]. Compounds used in these studies belong to two different classes in the isothiazole dioxide series. The first class includes 3-dimethylamino-4-(4-methoxyphenyl)-isothiazole *S,S*-dioxides **135**, unsubstituted or methyl substituted on C-5, respectively [43], or functionalised at C-5 with substituents ranging from alkenyl to aryl or heteroaryl groups. The second class consists of a series of 3-dimethylamino-4-(4-methoxyphenyl)-isothiazole *S,S*-dioxides and the corresponding 4,5-dihydro derivatives whose main feature is an *S*-atom as a linker between the isothiazole moiety and the substituent.

The 5-bromo derivative **136** was the key starting material for the preparation of both classes of compounds. It was obtained from **135** (R = H) by addition of bromine to the C4–C5 double bond followed by elimination of HBr, which can be spontaneous or induced by TEA. Stille reaction on **136** (Bu_3SnR, $(Ph_3P)_2BnClPd$) performed in toluene afforded the vinyl compound **137** (70%), the ethenyl derivative **138** (86%), and the aryl or heteroaryl compounds **139** (45–70%). Isothiazoles **140** functionalised at C-5 with an isoxazoline ring were prepared by a regioselective cycloaddition reaction starting from the vinyl derivative **137** and nitrile oxides [46].

The Michael addition of different mercaptans to isothiazoles **135** and **136** was regioselective and occurred at C-5 [47]. The addition to **135** of mercaptans gave a mixture of *trans* (major isomer) and *cis* diastereomers **141**

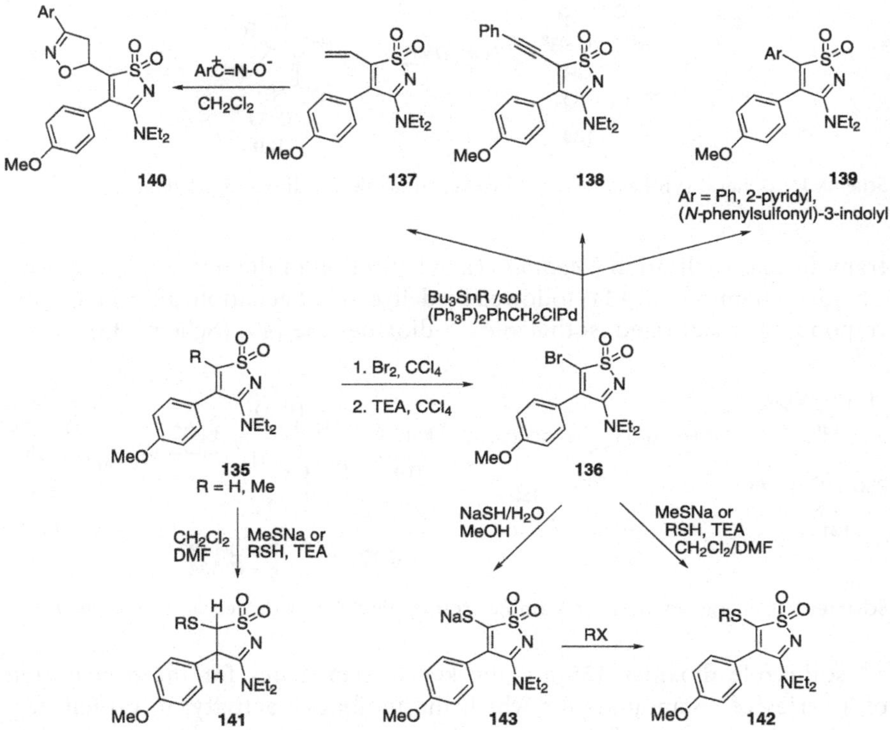

Scheme 36 General reactivity of 3-amino-5-bromo-isothiazole *S*,*S*-dioxide derivatives

(60–90%). Compound **136** gave the addition elimination products **142** with methyl thiolate (R = Me, 68%) or mercaptans (R = alkyl, aryl and heteroaryl, 31–88%) in CH_2Cl_2/DMF. The use of sodium thiolate in MeOH afforded **143**, which was alkylated at a sulfur atom to give **142** (R = Bn 79% yield, R = farnesyl 47% yield) using alkyl bromides and a base.

Analogues **145** and **146**, unsubstituted at C-4, were prepared as inhibitors of rat aortic myocyte proliferation [45] starting from isothiazoles **144** (R^1 = H, Cl) bearing different amino groups at C-3. Compounds **144** were synthesised from the corresponding isothiazolones prepared according to *Method C* followed by reaction with $POCl_3$ and NH_3 and oxidation of the sulfur atom with 3-chloroperbenzoic acid (Scheme 37) [48]. By reacting **144** (R^1 = H) with mercaptans, dihydro isothiazole derivatives **145** were formed in 40–91% yield. Through an addition–elimination process, compounds **146** (56–89%) were obtained from **144** (R^1 = Cl) and mercaptans.

According to *Method F* 4-amino-2,3-dihydroisothiazole *S*,*S*-dioxides **127** were synthesised through base-catalyzed ring closure starting from a variety of alkylsulfonamides **126** (Scheme 38). Several compounds belonging to this class have been studied for their potential anti-HIV activity [49–51].

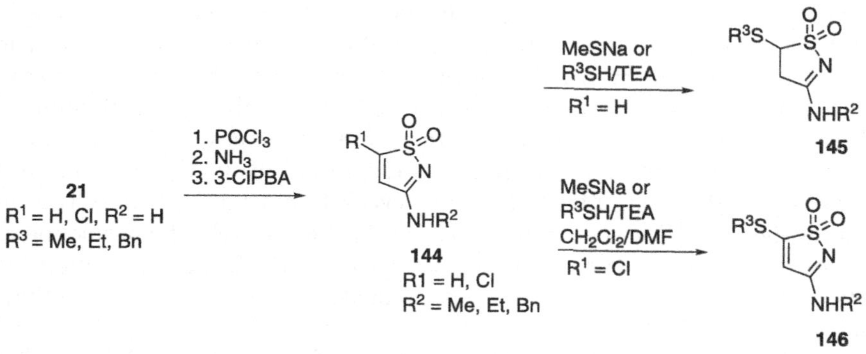

Scheme 37 Reactivity of 3-alkylamino-isothiazole *S,S*-dioxides with mercaptans

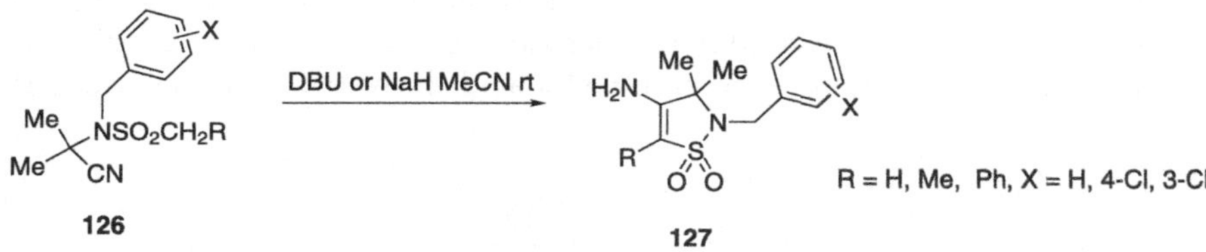

Scheme 38 Synthesis of 4-amino-2,3-dihydroisothiazole *S,S*-dioxides

More recently, several studies on alkanesulfonamides located on monosaccharide backbones have been performed due to the biological interest of spiro products like **150** (Scheme 39) [49]. These new bicyclic systems were used as glycone precursor for aza analogues of TSAO RT inhibitors. Several families of

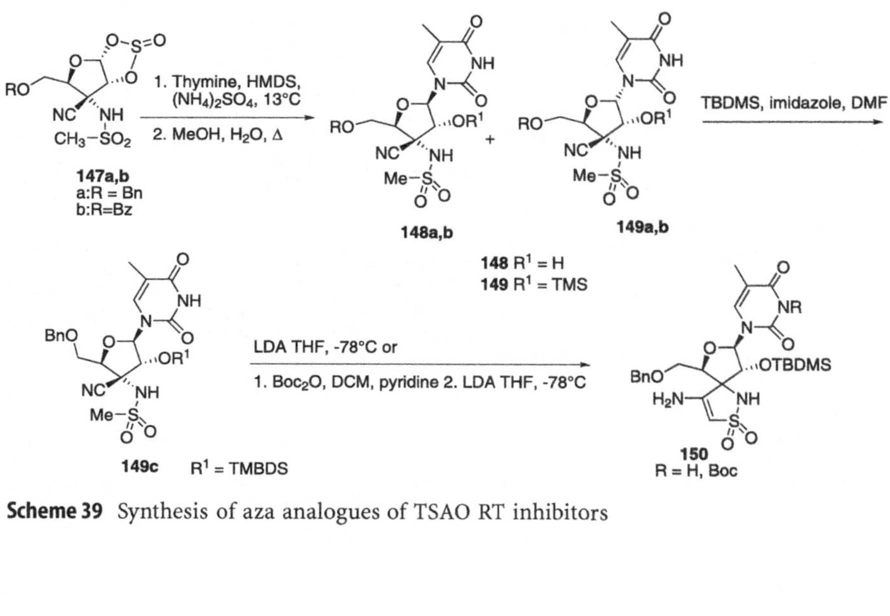

Scheme 39 Synthesis of aza analogues of TSAO RT inhibitors

compounds, depending on the substitution at both *N*-3 and *N*-2″, were synthesised. Sulfite derivatives **147** as *endo* and *exo* mixtures (*endo/exo* 1 : 1 to 3 : 2) were prepared by reaction of the corresponding dihydroxy derivatives with $SO(Im)_2$ in THF. Compounds **147** were treated with silylated thymine in dry conditions at 125 °C to give a mixture of regioisomeric 2′-*O*-silylated and 2′-hydroxy-5-methyluridine derivatives **148a,b** and **149a,b.** After protection of **148a** as TBDMS giving **148c** followed by cyclisation using LDA in THF, compound **150** (R = H) was obtained (*Method F*). Alternatively, the protection as *N*-Boc derivative of **148c** and subsequent cyclisation afforded compound **150** (R= Boc). Several aza derivatives were evaluated for their inhibitory activity against HIV-1 (III_B) and HIV-2(ROD).

2,3-Dihydro-4-ethylaminomethylisothiazole *S,S*-dioxide (**151**) is the key compound for the preparation of a large series of compounds like **156** designed as growth hormone secretagogues (GHSs) (Scheme 40) [52]. Their synthesis is based on the coupling of derivatives **151**, prepared according to *Method G*, with the opportune carboxylic acid derivatives and subsequent

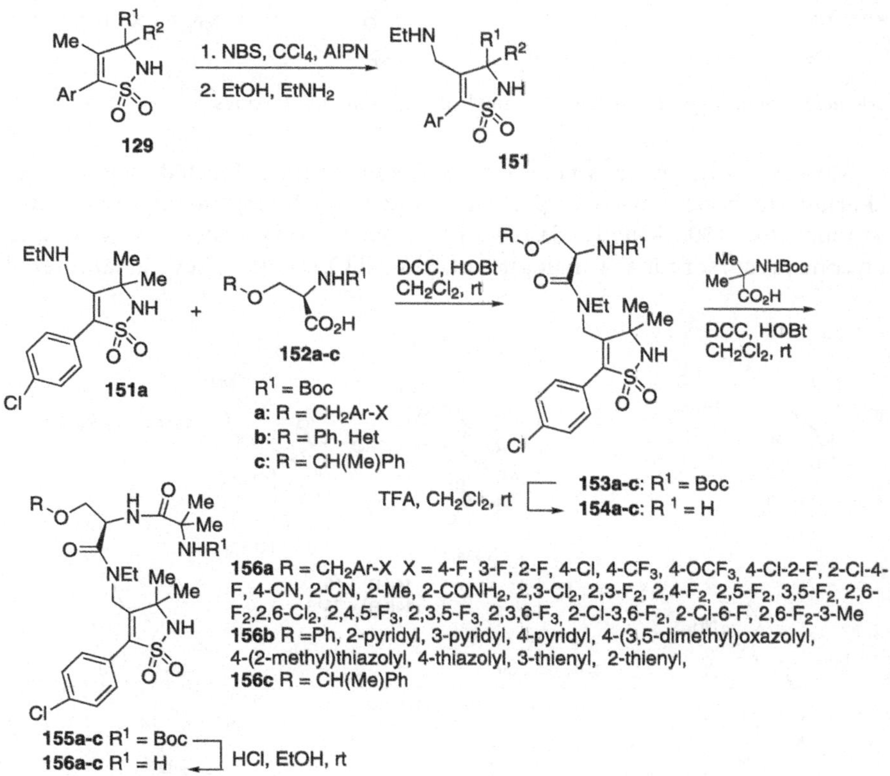

Scheme 40 Synthesis of growth hormone segretatgogues (GHSs) based on the 2,3-dihydroisothiazole nucleus

deprotection steps. Among the large number of differently substituted derivatives claimed in the patents [40–42], the synthesis of D-serine derivatives **156** are reported starting from **151a** as shown in Scheme 40.

2.4.2
Isothiazol- and Dihydroisothiazol-3-One *S,S*-Dioxides

Formation of the C – N bond through a cyclisation process is the key step for the synthesis of the isothiazolone *S,S*-oxide ring **158** from the corresponding **157** according to *Method I*, which is also general for dihydroderivatives and for benzisothiazolones [1, 2].

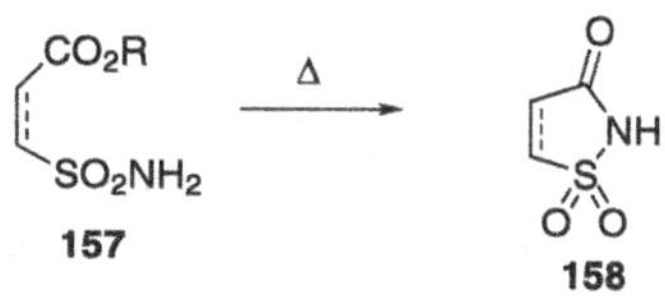

Scheme 41 Synthesis of isothiazol-and dihydro-isothiazol-3-one *S,S*-dioxides: *Method I*

This scaffold characterises several potent mechanism-based inhibitors of HLE (see Sect. 5.3) such as tetrahydrobenzisothiazolones **162** (R = alkyl, n = 1) [53], which were prepared according to *Method I* from **159**, oxidised with Cl_2 in AcOH to the corresponding intermediates, which were transformed into the sulfonamides **160**. Their cyclisation with MeONa afforded isothiazoles **161** (70–80% overall yield). The functionalisation of the nitrogen atom was performed using $PhSCH_2Cl$ and $(t\text{-Bu})_4NBr$ (70–80%) followed by treatment of the intermediate with SO_2Cl_2 in CH_2Cl_2 giving the *N*-chloromethyl intermediate (60–80%). Their reaction with the benzoic acid derivative in basic conditions afforded compounds **162** (50–60%) (Scheme 42). Compounds

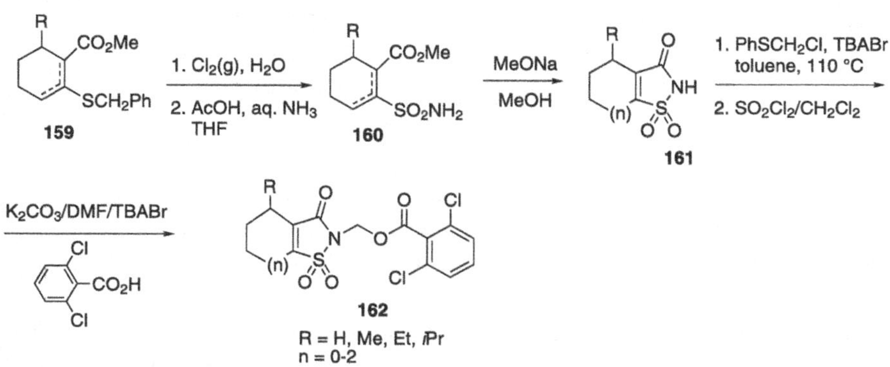

Scheme 42 Synthesis of 4,5,6,7-tetrahydrobenzo[*d*]isothiazolones

162 (R = H, $n = 0.2$) were also prepared from the available intermediates **161** ($n = 0.2$).

The analogue norbornane compound **165** was obtained from bromo isothiazolone dioxide **163** and cyclopentadiene, which gave the intermediate cycloadduct (95%), which was transformed into derivative **164** (40% overall yield) by reduction of the double bond followed by elimination of hydrogen bromide and deprotection (48%). The latter was converted to the target dichlorobenzoate **165** as described before for compounds **162** (Scheme 43) [53].

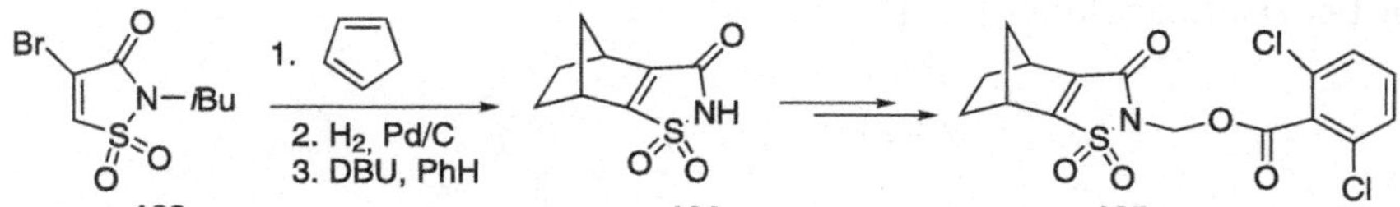

Scheme 43 Synthesis of norbornane-isothiazole derivative 165

Following *Method I* simple isothiazolidin-3-one *S*,*S*-dioxide derivatives of general formula **166** were synthesised from intermediates like **157** [54]. The functionalisation of nitrogen gave **167** (Scheme 44).

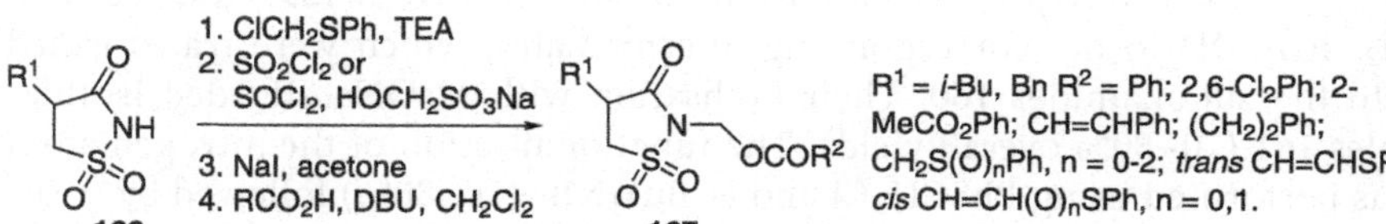

Scheme 44 Synthesis of isothiazolidin-3-one *S*,*S*-dioxide derivatives

Phosphate derivatives **168** (Fig. 1) were also prepared starting from racemic 4-*i*-propyl-2-bromomethyl-isothiazolidin-3-one *S*,*S*-dioxide and di *n*-butyl- and benzyl phosphate in the presence of DBU [55].

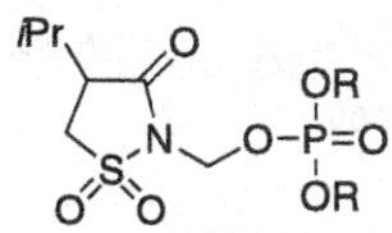

168
R = Bu, Bn

Fig. 1 *N*-Methylen-isothiazolidin-3-one *S*,*S*-dioxide phosphate

Functionalisation of the isothiazol-3-one *S*,*S*-dioxide system at C-5 provided several interesting derivatives. An example regarding the synthesis of some protein tyrosine phosphatase inhibitors [56] is reported below in

Scheme 45. Peptides containing the IZD heterocyclic pTyr mimetics were synthesised in 10–11 linear steps from readily available starting materials such as 5-chloro-isothiazol-3-one **169**, synthesised according to *Method C*, and amino acid derivatives **170**. The key synthetic reaction was a novel Suzuki coupling of chloroheterocycle **169** with 4-phenylalanineboronic acid **170** to afford the fully protected scaffold **171**. The *N*-terminus of **171** was subsequently elaborated via peptide coupling and the dipeptide **172** was deprotected to give inhibitor **173**. The 4,5 double bond of **172** was reduced, isomers were separated, and each of them was further elaborated to afford compounds **174** (Scheme 45).

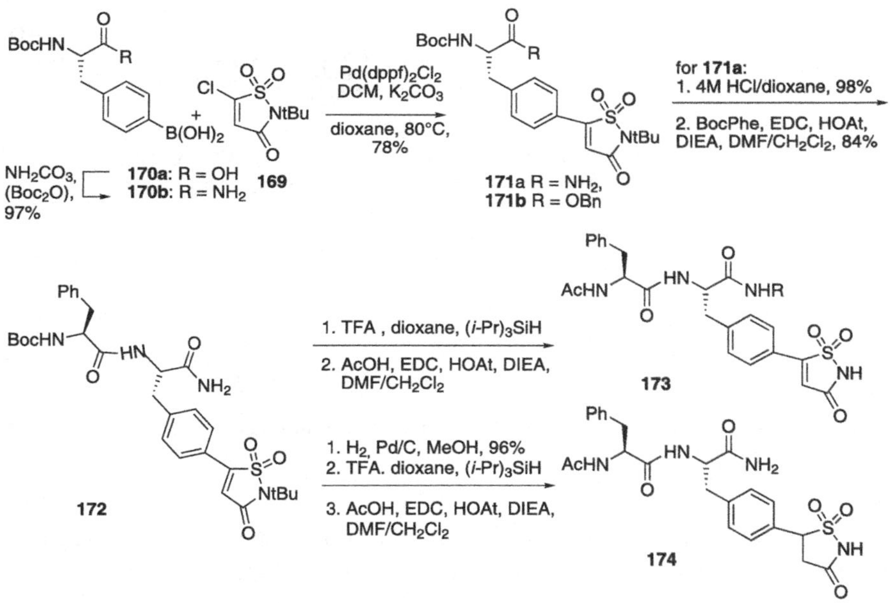

Scheme 45 Synthesis of peptidomimetics **173** and **174** containing the isothiazolidin-3-one *S,S*-dioxide scaffold

From the optimisation of the above compounds, potent nonpeptidic benzimidazole sulfonamide inhibitors were disclosed [57]. The synthesis followed the route previously depicted. The ester **171b** was reduced and deprotected giving **175** in high yield and its coupling with substituted phenylendiamine **176** followed by cyclisation under acidic conditions afforded the desired benzimidazole **177**. Careful control of the temperature was critical because ring closure at higher temperatures proceeded more rapidly but returned mixtures of diastereomers at the α-centre of the amino acid. The (*R*/*S*)-IZD diastereomers were easily separated by chiral HPLC to afford two discrete isomers. Both diastereomers were further elaborated to the final compounds providing

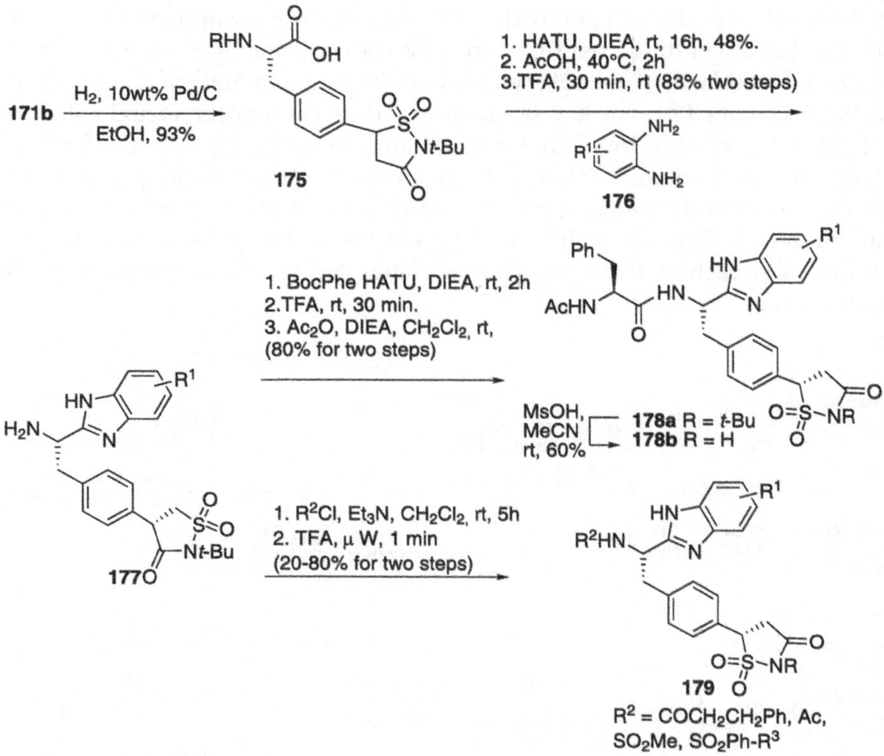

Scheme 46 Synthesis of peptidomimetics **178** and **179** containing the isothiazolidin-3-one *S,S*-dioxide scaffold

an active (*S*) and inactive pair. The removal of Boc group with TFA furnished the free amine, which was acylated with a variety of reagents under mild conditions to give amides, ureas, sulfonamides and carbamates **179** (Scheme 46).

3 Sultams

γ-Sultams are useful heterocycles for asymmetric synthesis and medicinal chemistry. Several compounds of biological interest containing the sultam moiety were synthesised and in many cases their preparation can be performed by simple functionalisation of the unsubstituted sultam or by modification of preexisting functional groups. Most compounds are *N*-substituted sultams but there are also several interesting derivatives substituted at C-2, C-3 or/and C-4. Some examples of the synthesis of sultams with a particular substitution pattern were reported.

3.1 Synthesis of Sultams

Various protocols for the synthesis of the substituted sultams have been developed using as the ring formation step either C – N, C – C or C – S bond formation [1, 2]. Very often the cyclisation of a γ-aminosulfonyl chloride deriving from nucleophilic substitution of an opportune amine on a γ-halosulfonyl chloride is used. In such a case the resulting sultam can be substituted or not at the ring nitrogen depending on the amine used in the process (*Method J1*). In the case of γ-aminopropanethiols oxidation of the thiol function and cyclisation was done in a one-step procedure (*Method J2*) (Scheme 47). A large number of different combination of reagents were used in the sultam ring synthesis based on this procedure and will be described for each significant compound.

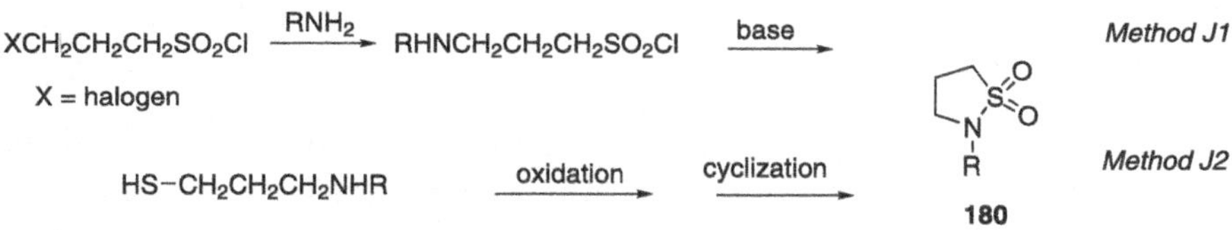

Scheme 47 Synthesis of sultams: *Method J*

Recently developed powerful methodologies for the generation of these cyclic sulfonamides include pericyclic reactions as the intramolecular Diels–Alder reaction affording bi- or polycyclic sultams (*Method K*) (Scheme 48).

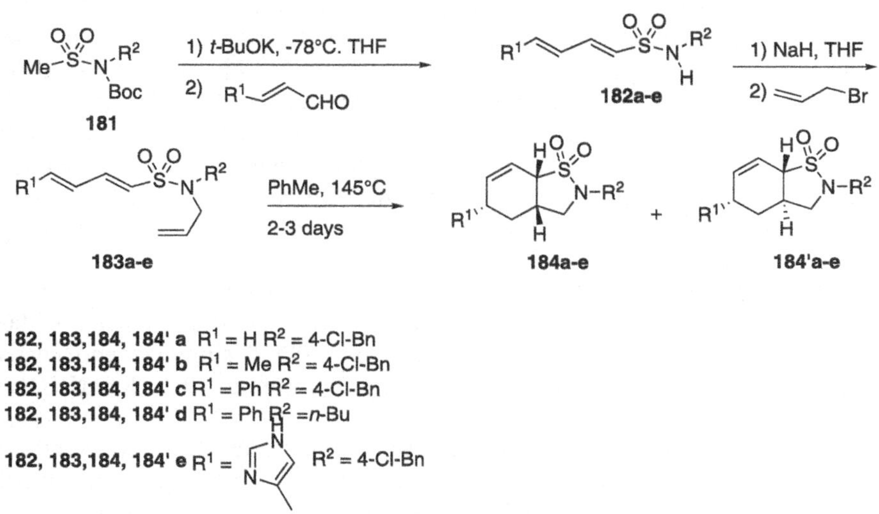

Scheme 48 Synthesis of polycyclic sultams via Diels–Alder reaction: *Method K*

Such a procedure has been exploited for the synthesis of several derivatives for which an anti-inflammatory activity has been claimed [58]. A derivative under study as the Histamine H3 antagonist was prepared by the thermal intramolecular Diels–Alder reaction of a triene derivative of buta-1,3-diene-1-sulfonic acid amide. 1,3-Butadiene sulfonamides **182** (**a**: 67%, **b**: 69%, **c**: 99%, **d**: 51%) were prepared by the base mediated condensation of *N*-Boc-methanesulfonamides (**181**) with a series of aldheydes. *N*-akylation of **182** to give trienes **183** (**a**: 69%, **b**: 76%, **c**: 82%, **d**: 59%) was achieved by reacting the sodium salts with allyl bromide in THF at reflux. The intramolecular Diels–Alder reactions of compounds **183** were performed at 145 °C in toluene in a sealed vessel under argon. Under these conditions compounds **184** and **184′** were obtained in good yields (**a**: 76% ratio 6 : 1, **b**: 71% ratio 6 : 1, **c**: 92% ratio 3 : 1, **d**: 87% ratio 3 : 1).

3.2 Reactivity of Sultams

N-Unsubstituted γ-sultams can be easily substituted at the nitrogen ring by alkylation, arylation or acylation. Arylation can be performed efficiently via copper promoted chemistry by using arylboronic acids. Alkylation is usually performed with halogen derivatives by using bases such as K_2CO_3, NaH, TEA. In some cases good results were obtained from hydroxy substituted compounds and DEAD/PPh_3 [59, 60].

Synthesis of sultam hydroxamates prepared as potential anti-inflammatory agents was accomplished as shown in Scheme 49 by applying the methodology already described (*Method C*). Racemic homocystine **185** was oxidised to the sulfonyl chloride and cyclised. The resulting sultam **186** was arylated or alkylated affording **187a–e** (**a**: R = 4-Ph – Ph, 13%; **b**: R = 4-PhCH$_2$O – Ph, 26%; **c**: R = Ph – PhCH$_2$CH$_2$, 88%; **d** R = PhCH$_2$OPhCH$_2$, 44%; **e**: R = PhCH$_2$OPhCH$_2$CH$_2$). Deprotection of oxygen in **187b,d,e** and transformation of the phenolic group of **188** into OTf followed by Suzuki reaction gave the bis-aryls **189a–c**. When **188** was treated with 4-(hydroxymethyl)- or with 4-(chloromethyl)-2-methylquinoline, **190** was formed (Scheme 49). Treatment of **187** and **190** with basic hydroxylamine gave the corresponding hydroxamates **191**.

Enantiopure homochiral sultams were prepared from chiral alcohol **192a** that was converted to the thioacetate **192b** (R = AcS) prior to chlorine oxidation, selective *N*-Boc removal and cyclisation to **193**. A similar chemistry as that reported in Scheme 49 afforded the target compound **194b** (R = NHOH) (Scheme 50).

A novel series of HIV protease inhibitors containing the sultam scaffold **197** has been synthesised [61]. Compound **197** was prepared (Scheme 50) starting from the thioacetate **195** prepared as described in the literature [61]. The usual oxidation/chlorination one-pot process (*Method J1*) with Cl_2 gas

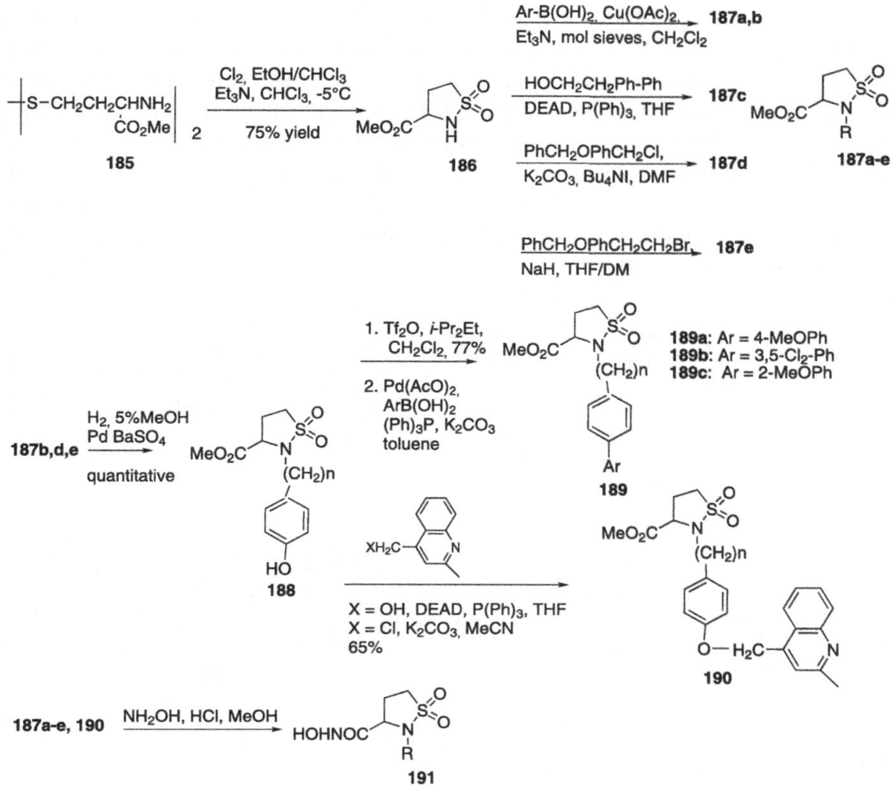

Scheme 49 Synthesis of sultam hydroxamates

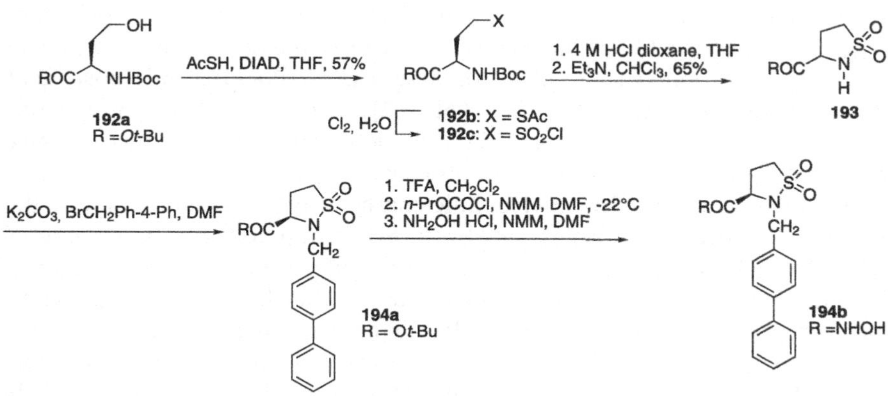

Scheme 50 Synthesis of homochiral sultams 194

in AcOH/aq. HCl produced sulfonyl chloride **195b** (95%). The Cbz group was removed (48% HBr/AcOH) and the cyclisation to **196** (58%) was performed using TEA. The benzyl substituent was introduced starting from the *N*-Boc derivative, obtained from **196** (Boc_2O/THF), which was treated with LDA and BnBr in THF. Finally, the nitrogen atom was deprotected (TFA/CH_2Cl_2) and the isothiazolidine *S,S*-dioxide **197** was isolated. Compound **197** was then coupled with epoxide **198** using NaH in DMF at 80 °C. A partial isomerisation occurred and a mixture of *trans/cis* epimers (3 : 1) was formed (the *trans* isomer **199** is shown in Scheme 51).

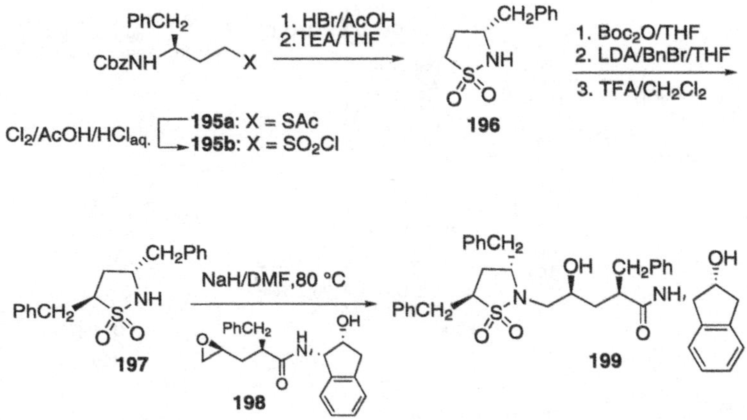

Scheme 51 Synthesis of 3,5-dibenzyl-*N*-substituted sultams

By the same chemistry based on the reaction of 3-chloropropanesulfonyl chloride with amines as shown in *Method J1*, compounds **201** [62] and **203** were prepared from amines **200** and **202**, respectively (Scheme 52). Further alkylation of **203** with methyl iodide resulted in the formation of **204**. Removal of the Boc group in **203** and **204** with HCl followed by coupling with *N*-Boc-D-(4-Cl)Phe under EDC conditions gave **205**. Removal of Boc protection followed by EDC coupling with *N*-Boc-Tic and subsequent Boc removal furnished **206a** and **206b** [63]. Compound **201** showed anti-inflammatory activity, whereas **206a** and **206b** were synthesised as selective human melanocortin subtype-4 receptor ligands.

Taking advantage of the nucleophilic character of the C-5 atom due to the electron-withdrawing effect exerted by the SO_2 group, condensation reactions with aldheydes are possible giving rise to the alkylidene derivatives [64, 65]. By this way, through an aldol condensation, antiarthritic drug candidates, were prepared. The *N*-substituted sultam ring, from the cyclisation process, was condensed with 3,5-di-*t*-butyl-4-hydroxybenzaldehyde affording the corresponding adducts **207** as diastereomeric mixture. Treatment of the crude aldol adducts with a catalytic amount of *p*TsOH resulted in dehydration and

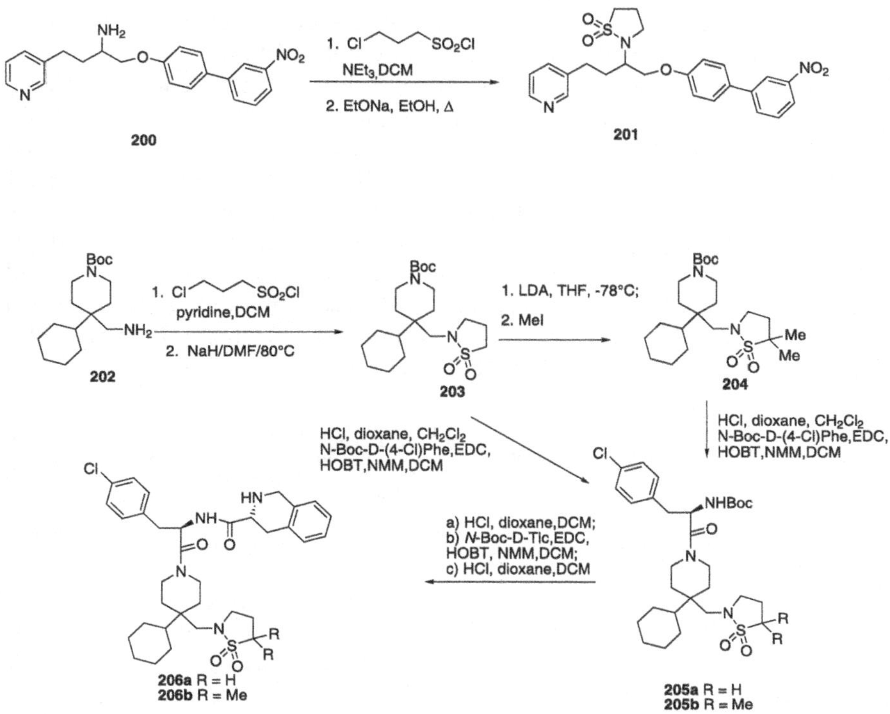

Scheme 52 Synthesis of *N*-functionalised sultams as selective human melanocortin subtype-4 receptor ligands

removal of the MOM group yielded an *E*/*Z* mixture of 5-benzylidene-γ-sultam derivatives **208**. Starting from the *N*-unprotected sultam **208**, several alkylated or acylated derivatives were formed by introduction of an alkyl or acyl substituent at the nitrogen atom. The use of the *p*-quinone methide

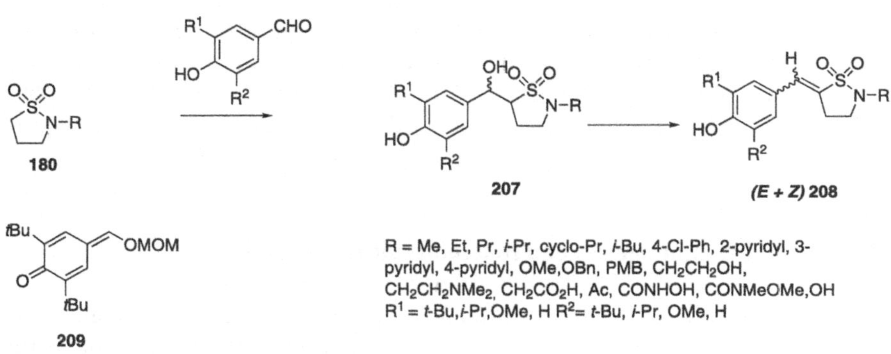

Scheme 53 Synthesis of 5-benzylidene-γ-sultam derivatives

derivative **209**, generated from the corresponding benzaldehyde, allows to obtain a single *E*-isomer (R = Et), via a 1,6-Michael addition (Scheme 53).

By a Wittig-Horner reaction, alkylidene derivatives **212** were similarly prepared and tested on *Plasmodium falciparum*. Phosphonates **210** were obtained from sultams **180** by using $(EtO)_2POCl$ and LDA at – 78 °C and used in a Wittig-Horner reaction with Boc-Phe-CHO (*n*BuLi in THF) affording **211** (*E*: 60%, *Z*: 16–29%) after deprotection of the nitrogen atom (TFA in CH_2Cl_2). The dipeptides **212** (56–88%) were obtained by condensation of **211** with *N*-Boc-L-leucine-*N*-hydroxysuccinimide ester in presence of TEA in THF and deprotection of nitrogen with TFA in CH_2Cl_2. Carbamates **212** (R = $PhCH_2$) were obtained in low yield from **212** (R = H) and 4-morpholinecarbonyl chloride in presence of TEA in THF (Scheme 54) [66].

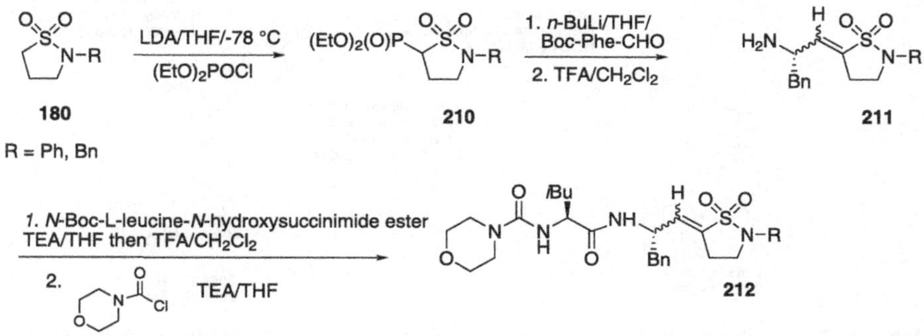

Scheme 54 Synthesis of 5-alkylidene-γ-sultam dipeptides

4
Benzisothiazoles

Examples of new synthetic approaches to the benzisothiazole ring are related to compounds characterised by a particular substitution pattern or to heterobicyclic derivatives. Instead, the preformed benzisothiazole ring was extensively used for the preparation of biological compounds functionalised with several heterosubstituted chains at C-3 or at the nitrogen atom. Naphtho[1,8-*c,d*]isothiazole (naphtosultam) represents the most used ring when functionalisation at the nitrogen atom with pharmacophores was done.

4.1
Synthesis of Benzisothiazoles and Heterobicyclic Isothiazoles

S,S-Dioxo-2,3-dihydrobenzo[*d*]isothiazol-5-yl-pyrazole (**215**) [67], an anti-inflammatory agent, was synthesised from compound **214** by treatment with CBr_4, Ph_3P and aqueous $NaHCO_3$. Preparation of the reagent **214** was ac-

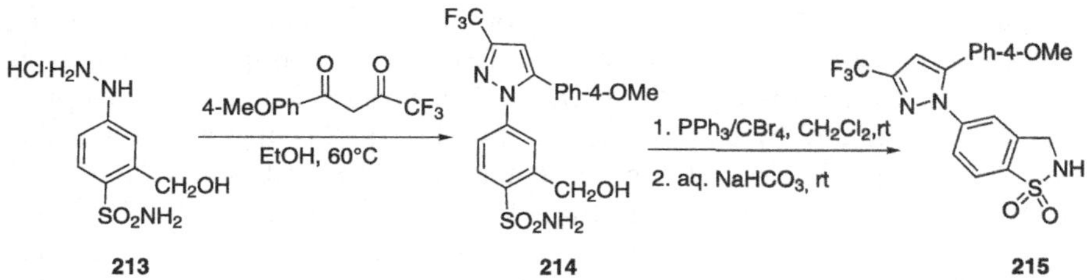

Scheme 55 Synthesis of *S*,*S*-dioxo-2,3-dihydrobenzo[*d*]isothiazol-5-yl-pyrazole

complished as shown in Scheme 55 by simple coupling of phenylhydrazine hydrochloride **213** with 1-(4-methoxyphenyl)-3-trifluoromethyl-1,3-propane dione in EtOH.

Benzo[*d*]isothiazoles **220** and **221** (Scheme 56) were synthesised as oxidosqualene cyclase inhibitors. By reaction of fluorobenzophenones **216** with potassium benzylthiolate followed by *S*-chlorination, cleavage of the benzyl group with SO_2Cl_2 and reaction with NH_3, compound **217** was formed. Methoxy deprotection and alkylation with 1,6-dibromohexane gave intermediate **218**, which was converted to **220** with *N*-allylmethylamine. Intermediate **218** was oxidised to the *S*,*S*-dioxide **219** and then functionalised at C-6 affording **221**. Similarly, benzo[*d*]isothiazole **223** was prepared from benzophenone **222** [68].

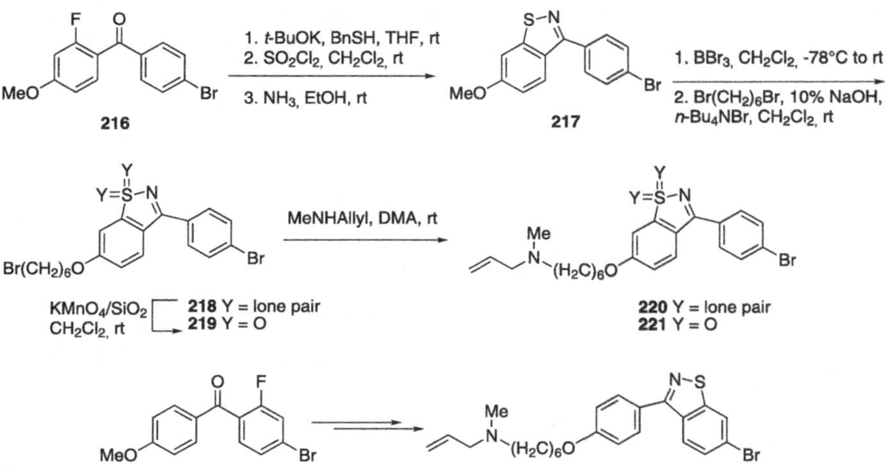

Scheme 56 Synthesis of 3-aryl substituted benzisothiazoles as oxidosqualene cyclase inhibitors

A series of benzisothiazoles *S*,*S*-dioxides functionalised with a nitrogen-linked chain at C-3 and with different substituents on the benzene ring, which are particularly useful in the treatment of pain, inflammatory hyperalgesia, and urinary disfunctions, were patented [69]. As an example, the synthesis of compounds **226** was reported (Scheme 57). From substituted chloropyridine derivatives and by using a Suzuki reaction, the *bis*-aryl derivatives **224** were synthesised. Their treatment, in sequence, with SO_2Cl_2, aq. NH_4OH, 1 M NaOH and with $KMnO_4$ afforded the 3-hydroxy-benzisothiazoles **225**. Their reaction with $POCl_3$ followed by treatment of the corresponding 3-chloro intermediates with a proper amine gave **226**.

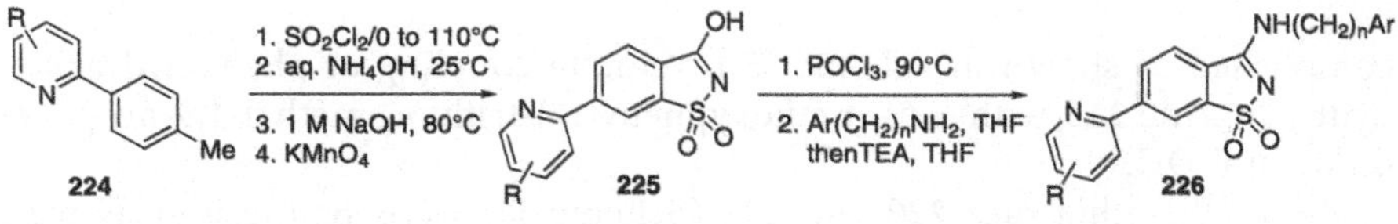

Scheme 57 Synthesis of 3-aminosubstituted benzisothiazoles *S,S*-dioxides

Potential new inhibitors of *Leptospheria maculansi* mediated detoxification of phytoalexin brassilexin were designed and synthesised by analogy with the heteroaromatic structure of isothiazolo[5,4-*b*]indole (brassilexin) [9]. The above ring was replaced by quinolino[5,4-*b*]-isothiazole **227**, benzothiophene[5,4-*b*]-isothiazole **230** and the simple benzoisothiazole **232** rings. In addition, 4-arylisothiazoles **7** resulting from disconnecting the "*a*" bond of the indole ring of brassilexin and replacing the NH group *ortho* to isothiazole with different substituents were prepared (see Sect. 2.1). Even if the synthesis of the above ring is known in general, some changes in the

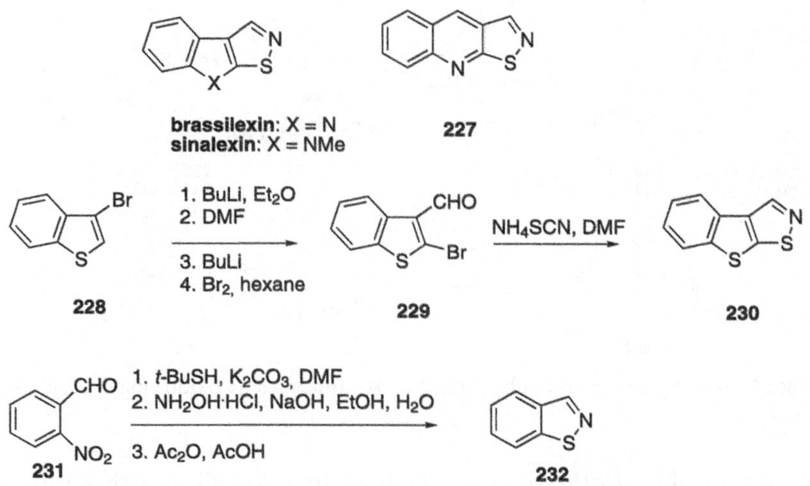

Scheme 58 Synthesis of benzothiopheneisothiazole and benzisothiazole

use of the reagents were made aiming to improve the yield or to find less harsh conditions or to simplify the purification processes. A new synthetic approach was adopted for the preparation of **230** starting from 3-bromo-benzo[*b*]thiophene **228** was treated with *t*-BuLi then quenched with DMF. After bromination in basic conditions, **229** (35% overall yield) was obtained and then heated with NH_4SCN affording **230** (11%). Aiming to avoid several chromatographic steps and to increase the yields, the synthesis of unsubstituted 1,2-benzoisothiazole **232** (64%, overall yield), starting from aldehyde **231**, was revisited. New conditions are reported in Scheme 58.

4.2
Reactivity of Benzisothiazoles and Heterobicyclic Isothiazoles

Several compounds of biological interest containing the benzisothiazole ring were prepared and, in many cases, well-known starting materials were used. In general, their chemistry is related to the simple functionalisation of preexisting substituents. They can be divided into two main classes, i.e. benzisothiazoles functionalised at C-3 or at nitrogen. The majority of C-3 functionalised compounds bears a nitrogen atom directly linked to C-3 or a carbon chain containing a nitrogen. Other heteroatoms can be directly linked to C-3, such as oxygen and sulphur.

4.2.1
3-Carbon Linked Benzisothiazoles

Vicini et al. reported on the preparation of hydrazones **234** obtained from the condensation of a proper aldehyde with 1,2-benzisothiazole hydrazide derivatives **233** characterised by a different length of the linker chain (Scheme 59) [70] and their antimicrobial [71, 72] and anti-inflammatory [73] activities were evaluated. Both theoretical and experimental lipophilic indices were calculated and QSAR studies were also reported [70].

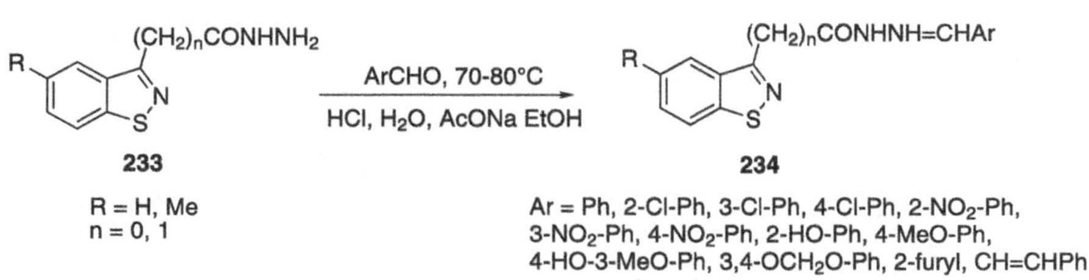

Scheme 59 Synthesis of hydrazones of hydrazido benzisothiazole derivatives

Below two examples of benzisothiazole derivatives characterised by anti-inflammatory activity functionalised at C-3 with a heterocyclic ring, which was built starting from homologous nitrile compounds **238** and **242b**, are given. Compound **238** was prepared by using two different synthetic strategies starting from the available 3-chlorobenzisothiazole (see Sect. 4.2.2) condensed with both ethyl cyanoacetate and phenylacetonitrile in EtOH/Na affording **235** and **238** (R^1 = Ph, R = H), respectively [74]. The alkylation of **235** with MeI gave **236** hydrolyzed and esterified to **237** (R^1 = Me), then transformed into nitrile **238** (R^1 = Me), by treatment first with NH_4OH and then with P_2O_5. Compound **235** was directly hydrolyzed in aqueous DMSO to **238** (R^1 = H). Tetrazolyl derivatives **239a,b** were prepared by treatment of **238** with NaN_3 in DMF in the presence of NH_4Cl. The benzisothiazolylalkanoic acids **240** were simply prepared by alkaline hydrolysis of the corresponding nitriles. From ethyl 2-(1,2-benzisothiazol-3-yl)-acetate (**237**, R^1 = H) some new 2-(1,2-benzisothiazol-3-yl)ethylamine derivatives **241** were synthesised and their putative histaminergic activity was investigated [75]. 2-(1,2-Benzisothiazol-3-yl)ethanol was obtained by reduction of ester **237** ($LiAlH_4$, Et_2O), which was treated with $SOCl_2$ to obtain the chloro derivative, and then reacted with a suitable amine in EtOH affording the targeted compounds **241** (Scheme 60).

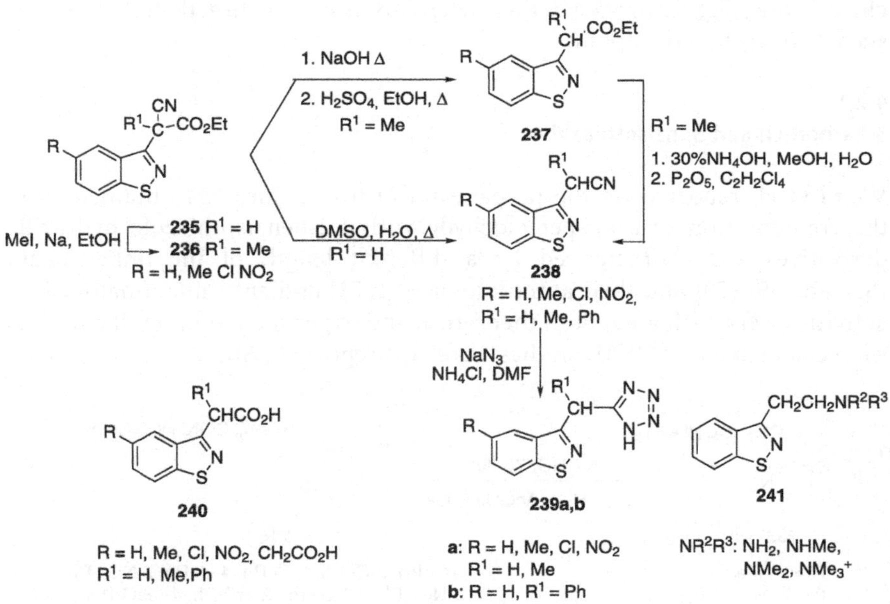

Scheme 60 Synthesis of 3-tetrazolyl- and 3-carboxymethyl-benzisothiazoles

Benzisothiazole 3-carboxamide **242a** was converted, through the intermediate nitrile **242b**, into the thioamide **242c** ($POCl_3$ then NH_3/H_2S), the key starting material for the preparation of several 2-(1,2-benzisothiazol-

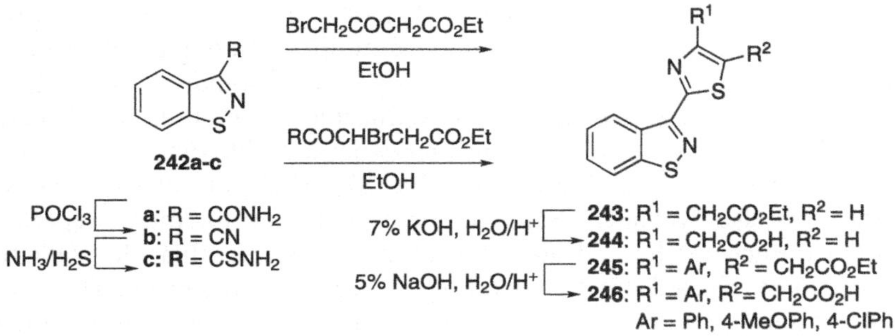

Scheme 61 Synthesis 2-(1,2-benzisothiazol-3-yl)-thiazolyl-4- or 5-acetic acid derivatives

3-yl)-thiazolyl-4- or 5-acetic esters **243** and **245** obtained by reaction with bromo ketoesters. The hydrolysis of the ester function gave acids **244** and **246** (Scheme 61) [76].

4.2.2 3-Heterosubstituted Benzisothiazoles

A large number of biologically interesting compounds were synthesised from two key reagents represented by 3-chloro-1,2-benzisothiazole (**248**) and 3-amino-2-benzisothiazole (**249**) (Scheme 62). The chloro derivative was prepared from 1,2-benzisothiazolone (**247**) obtained according to *Method C* from 2,2′-dithiosalicyclic acid ($SOCl_2$, DMF, toluene, 75 °C; Cl_2, CH_2Cl_2, then NH_4OH). By adding $POCl_3$ to **247** and heating it gradually to 120 °C, the 3-chloro derivative **248** was obtained (77% yield) [77]. Other recent preparations of **248** are claimed in several patents [78, 79].

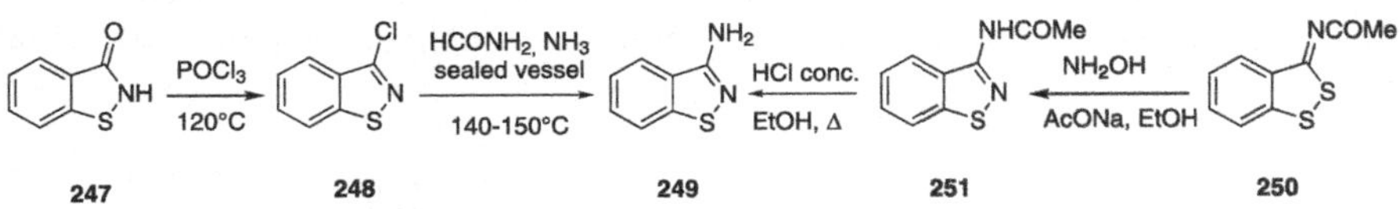

Scheme 62 Synthesis of 3-chloro- and 3-amino-1,2-benzisothiazole

The preparation of 3-amino-1,2-benzisothiazole **249** can be done in different ways. A well-known method started from **248**, which was treated with a formamide solution containing NH_3 [80] (Scheme 62). An alternative method started from hydroxylamine and *N*-(3*H*-benzo[*c*]1,2-dithiol-3-ylidene) acetamides (**250**) (obtained from **248** and thioacetic acid), and gave *N*-(3-benzisothiazolyl)acetamide (**251**) (52–62%) hydrolyzed to amine **249** (48–88%) [81].

Several 3-amino substituted benzisothiazoles, in which the amino group is involved in the formation of an amide, amidino, imino, guanidino groups or is a part of a cyclic amine (e.g. piperazine) were prepared and evaluated for different activities. 3-Amidinobenzisothiazole compounds **252** displayed remarkable analgesic action and an interesting antiphlogistic action that is often dissociated from antipyretic activity [82]. Their antimicrobial activity in vitro was also evaluated [83]. The main procedure for their preparation involved a nucleophilic addition of 3-amino-1,2-benzisothiazoles **249a,b** to the carbon of selected nitriles used both as reagent and solvent. The majority of the target compounds were obtained by heating the amine with the suitable cyanide and acidic catalyst enhancing the CN group reactivity (i.e. $SnCl_4$ or $AlCl_3$). Guanidino derivatives **253** and formamidino compounds **254** were prepared by reacting **249a,b** with cyanamine in HCl and *N,N*-dimethylformamide dimethylacetal in benzene, respectively (Scheme 63).

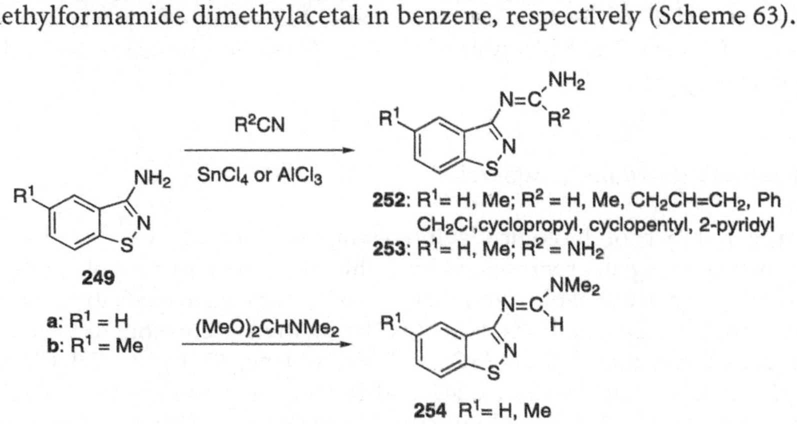

Scheme 63 Synthesis of 3-amidino- and 3-guanidino-1,2-benzisothiazoles

3-Aminoarylidene-1,2-benzisothiazol **255** were also prepared from 3-aminobenzisothiazole **249** and an opportune aldehyde (Scheme 64). This

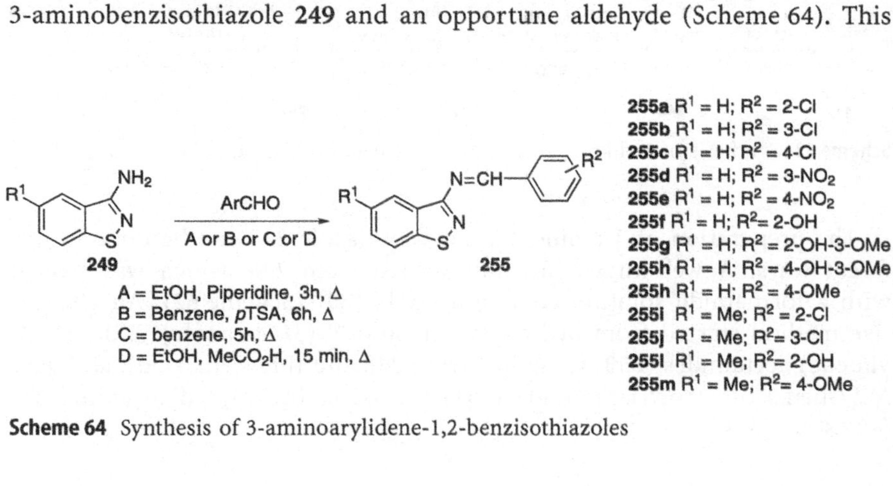

Scheme 64 Synthesis of 3-aminoarylidene-1,2-benzisothiazoles

reaction was not always straightforward and different experimental conditions (A, B, C or D) were developed [73].

There is a large literature concerning the preparation and the evaluation of the 3-(piperazinyl)-1,2-benzisothiazole derivatives **258** (Tables 2, 3) substituted at nitrogen atom with different oxygen functionalized chains [84–88] as antipsychotic agents. Three main synthetic strategies were adopted for their preparation (Scheme 65). The first one used as key reagent the 3-(piperazinyl)-1,2-benzisothiazole (**256**), which was appropriately alkylated to introduce a chain of suitable length ($n = 2, 3, 4$) affording the intermediate **257** and then made to react with the characteristic terminal group (reagent RZ) affording **258**. Some examples of the very large number of derivatives that have been synthesised are given in Table 2 (entries 1–9). The second strategy took advantage of reagents **259a,b** from which compounds of type **258** (Table 2, entries 10–12) were prepared. The use of reagent **259b** gave a mixture of two regioisomers (entries 11 and 12).

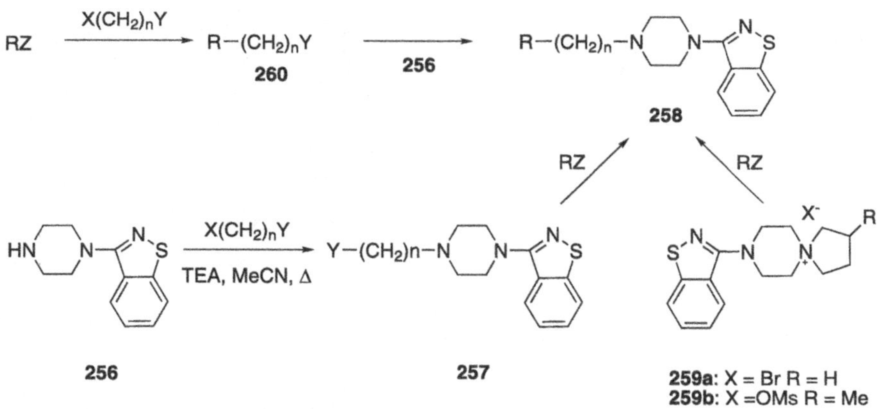

Scheme 65 Synthesis of 3-(piperazinyl)-1,2-benzisothiazole derivatives **258**

According to the third strategy, the chain was first linked to the terminal characteristic group (intermediate **260**) and then the piperazinylbenzisothiazole (**256**) was alkylated with the whole substituent **260** affording **258** (Table 3 entries 1–5).

3-Phenylpropionic acid derivatives of general formula **261**, functionalised at C-2 with a 3-heterosubstituted benzisothiazole, were patented for use in the treatment and/or prevention of peroxysome proliferator-activated receptor gamma (PPARgamma) mediated diseases [89]. Only the synthesis of the 3-sulfur linked compound **264** (40%) was extensively described starting from methyl 3-[4-(benzyloxy)phenyl]-2-chloropropionate (**263**) and benzoisothiazol-3(2*H*)-thione (**262**) operating in the presence of MeONa/MeOH (Scheme 66).

Table 2 Antipsychotic 3-(piperazinyl)-1,2-benzisothiazole derivatives **258** (part I)

Entry	n	Y	RZ	**258**, R:	Refs.
1	4	NH_2	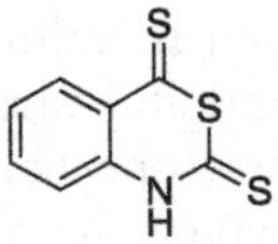	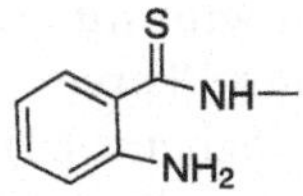	[85]
2	4	NH_2	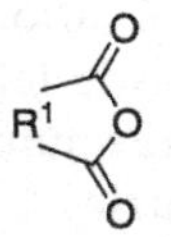	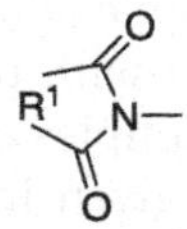	[86]
3	4	OH	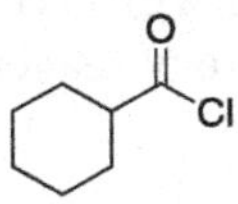	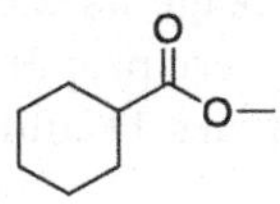	[86]
4	4	OH	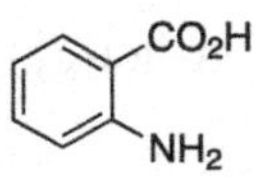	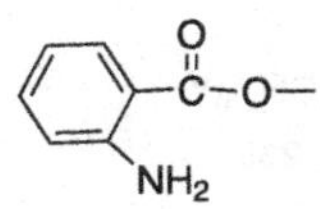	[85]
5	2		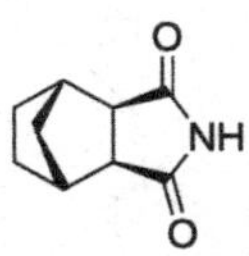	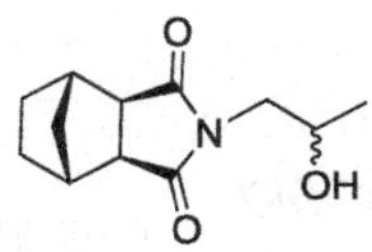	[86]
6	3	CO_2Et	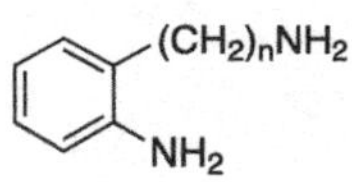	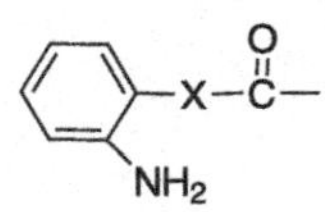 X = CH_2NH, NH	[85]
7	3	ONH_2	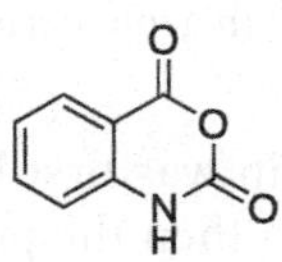	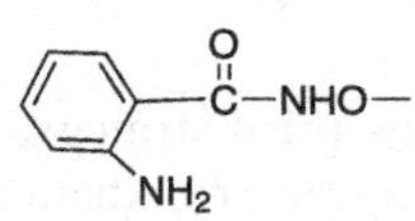	[85]
8	3	Cl	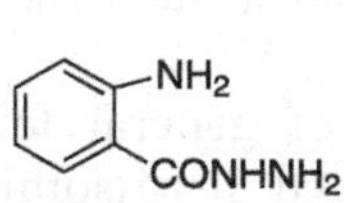	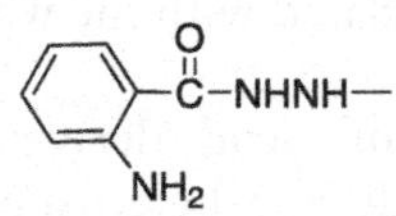	[85]
9	2	Cl	Het	Het	[87]
10	4	–	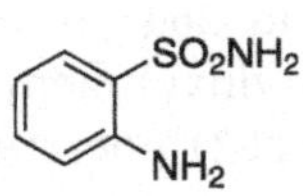	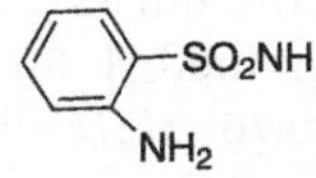	[85]

Table 2 (continued)

Entry	n	Y	RZ	**258**, R:	Refs.
12	4	–	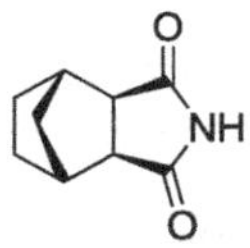	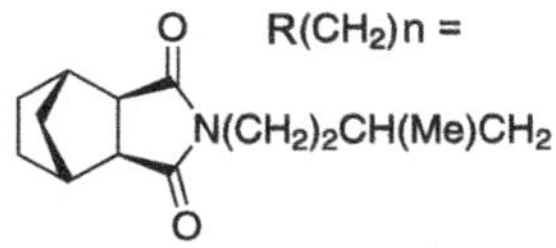	[85]

Table 3 Antipsychotic 3-(piperazinyl)-1,2-benzisothiazole derivatives **258** (part II)

Entry	R $(CH_2)_nY$ **260**	**258**, R:	Refs.
1	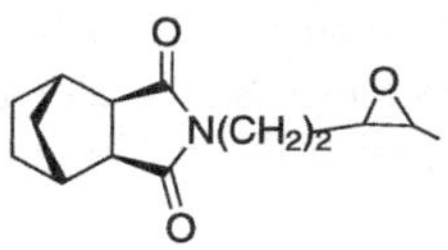	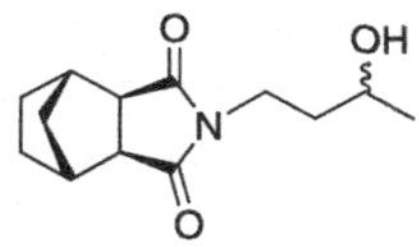	[86]
2	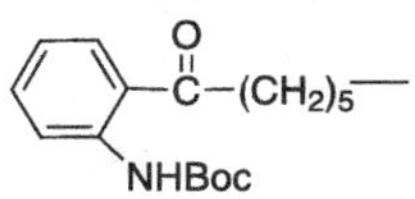	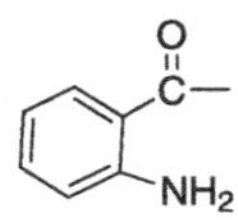	[85]
3	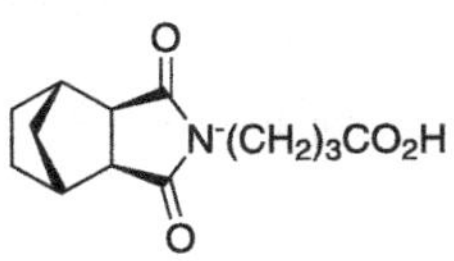	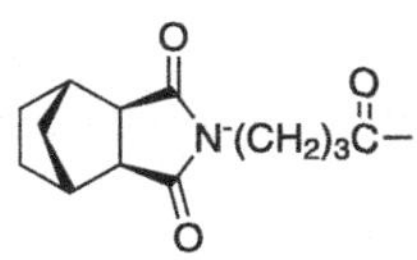	[86]
4	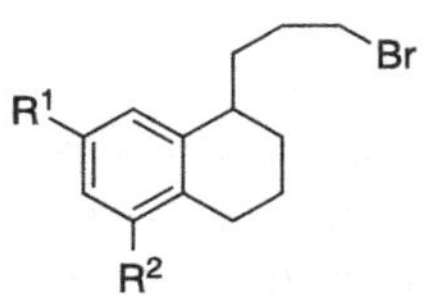 $R^1 = H$, $R^2 = OMe$; $R^1 = OMe$, $R^2 = H$	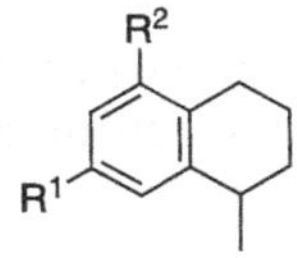	[88]
5	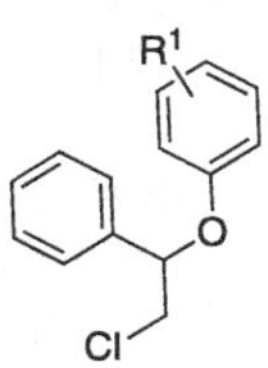	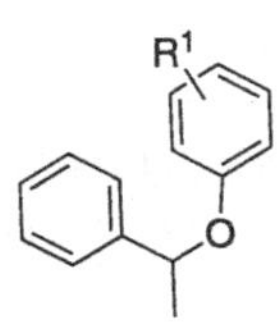 $R^1 = 2$-F; $R^1 = 3$-CF_3; $R^1 = 4$-OMe; $R^1 = 2$-CF_3; $R^1 = 4$-CF_3	[89]

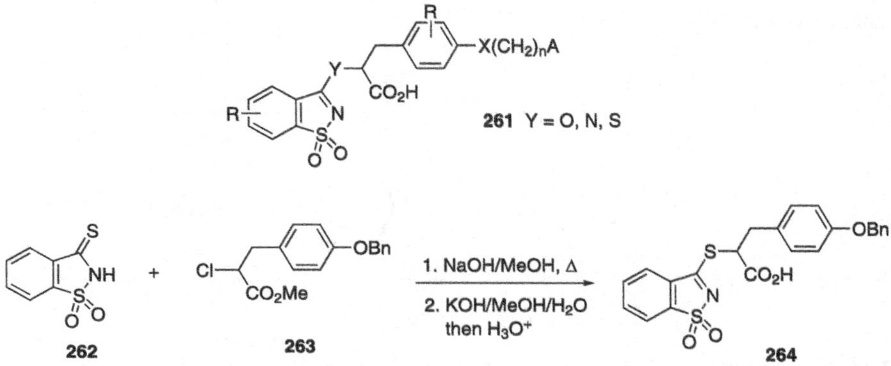

Scheme 66 Synthesis of 2-hetero-(benzisothiazol-3-yl)(3-phenyl)propionic acids

Benzisothiazole moieties substituted at C-3 with an oxygen are also present in compounds studied for β-adrenoceptor blocking activity such as **266** [90]. These compounds were prepared as racemic mixtures from **265** by a coupling reaction of the epoxide function of the benzene ring with the various amine derivatives (Scheme 67).

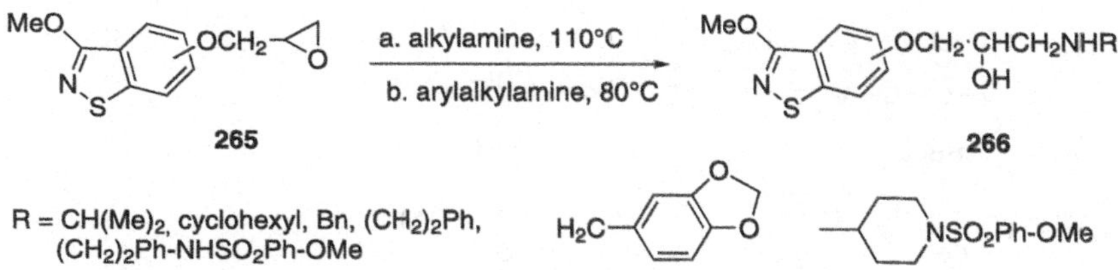

Scheme 67 Synthesis of 3-methoxybenzisothiazolyl substituted propanolamines

4.2.3
2-Substituted Benzisothiazoles and Naphtosultams

S,S-Dioxo-2*H*-naphtho[1,8-*c,d*]isothiazole (naphtosultam) is in general the more represented ring of this class of nitrogen functionalised compounds. A series of naphtosultams **268** substituted at nitrogen with a functionalised C-7 and C-8 chain were patented for their anti-cell-proliferation activity (Scheme 68) [91]. They were prepared from naphthalenesultam **267** and a ω-bromo-hexanoic or -heptanoic acid benzyloxy-amides in the presence of K_2CO_3. The resulting intermediates were reduced with H_2 and Pd/$BaSO_4$ giving **268** (60–70%).

The naphtosultam derivatives **270** characterised by anti-inflammatory activity are strictly correlated to the above compounds. They were prepared via intermediate **269**, obtained by reaction of **267** with dibromoalkyl compounds followed by treatment with the proper amine (Scheme 68) [92].

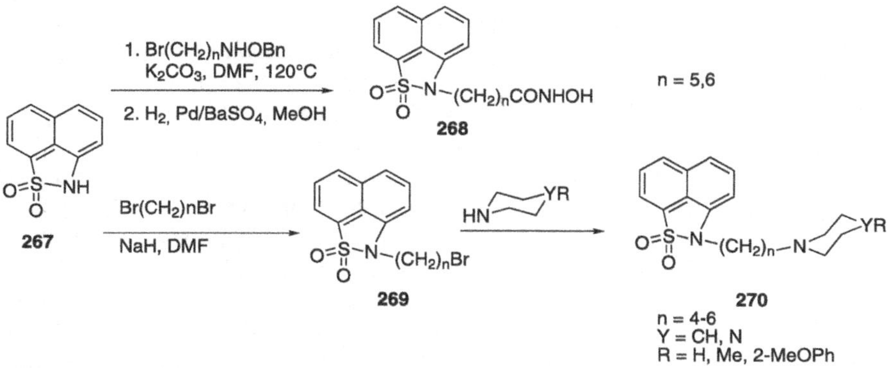

Scheme 68 Synthesis *N*-naphtosultam-substituted alkanoic acid hydroxyamides and alkylamines

Aryl sulfonamides were targeted as potential anti-MRS pharmacophores of 1-β-methylcarbapenems characterised by antimicrobic activity (Scheme 69). The functionalisation of carbapenem **272** with naphtosultam as well as with other isothiazole derivatives was achieved through a Mitsunobu reaction, taking advantage of the acidity of the arylsulfonamide function. The reaction of *bis*(allyl)protected hydroxymethylcarbapenem **271** and the proper sulfonamido compound in THF and in the presence of DEAD or DIAD and Ph_3P afforded **272** (average yield: 49%). The deprotection of oxygen atoms performed with a mixture of $Pd(Ph_3P)_3$, Ph_3P, $BuCHEtCO_2H$, $BuCHEtCO_2Na$ afforded **273** (average yield: 67%) [93].

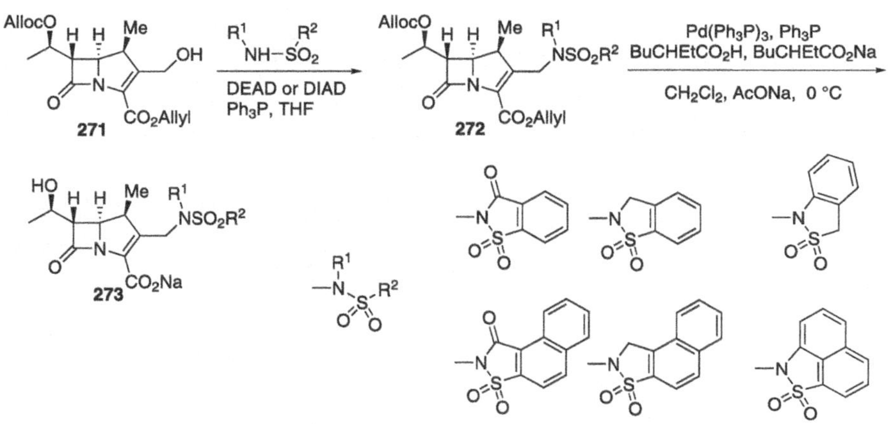

Scheme 69 General synthesis of carbapenems functionalized with benzisothiazole *S*,*S*-dioxide rings

In this case, the naphthosultamyl-methyl group was also selected as a good candidate and the structure was optimised to improve water solubility, pharmacokinetics and chemical stability. The Mitsunobu reaction of carbapenem **271** with a series of homologous silyloxyalkyl-1,8-naphtosultams **274** produced compounds **275** in good yields. Deprotection of the *O*-silyl group (TfOH, THF/H_2O, 55–65% from **271**) followed by activation of the hydroxy group as mesylate ($n = 1$: MsCl, TEA then NaI) or triflate ($n = 2,3$: Tf_2O, 2,6-lutidine) gave compounds **276**. Their reaction with substituted DABCO salts ($n = 1$: AgOTf, MeCN; $n = 2,3$: MeCN) followed by deprotection of oxygen atoms using the above reported McCombi procedure or, preferably, using dimedone as allyl scavenger ($Pd(Ph_3P)_3/Ph_3P$, dimedone, *i*Pr_2NEt, DMF) gave the cationic zwitterion products **277** (24–72%) purified by tandem ion-exchange and reverse-phase chromatography [94]. Being biologically optimised the linker between naphtosultam and amino group, and aiming to

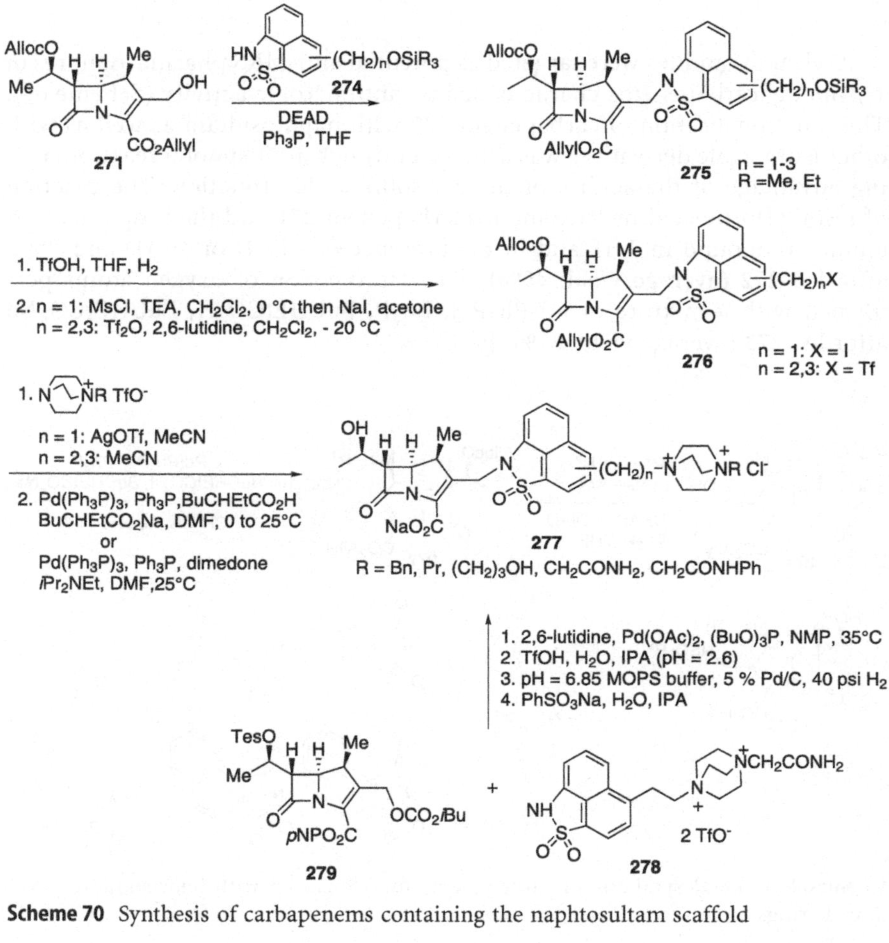

Scheme 70 Synthesis of carbapenems containing the naphtosultam scaffold

minimise manipulations with the sensitive carbapenem, the next synthetic challenge was the coupling of the preformed naphthosultam intermediate **278**, prepared as described in the literature, to carbapenem **279**. Compound **277** (**L-786, 392**, $R^1 = CH_2CONH_2$) was obtained in 97% yield [95] (Scheme 70).

4.2.4
Reactivity of Heterobicyclic Isothiazoles

Nucleosides **282–293** functionalised with the imidazo[4,5-*d*]isothiazole ring were prepared by reaction with different glycosides and their cytotoxic activity was evaluated (Scheme 71) [96]. Studies concerning the regiochemistry of the formation of the *β*-*N*-glycosyl bond between the sugar and the nitrogen atom of the imidazole ring were performed. The sodium salts of imidazo[4,5-*d*]isothiazoles **280a–e**, generated in situ using NaH, were condensed with different chloro sugars **281a,b**. From the *α*-chlorodesoxyribose derivative

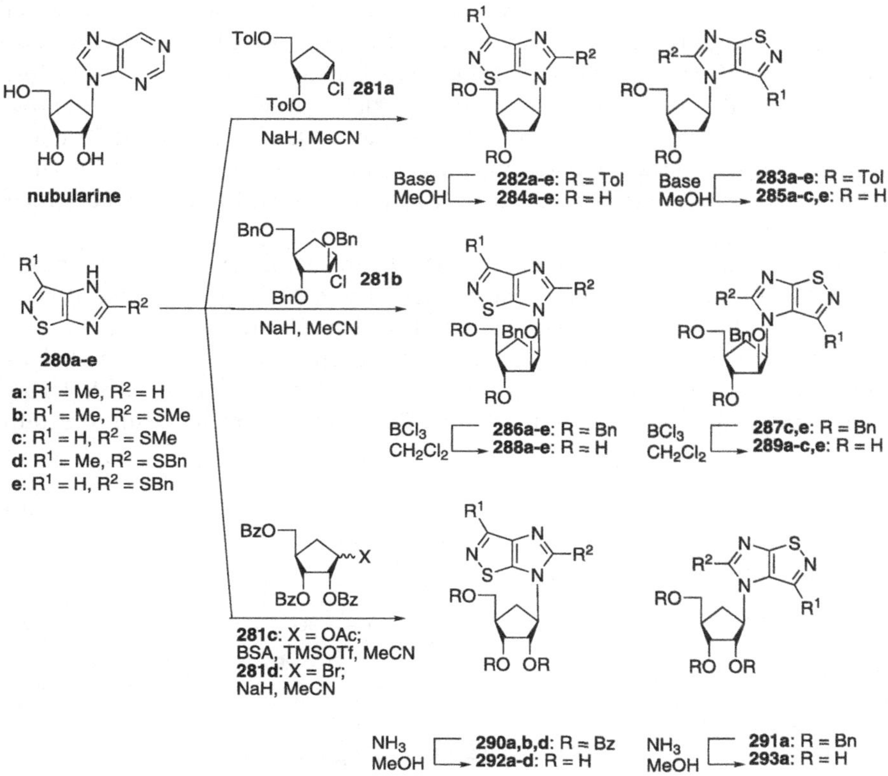

Scheme 71 Synthesis of imidazo[4,5-d]isothiazole-containing nucleosides

281a, a mixture of *N*-6 and *N*-4 *β*-nucleosides **282** and **283** (63–88%) then deprotected affording **284** and **285** (27–94%). The regiochemistry was governed by steric principles, as heterocycles bearing both 3- and 5-substituents gave primarily the *N*-6 nucleosides, while less hindered derivatives afforded the *N*-4 and *N*-6 products approximately in 1 : 1 ratio. The glycosylation of **280** with the more hindered *α*-chloroarabinose derivative **281b** is more regioselective and the *N*-6 nucleoside derivatives **286** were formed (46–68%) together with lesser amounts of the *N*-4 isomers **287** (4–30%). In particular, starting from **280c,e**, which do not have a substituent at C-3, compounds **287c,e** were obtained in 24% and 30% yields, respectively. A small amount of the corresponding *α*-anomers of **286** (3–5%) and **287c** (20%) and **288e** (7%) were detected (^{1}H NMR). The benzyl group was then removed affording the nucleosides **288a–e** and **289c,e**. Apart from compound **289a**, the 5-substituted compounds were subject to decomposition and were isolated in very low yields. The glycosylation of **280a,b,d** with *β*-D-ribofuranose **281c** was performed by in situ generation of the *N*-silyl derivatives with *N,O*-*bis*(trimethylsilylacetamide) ion followed by glycosylate with TMSOTf as catalyst and compounds **290a** (37%) and **291a** (23%) were formed. The reaction was not successful starting from 5-substituted **280b–d**. When the reaction of **280b,d** and **281c** was directly performed in presence of TMSOTf, the single *N*-6 regioisomers **290b** and **290d** were isolated in 52 and 21% yields, respectively. Deprotection of compounds **290ab,d** and **291a** with methanolic ammonia provided the nucleosides **292a,b,d** and **293a** in good yield. The *N*-6 isomer **292c** (11%) was prepared via the sodium salt of **280c** (MeONa/MeOH), which was treated with bromo sugar **281d** (large excess) to give **290c**, which was then deprotected.

To synthesise potent inhibitors of MMP-13 and MMP-9 with selectivity versus MMP-1 and TACE, the isothiazole derivatives **297** were prepared [97] as outlined in the Scheme 72. The bicyclic heteroaryl system of **297** was accessible by condensation of the isothiazole **109a** with diethyl ethoxymethylenemalonate followed by thermally induced cyclisation to give alcohol **294**. Treatment with $POCl_3$ converted the hydroxy substituent to a chloro group

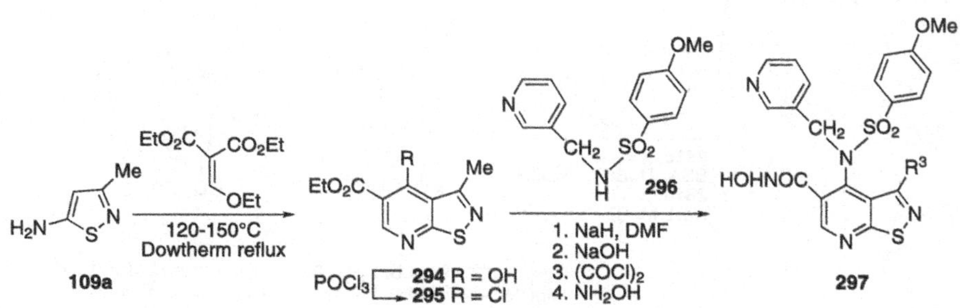

Scheme 72 Synthesis of isothiazolopyridine hydroxamic acids

and **295** was formed. The anion of the sulfonamide **296** displaced chlorine from the pyridine ring. Hydrolysis of the ester with NaOH followed by activation of the acid with oxalyl chloride and reaction of the intermediate acid chloride with hydroxylamine gave the hydroxamic acids **297**.

5
Benzisothiazolones

In this chapter benzisothiazolones and benzisothiazolone *S*,*S*-dioxides are separately treated. According to the large amount of literature already available on the second group of compounds (saccharin derivatives), most of all for their interest as sweetener, here only some examples of synthesis or reactivity that appear particularly appealing are considered. Regarding the benzisothiazolones, not oxidised at the sulphur atom, several synthetic methods were described and substitutions with different alkyl or heteroalkyl chain at the nitrogen atom considered. In many cases competitive *N*- and *O*-alkylation was observed. 3-Oxoisothiazolo[5,4-*b*]pyridines were also considered.

5.1
Synthesis of Benzisothiazol-3-One *S*,*S*-Dioxides

Methods L and *M* were the common methods used to prepare NH saccharins **300** ($R^2 = H$). Starting from substituted diethylbenzamides **298** (*Methods L*) (Scheme 73) bearing at the benzene ring a 4-*iso*-propyl or a 4-methoxy group or alkoxy groups in one of the four positions of the ring, intermediates **299** (80–84%) were prepared by introducing a sulfonamido group at the *or*-

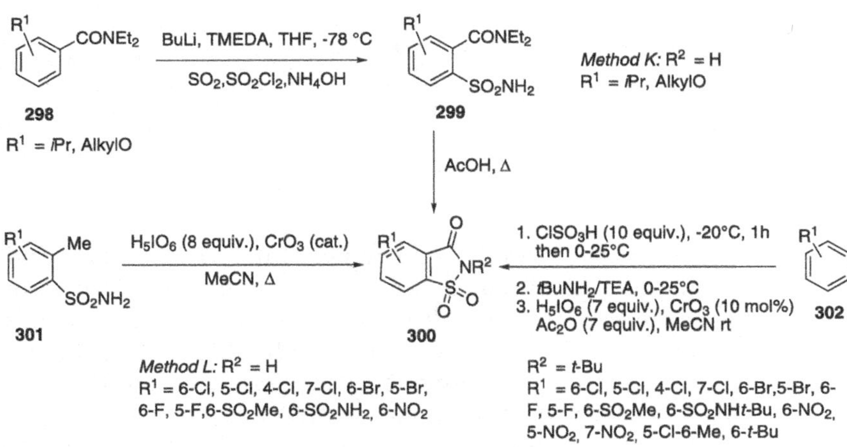

Scheme 73 Synthesis of benzisothiazolone *S*,*S*-dioxides: *Methods L* and *M*

tho position (BuLi in TMEDA then, in sequence, SO_2, SO_2Cl_2 and NH_4OH). By heating of **299** in acetic acid, **300** (R^2 = H) was obtained in 90–95% yield (Scheme 73) [98].

The practical and general *Method M* (Scheme 73) starts from various substituted *o*-toluenesulfonamides **301**, which were oxidised to the corresponding substituted saccharin derivatives **300** (R^2 = H) by refluxing with 8 equiv. of H_5IO_6 and a catalytic amount of CrO_3 in MeCN. A higher catalyst loading of CrO_3 was required for complete oxidation of substrates with strong electron-withdrawing groups.

In some cases, starting from the substituted toluene **302**, it is possible to perform a "one-pot reaction" (Scheme 73) with formation of the *N*-*t*-butyl saccharine derivatives, the *N*-*t*-butyl group being very useful as an *N*-protecting group for the preparation of other protected saccharin derivatives. *N*-*t*-Butyl-*o*-methyl arenesulfonamides were easily prepared by chlorosulfonation of substituted toluene derivatives with chlorosulfonic acid. The resulting sulfonyl chlorides were treated with *t*-$BuNH_2$ to afford the *N*-*t*-butyl-*o*-toluenesulfonamide derivatives, which were subjected to the oxidation step without purification, affording **300** (R^2 = *t*-Bu) [99].

The preparation of the saccharin derivative **305** (Scheme 74), substituted with thiomethyl group at C-5 and directly functionalised at nitrogen with a chain containing the bromo atom at C-4′, was prepared according to *Method N*. The substituted sulfonamide **304** was prepared by reaction of **303** in sequence with $ClSO_3H$, 4-aminobutanol and 2,3-dihydropyran. **304** was *ortho*-lithiated and carbonylated and the corresponding intermediate was finally cyclised, deprotected at the oxygen atom and transformed into the *N*-bromobutylsaccharin **305** [100].

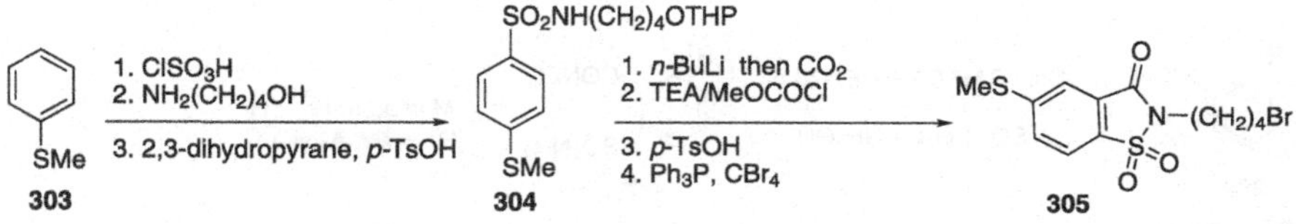

Scheme 74 Synthesis of *N*-4′-bromobutylsaccharin: *Method N*

Method N was also applied to the preparation of the naphto[1,2-*d*]- and naphto[2,3-*d*]isothiazole *S*,*S*-dioxide nuclei (NiT) by *ortho*-carbonylation of 2-(*N*-*t*-butyl)naphtalenesulfonamide (**306**). A mixture of three compounds **307a,b** and **308** was obtained, which were separated by chromatography and treated with PPA resulting in the corresponding naphtoisothiazoles. The compounds were then reacted with ethyl bromo acetate and hydrolyzed to esters **309a,b** and **310** as shown in Scheme 75. The nitration of ethyl ester of **309** (fuming HNO_3) gave regioisomeric esters hydrolyzed to **311a,c**. Alternatively,

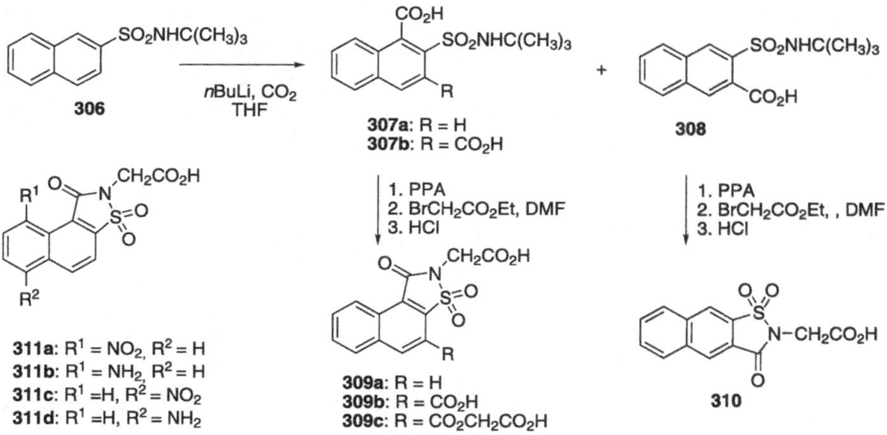

Scheme 75 Synthesis of naphto[1,2-*d*]- and naphto[2,3-*d*]isothiazolone *S,S*-dioxide derivatives

the nitrocompounds were first reduced to the corresponding amines and then hydrolyzed to **311b,d** (H_2, Pd/C) [101].

Another two methods (*Methods O, P*) were reported to obtain saccharin derivatives from heterocylic starting materials.

The preparation of all four mononitro aromatic derivatives of 1,2-benzisothiazol-3-one *S,S*-dioxide was reported. As an example, the preparation of the 5-nitrosaccharin **314** (*Method O*) from nitro-1*H*-benzo[*d*][1,3]oxazine-2,4-dione **312** is outlined in the Scheme 76. Starting from **312**, which was treated with K_2CO_3 in MeOH, the methyl 2-amino-5-nitrobenzoate **313** was formed. The amino group was transformed into the diazonium salt ($NaNO_2$ in AcOH/HCl). Its reaction with $CuCl_2$ and sulphur dioxide afforded the corresponding sulfonyl chloride, which was added to cold concentrated ammonium hydroxide and **314** was isolated in 74% yield [100].

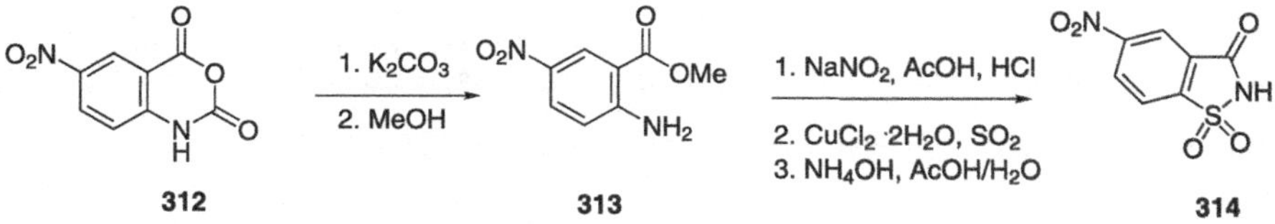

Scheme 76 Synthesis of nitro-substituted benzisothiazolone *S,S*-dioxides: *Method O*

Saccharin substituted at nitrogen with an OR group can be obtained in good yield starting from *O*-sulfobenzoic anhydride **315** according to *Method P*, via the synthon *O*-carbethoxybenzenesulfonyl chloride **316**, which was condensed with the corresponding *O*-alkylhydroxylamine to give intermediates **317a,b,c**. Thermal cyclisation, either neat or in refluxing AcOH, afforded **319a,b** from **317a,b**. The synthesis of *N*-hydroxysaccharin **319c** could not be done by deprotection of **317a,b** but through protection of compound **317c** as the dihydropyranyl derivative **318**. Its cyclisation in smooth conditions (*i*-butyl chloroformate, TEA) followed by acid-catalyzed removal of the THP group afforded **319c** (61% yield) (Scheme 77) [102].

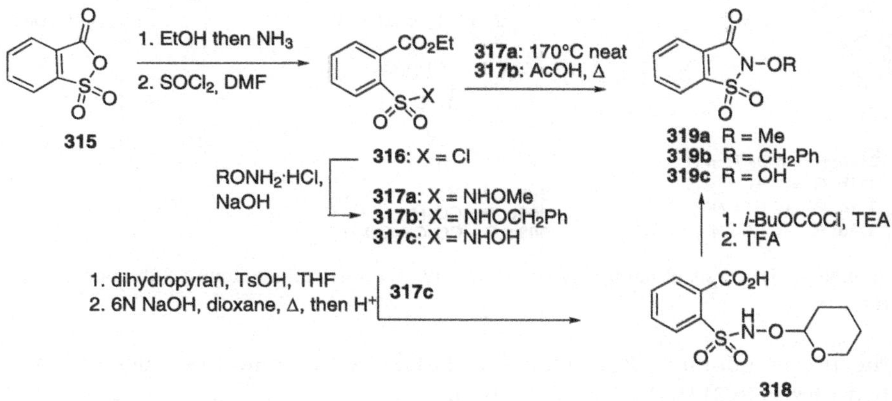

Scheme 77 Synthesis of *N*-hydroxy-benzisothiazolone *S,S*-dioxides: *Method P*

Compounds **321** (Scheme 78) were prepared as potential protease inhibitors [103] according to *Method Q* using the 2-sulfobenzoic anhydride **315**, which by reaction with alanine or *S*-benzyl-L-cysteine methyl esters gave intermediates **320** directly treated with PCl_5 at 80 °C. Compounds **321a** (16%, racemic) and **321b** (11%) were formed, respectively. As a by-product in the case of the cysteine starting material, compound **321c** was formed.

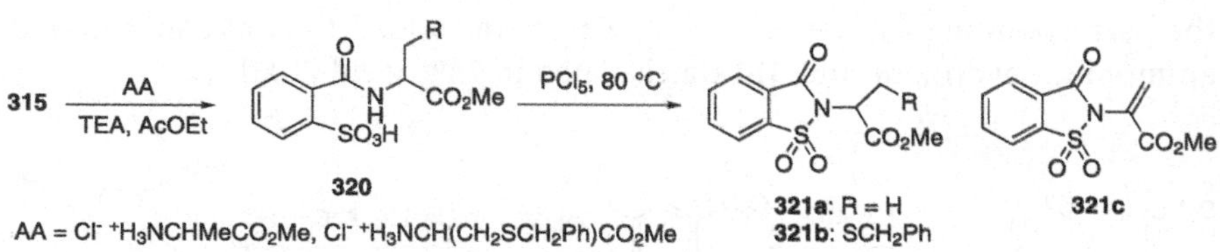

Scheme 78 Synthesis of methyl *N*-saccarinyl acetate derivatives: *Method Q*

5.2 Synthesis of Benzisothiazol-3-Ones and Heterobicyclic Isothiazolones

1,2-Benzisothiazol-3-ones **323** and 3-oxoisothiazolo[5,4-*b*]pyridines **324** were prepared by modification of known synthetic protocols and, usually, the *N*-substituted compounds were obtained directly. Different reagents were used to achieve this synthetic target according to *Methods C, M-T*. The main synthetic procedure follows the general *Method C* using the 2,2′-dithiobis(benzoic acid) (**322**) as building block, which after activation of the carboxylic function and reaction with different nitrogen donors gave the 1,2-benzisothiazol-3-one derivatives **323** (Scheme 79). 3*H*-1,2-Benzodithiole-3-thione **325** and pyrido derivative **326** are the starting materials for the preparation of 1,2-benzisothiazol-3(2H)-thione **327** and 3-oxoisothiazolo [5,4-*b*]pyridine **328**, respectively, by reaction with a proper amino derivative (*Method R*). The thione derivatives are not stable and can equilibrate in solution affording 3-imino compounds **327′** and **328′**. The transformation of **323** into the corresponding thione **327** was possible by using the Lawesson's reagent (Scheme 79).

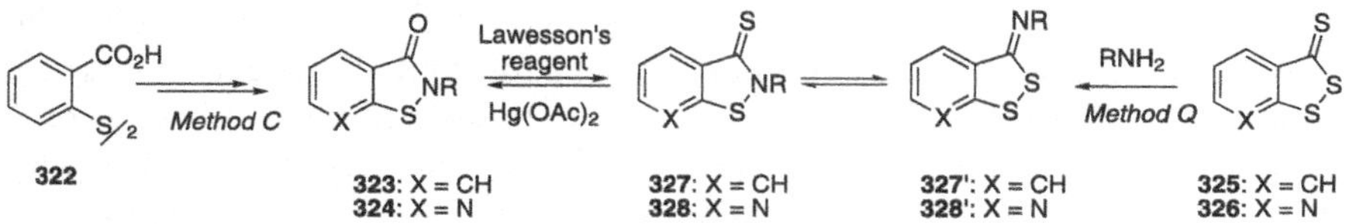

Scheme 79 Synthesis of 1,2-benzoisothiazol-3(2*H*)-ones and 3-oxoisothiazolo[5,4-*b*]pyridines: *Methods C* and *R*

According to both *Methods C* and *R*, a series of 1,2-benzisothiazol-3(2H)-ones **329** and the corresponding thiones **330**, substituted at the nitrogen atom with a chain having a hydroxy group, were prepared and tested as antimicrobics (Scheme 80). Compounds **329** were prepared through a "one-pot" procedure starting from **322** [104, 105]. The reaction of thione **325** with hydroxyamines gave compounds **330**, which are in equilibrium with **330′** [105]. Depending on the nature of the amine, the solvent polarity and the temperature, different ratios of the two isomers were obtained. Using ethanolamine as coupling reagent a dynamic equilibrium occurs in solution, making the isomers inseparable. This last synthetic approach was unsatisfactory in the case of compounds **330**, where X = $(Me)_2CCH_2$, $Me(CH_2OH)CCH$, which were prepared in better yields by treating the corresponding compounds **329** with the Lawesson's reagent.

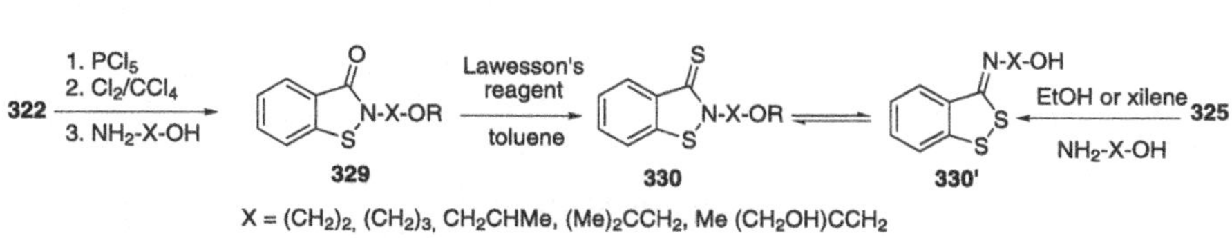

Scheme 80 Synthesis of 2-hydroxyalkyl-1,2-benzoisothiazol-3(2*H*)-ones and thiones

(*S*-(*R**, *R**))-3-Methyl-2-(oxo-3*H*-benzo[*d*]isothiazol-2-yl)pentanoic acid (**332**) (Scheme 81) was prepared by using *Method C* on a pilot scale (amount of **332** produced: 37 Kg) from **322**, which was transformed into the dichloride

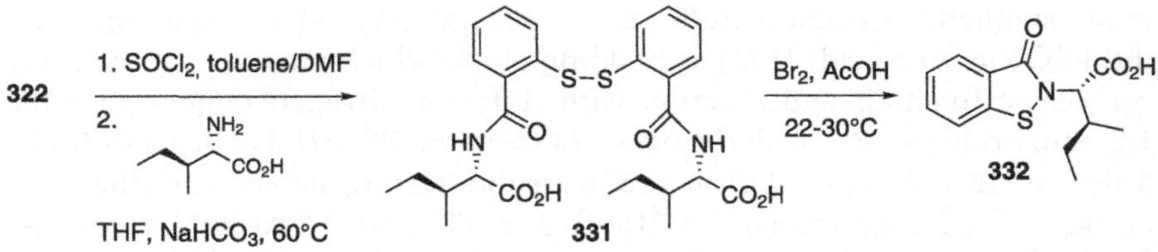

Scheme 81 Synthesis of (*S*-(*R**, *R**))-3-Methyl-2-(oxo-3*H*-benzo[*d*]isothiazol-2-yl)pentanoic acid

(toluene/DMF/$SOCl_2$, 83%). Several esters of leucine were used and reacted with the dichloride, but the best result was found using L-leucine itself. The choice of the reaction solvent was critical and THF was found to give the product in satisfactory yield and purity. The addition to the reaction mixture of a base such as $NaHCO_3$ at 60 °C improved the yield and purity: compound **331** was obtained in 75% yield, without epimerisation at the amino acid stereocentre. The oxidative cyclisation of **331** afforded compound **332** in 74% yield [106].

From **322** and using *Method C*, a large class of benzisothiazolones, named **BITA1–3**, substituted at nitrogen atom with an aryl or heteroaryl nucleus bearing a SO_2 group at the *para* position were prepared and tested for anti-HIV activity (Fig. 2) [107].

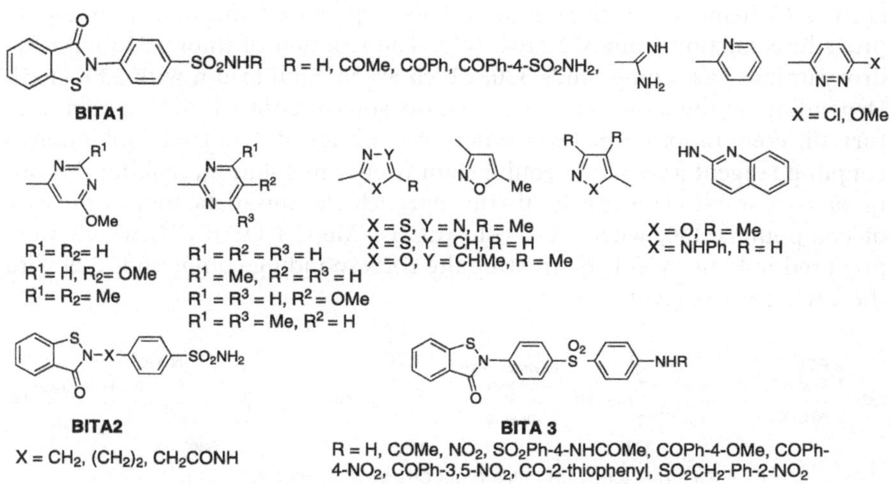

Fig. 2 Synthesis of **BITA** compounds

A different way was used to build the *N*-amino substituted benzisothiazolone nucleus from chlorocarbonylphenylsulfenylchlorides (**333**) (*Method S*) which, on reaction with *N*-Boc-hydrazine in pyridine and ether gave compounds **334**. The deprotection was achieved using TFA and *N*-amino-benzisothiazol-3-ones **335** were isolated in 70–75% yields (Scheme 82) [71–73].

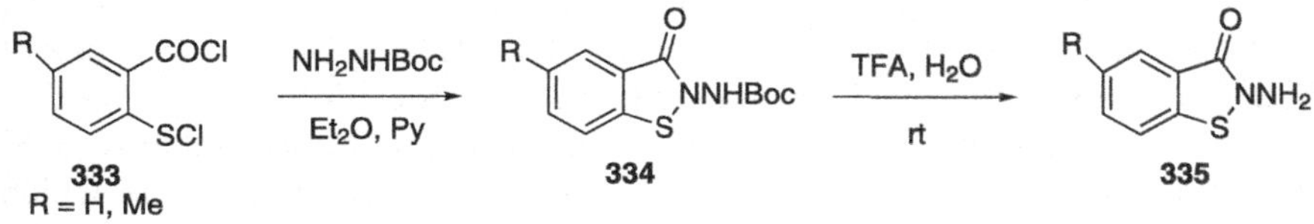

Scheme 82 Synthesis of 2-*N*-amino-benzisothiazolones: *Method S*

Pyridoisothiazolones and the corresponding thiones were prepared according to *Method R* and used for evaluations in vitro of anti-*micobacterium* activity. 3*H*-1,2-Dithiol-[3,4-*b*]pyridin-3-thione (**326**) was transformed into the corresponding oxo compound **337** (98%) with $Hg(OAc)_2$ in AcOH [108]. The reaction of **336** with hexylamine gave the thioamide intermediate, which was then oxidised with I_2 in basic conditions affording **338**, which equilibrates to **338′** in different solvents such as DMSO, DMF, acetone and H_2O. From **337**, *N*-hexylisothiazolo[5,4-*b*]pyridine-3-(2H)-one (**339**) (36%) was prepared using the sequence shown above (Scheme 83) [109].

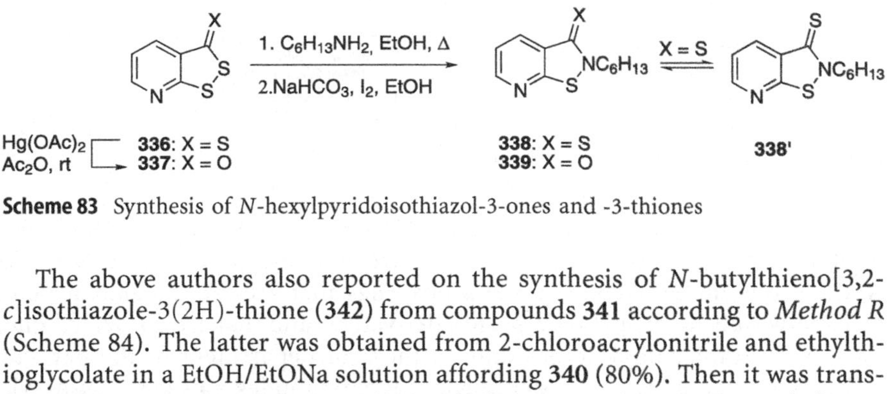

Scheme 83 Synthesis of *N*-hexylpyridoisothiazol-3-ones and -3-thiones

The above authors also reported on the synthesis of *N*-butylthieno[3,2-*c*]isothiazole-3(2H)-thione (**342**) from compounds **341** according to *Method R* (Scheme 84). The latter was obtained from 2-chloroacrylonitrile and ethylthioglycolate in a EtOH/EtONa solution affording **340** (80%). Then it was transformed into **341** (32%) by reaction with i) $NaNO_2$, ii) potassium ethyl xantogenate and iii) P_4S_{10}. The reaction of **341** with *n*-butylamine followed by oxidation with I_2, afforded a mixture of **342** and **343** (57%) [110].

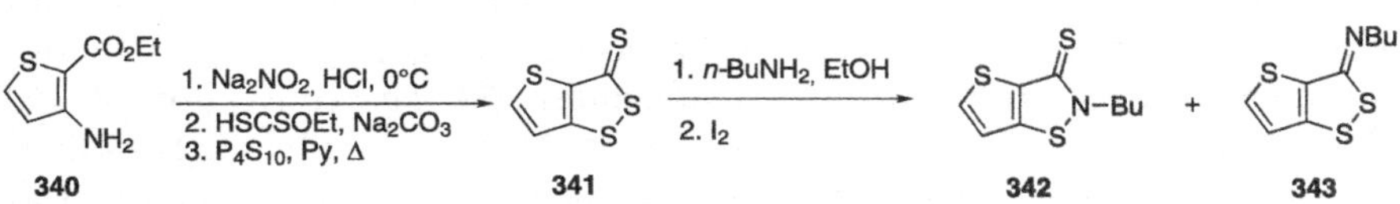

Scheme 84 Synthesis of butylthieno[3,2-*c*]isothiazole-3(2*H*)-thione

An efficient method to prepare *N*-substituted pyridoisothiazol-3-ones is shown in Scheme 85 (*Method R*). It employs a readily available starting material and inexpensive reagents, and can be carried out on a large scale without purification of the intermediate. Commercially available mercaptonicotinic acid (**344**) was converted into mercaptonicotinamide by sequential treatment with thionyl chloride, buffered NH_4OH and $NaBH_4$. The use of $NaBH_4$ improved the yield by reducing the disulfide *bis*-amide preventing its disproportionation in the reaction mixture. 2-Mercaptonicotinamide (**345**) was oxidatively cyclised to the 2H-pyridoisothiazolone **347a** in H_2SO_4, which acts as oxidant and solvent. This avoids use of chlorine, hydrogen peroxide or peracids to effect this transformation [111]. Similarly, a large number of pyridoisothiazol-3-ones **347b**, substituted with a functionalised benzyl group at the nitrogen atom, can be prepared from the 2-(benzylthio)nicotinamides **346**, which were synthesised by standard methods [112]. This method is particularly useful when the substituted benzylamine is readily available and the alkylation at nitrogen of compound **347** is not feasible because of the instability of the benzyl halide (*Method D1*). The oxidative cyclisation of **346b** to **347b** was carried out in one step by heating with sulfuryl chloride. Some selected examples are listed in Scheme 85.

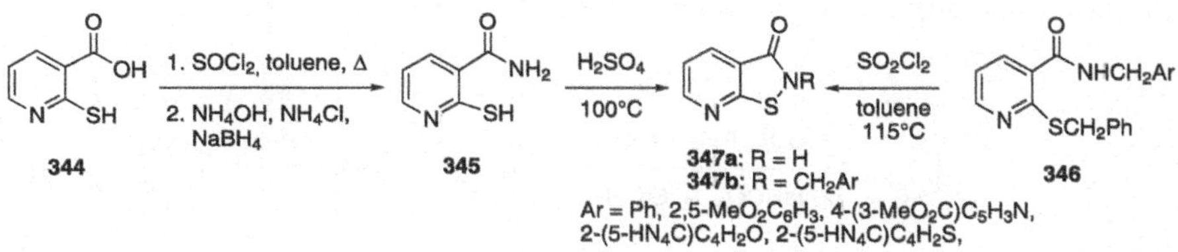

Scheme 85 Synthesis of pyridoisothiazol-3-ones: *Method R*

N-Substituted pyridoisothiazolones **351** and **352** displayed a high inhibitory activity of H+/K+ ATPase in vitro but non in vivo (see Sect. 6). A study was undertaken on finding prodrugs that are more stable at neutral and weakly acidic pH than pyrido derivatives and that are converted into the active isothiazolopyridines only in the acid department of the parietal cells. These prodrugs were identified in nicotinamides **349** [113]. *S*-Alkylnicotinic acids **348** were prepared by condensation of 2-mercaptonicotinic acid **344** with the corresponding benzyl chlorides operating in DMF or with the corresponding benzyl alcohols under acidic conditions (HCl/acetone). Their reaction with *i*-$BuNH_2$ by the use of EDC (procedure A) or with 4-aminopyridine and oxalyl chloride (procedure B) gave the nicotinamides **349** then oxidised with *m*-CPBA to sulfoxides **350**. The conversion of **350** ($R^1 = i$-Bu, Py) into the respective *N*-*i*-butyl (**351**) or -pyridyl (**352**) isothiazolopyridines was done and a study on the efficacy of R as a leaving group was performed (Scheme 86).

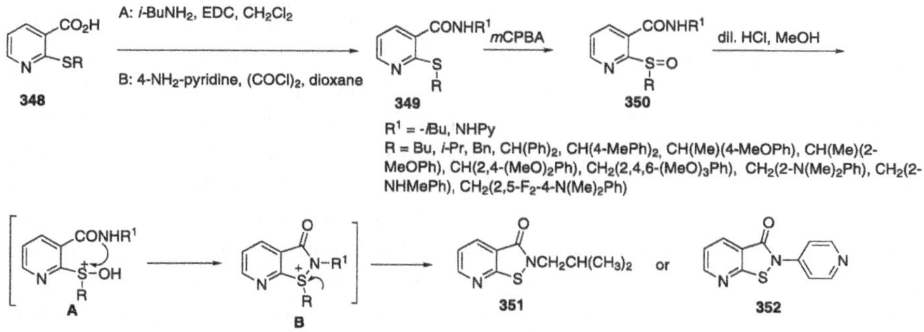

Scheme 86 Synthesis of pyridoisothiazol-3-ones: *Method T*

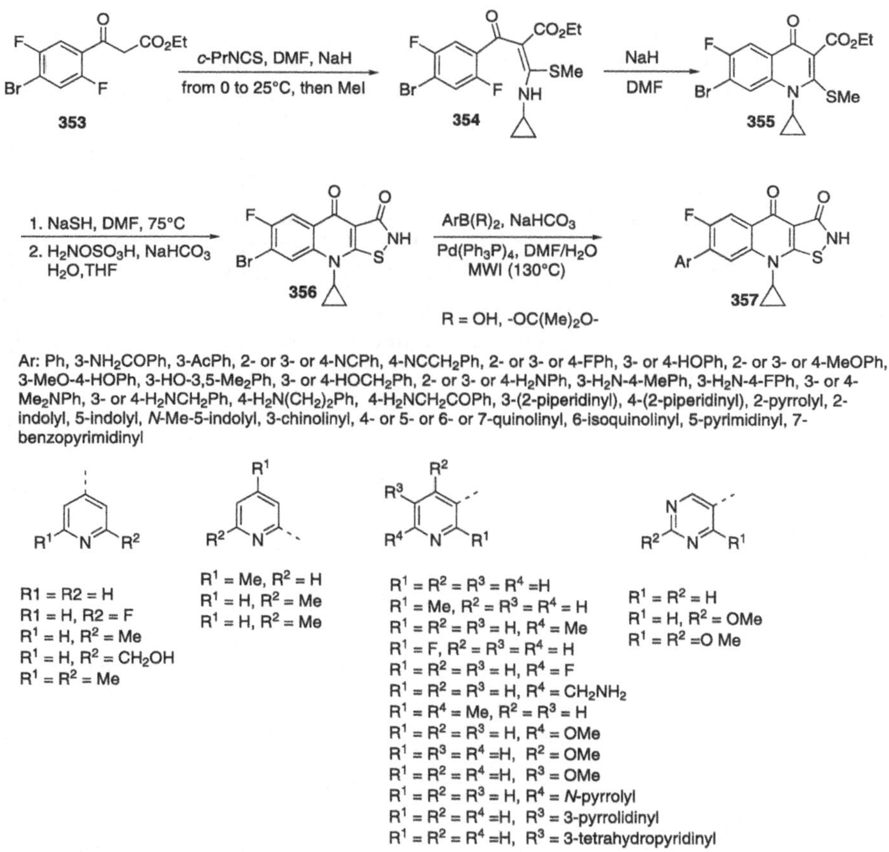

Scheme 87 Synthesis of isothiazoloquinolone derivatives

SAR studies were performed on compounds containing the 9*H*-isothiazolo[5,4-*b*]quinoline-3,4-dione (ITQ) nucleus and it was found that some of them are potent antibacterial agents (see Scheme 101) [114, 115]. They were prepared from compound **353**, which was treated with cyclopropyl isothiocyanate in DMF and then with MeI. Compound **354** (94%) was obtained and treated with in NaH in DMF to give the isothiazolo[5,4-*b*]quinoline compound **355** (93%). Its treatment with anhydrous NaSH gave the corresponding mercaptan (84%), which was directly cyclised without purification to **356** (85%) in the presence of hydroxylamine-*O*-sulfonic acid. Microwave-assisted Suzuki–Miyaura cross-coupling of the ITQ nucleus **356** with the desired arylboronic esters or acids afforded derivatives **357**, typically, in 30–50% yield after HPLC purification (Scheme 87).

5.3 Reactivity of Benzisothiazol-3-Ones and Benzisothiazol-3-One *S,S*-Dioxides

Several functionalisations were introduced at the nitrogen atom of benzisothiazol-3-ones and 3-oxo-isothiazolo[5,4-*b*]pyridines of general formula **358**, in which $n = 0$ or 2, adopting two main synthetic strategies consisting in i) the formation of the sodium salt, which was then reacted with electrophiles to directly give the compound **359** (*Method U*), or ii) the preparation of key reagents in which the nitrogen atom of saccharin was substituted with an activated methylgroup (intermediate **361**), which was then reacted with a proper nucleophile affording **362** (*Method V*). The choice of the above methods is strictly dependent on the kind of chain. The main limitation of *Method U* is related to the possibility to obtain the product of *N*-alkylation **359** and of the *O*-alkylation **360**. This occurs particularly for non-oxidised compounds ($n = 0$) (Scheme 88).

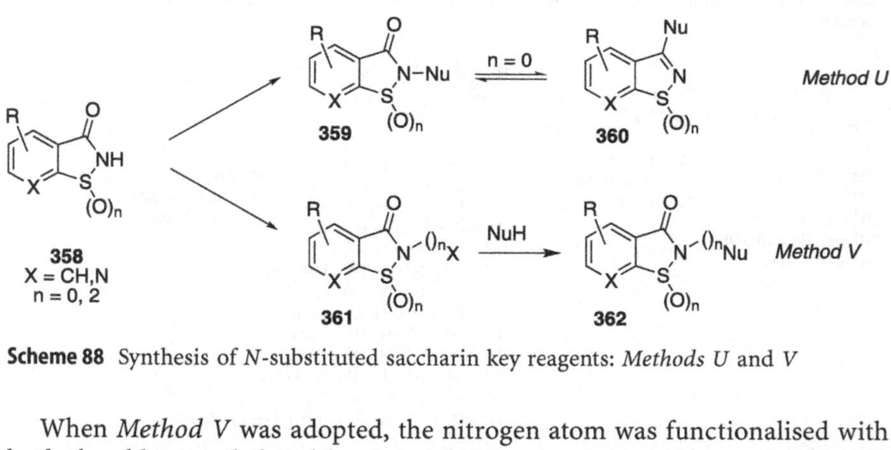

Scheme 88 Synthesis of *N*-substituted saccharin key reagents: *Methods U* and *V*

When *Method V* was adopted, the nitrogen atom was functionalised with both the chloromethyl and bromomethyl groups. By treating saccharins **363** with $PhSCH_2Cl$ and TEBABr [116] or i-Pr_2NEt [117] in toluene at reflux, compounds **364** were obtained, which were then treated with SO_2Cl_2 in CH_2Cl_2, and the chloromethyl compounds **365** (80–90%) were formed. Alternatively, **365** were prepared from the corresponding alcohols **366** with $SOCl_2$ [118]. The bromomethyl derivatives **368** (95% overall yield) were obtained by treating **363** with chloromethyl pivalate and Hünig's base affording **367**, then treated with HBr in acetic acid (Scheme 89) [116].

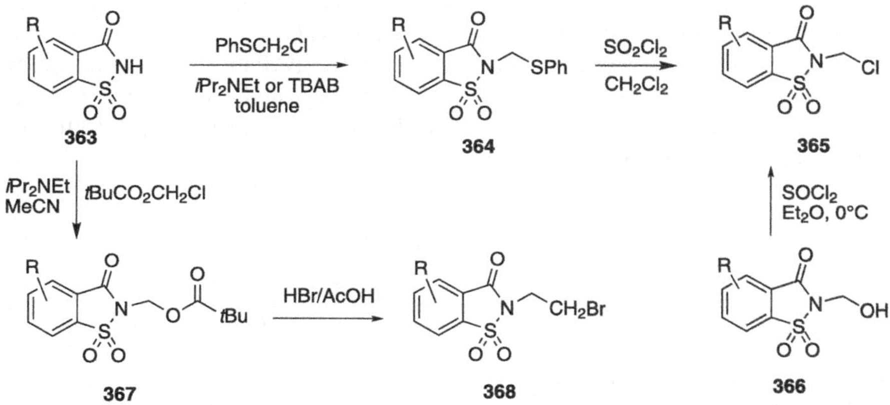

Scheme 89 Synthesis of saccharin key reagents

A large number of molecules containing the benzisothiazolone *S*,*S*-dioxide scaffold (BIT) of the general formula **369** were prepared to find molecules characterised by protease inhibition activity (Fig. 3). Analogous isothiazole derivatives are reported in Sect. 2.4.2. SAR studies were done both considering the substitution pattern of the benzene ring and, most of all, the substituent on the nitrogen atom represented by the CH_2LG group, in which LG is a different leaving group such as substituted benzoic acids, amino

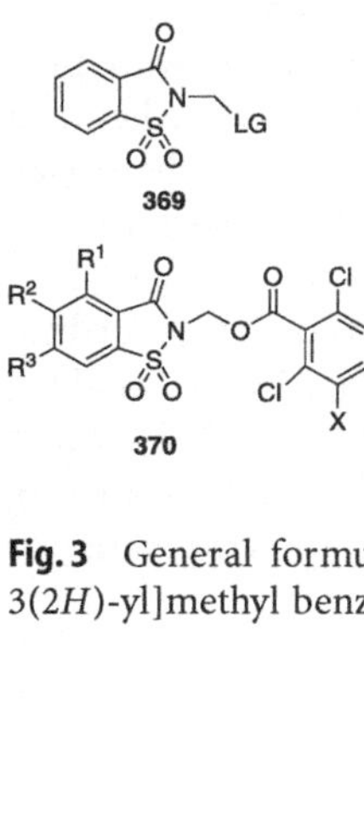

LG = OR, OHet, OCOR, $OPO_2(R)$, SO_2R

[115]
R^1 = iPr, R^2 = R^3 = H, X = O-$(CH_2)_2$-4-morpholinyl
R^1 = iPr, R^2 = H, R^3 = OMe, X = O-$(CH_2)_2$-1-pyrrolidinyl
R^1 = iPr, R^2 = H, R^3 = OMe, X = OCH_2CO_2H
R^1 = iPr, R^2 = H, R^3 = OMe, X = $SO_2NMe(CH_2)_2NMe_2$
R^1 = iPr, R^2 = H, R^3 = OMe, X = SO_2-1-piperazinyl-4-Me
R^1 = iPr, R^2 = H, R^3 = OH, X = SO_2-1-piperazinyl-4-Me
[96]
R^1 = R^2 = OEt, R^3 = H, X = O-$(CH_2)_2$-4-morpholinyl
R^1 = R^2 = OEt, R^3 = H, X = O-$(CH_2)_2$-1-pyrrolidinyl
R^1 = OEt, R^2 = H, R^3 = OMeX = O-$(CH_2)_2$-4-morpholinyl

[96]
X = H
R^1 = OEt, R^2 = R^3 = H
R^1 = iPr, R^2 = R^3 = H
R^1 = R^2 = OMe, R^3 = H
R^1 = R^3 = OMe, R^2 = H
R^1 = OEt, R^2 = H, R^3 = OMe
R^1 = OPr, R^2 = H, R^3 = OMe
R^1 = R^3 = OMe, R^2 = H
R^1 = R^2 = H, R^3 = OMe
[117]
R^1 = iPr, R^2 = H, R^3 = F
R^1 = iPr, R^2 = H, R^3 = OMe
R^1 = iPr, R^2 = H, R^3 = NMe_2
R^1 = iPr, R^2 = H, R^3 = 1-Me-4piperazinyl

Fig. 3 General formula of several protease inhibitors and 3-oxo-1,2-benzisothiazolone-3(2*H*)-yl]methyl benzoate *S*,*S*-dioxide derivatives

acids, hydroxy-heterocycles or cyclic enolates, phosphinates or phosphonates and sulfinates. The above compounds were prepared directly from saccharin derivatives or, generally, from the key reagents **365** and **368** by reaction with nucleophiles. The reaction of chloroderivatives **365** with 2,6-dichlorobenzoic acids, having a further *meta*-substituent, in the presence of K_2CO_3/TEBABr in DMF afforded compounds **370** (75–95%) [98, 117, 119].

A library of *N*-(acyloxymethyl)benzisothiazolone *S*,*S*-dioxide derivatives **371** and **372**, in which the acyl group represents a *para* substituted benzoate or an amino acid or a dipeptide, was prepared using parallel synthetic methods and automated purification, when possible, starting from bromo compound **368** (R = H) and 300 commercially available carboxylic acids using, as reaction conditions, K_2CO_3/DMF (23 °C) or K_2CO_3/MeCN (60 °C) or *i*-Pr_2EtN/DMF (23 °C), depending on the solubility of the acid [120]. Selected compounds are shown in Fig. 4.

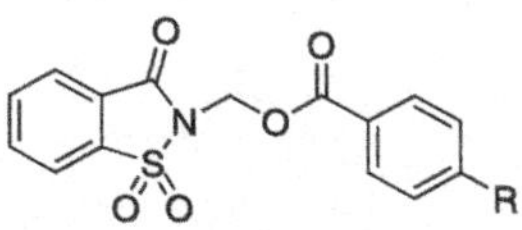

371

R = NHCbz, CONH$(CH_2)_2$Ph, NHCO$(CH_2)_2$Ph, CH_2NHCONHPh, CH_2NHCO_2Ph, NHCONHCH_2Ph, CH_2NHCOCH_2Ph, NH_2, $CH_2NHCO^2CH_2$Ph, O$(CH_2)_2$OPh

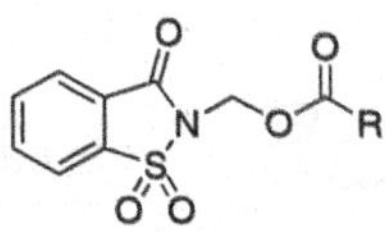

372

R = $(CH_2)_n$NHR^1, n = 1-5, R^1 = Cbz, Boc
R = L-Ala-Gly-*N*-Cbz, -Gly-Gly-*N*-Cbz, *N*$_\alpha$-Boc-*N*$_\delta$Cbz-L-lysine, *N*$_\alpha$-Boc-N$_\delta$-Cbz-D-lysine Gly-β-Ala-*N*Cbz, β-Ala-β-Ala-*N*Cbz

Fig. 4 Selected compounds of a library of *N*-(acyloxymethyl)benzisothiazolone *S*,*S*-oxide derivatives

Aiming to avoid basic treatment, which causes instability of these derivatives, an alternative synthetic approach was found that operates in the solid phase [121]. To prepare libraries of acyl *N*-(acyloxymethyl)benzisothiazolone *S*,*S*-dioxides, those authors applied split/pool methodology. Two different series of compounds were prepared such as compounds **376** containing an amino acid group acylated at the nitrogen atom. According to strategy I shown in Scheme 90, a series of structurally diverse carboxylic acid derivatives were prepared from supported aldehyde **373** condensed with different protected amino acids in the presence of NaBH$(OAc)_3$ to afford intermediate **374**, which was then acylated affording **375**. After hydrolysis of ester and reaction with bromo derivative **368** (R = H) in PTC conditions, compounds **376** were obtained.

A different series of carboxylic acid derivatives **379** was prepared according to strategy II (Scheme 91) starting from **373**, which was treated with an alkylamine in reductive conditions. The acylation of amine **377** gave **378**, which was transformed into amide **379** by condensation with **368** (R = H). According to the protocol depicted in Scheme 91, the elaboration of the intermediated so formed gave **379**.

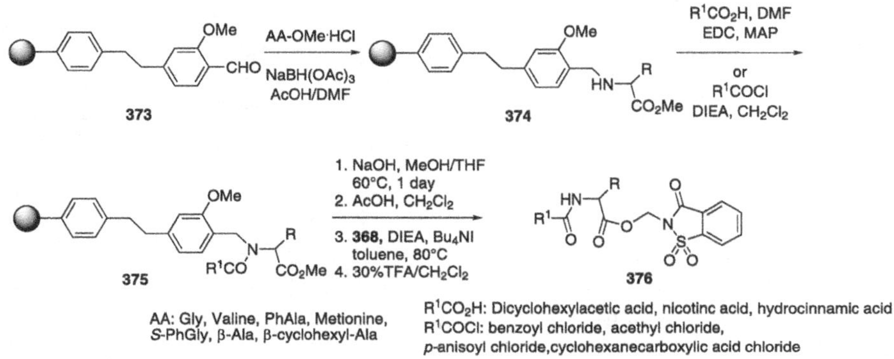

Scheme 90 Solid phase synthesis of *N*-(acyloxymethyl)benzisothiazolone *S*,*S*-dioxide derivatives: strategy I

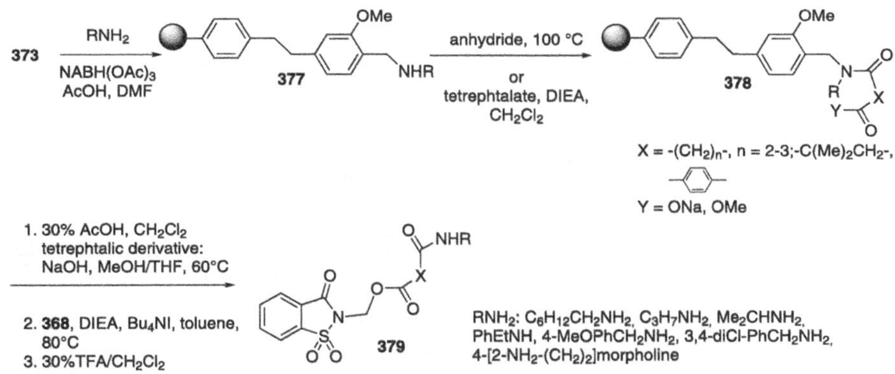

Scheme 91 Solid phase synthesis of *N*-(acyloxymethyl)benzisothiazolone *S*,*S*-dioxide derivatives: strategy II

Both chloro- and bromomethyl aryl substituted derivatives **365** and **368** were used for the preparation of *N*-(heteroaryloxymethyl)benzisothiazolone *S*,*S*-dioxide derivatives **380**, obtained in low to good yields, using different bases (NaH/DMF, $CsCO_3$/DMF, *i*-Pr_2NEt/DMF, MTBD/MeCN). This way, compounds containing cyclic enolates were also prepared (Fig. 5) [116, 122].

Benzisothiazolones with a phosphonate leaving group were also prepared. Substituted chloro derivatives **365** were used to prepare the dialkyl phosphates **381a** and dialkyl phosphinate **381b** by using a proper dialkyl phosphate or dialkyl phosphinate in CH_2Cl_2 in the presence of TEA. Only the yield (66%) of compound **381** (R^1 = 4-*i*-Pr, 6-OMe; R^2 = OEt) is reported (Fig. 6) [123].

SAR studies were performed to find the more active tetrazolyloxy substituted compounds **382**, and the 4-*i*-propyl-6-methoxy derivative showed an improved activity and blood stability (Fig. 6) [119].

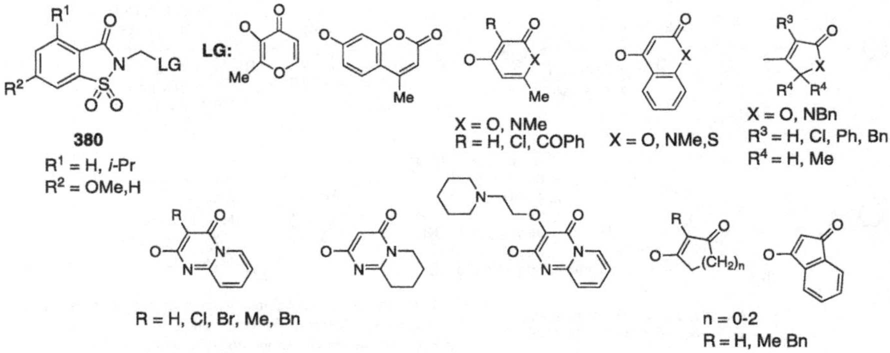

Fig. 5 *N*-(heteroaryloxyymethyl)benzisothiazolone *S*,*S*-oxide

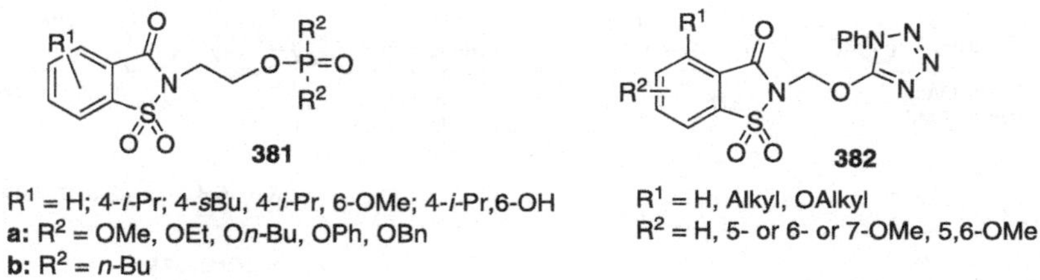

Fig. 6 *N*-(Phosphoxymethyl)- and *N*-(tetrazolylmethyl)benzisothiazolone *S*,*S*-oxide derivatives

Benzisothiazolones **385** containing sulphur leaving groups were prepared via thioderivatives **383** [103, 124]. The functionalisation of the nitrogen atom with a sulphur containing chain was achieved in two different ways depending on the substituent linked to the sulphur atom. The first one consists in the reaction of sulfides **384** ($R^2 = $ Me, Ph) with the sodium salt of saccharin **363** (R = H) in DMF affording **383**. Alternatively, compounds **383** (25–60 overall yields) were prepared from the chloromethyl derivative **365** (R = H) and the appropriate mercaptan in the presence of DBU/MeCN or TEA/THF. Thiophenol reacted with bromo derivative **368** (R = H) in the presence of TEA. Sulfones ($n = 2$) or sulfoxides ($n = 1$) **385** were prepared with *m*-chloroperbenzoic acid and their distribution depended on the stoichiometry of the oxidant and on the kind of R^1 and R^2 on **383** (Scheme 92).

Dithiocarbamates **386a** (10–62%) or *O*-alkylthiocarbonates **386b** (37–46%) were prepared from potassium *N*,*N*-disubstituted dithiocarbamates or potassium *O*-alkyl dithiocarbonates and **365** (Scheme 92). Their antimicobacterial activity was checked [118].

Several *N*-(4-substituted piperazin-1-yl-alkyl)benzisothiazol-3-one *S*,*S*-dioxides of general formula **387**, which were studied as 5-HT_{1A} receptor ligands, were prepared (Scheme 93). They are described in a patent applica-

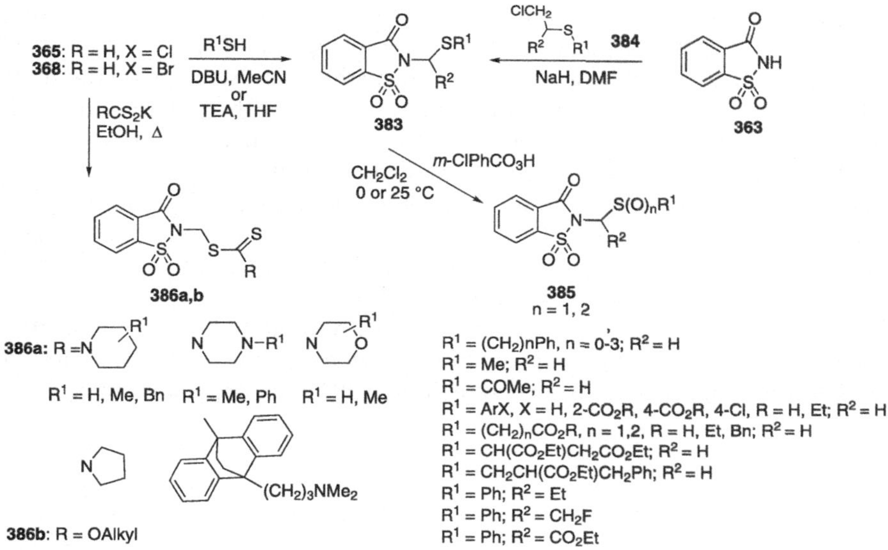

Scheme 92 Synthesis of *N*-(thiomethyl)-benzisothiazolone *S*,*S*-dioxide derivatives

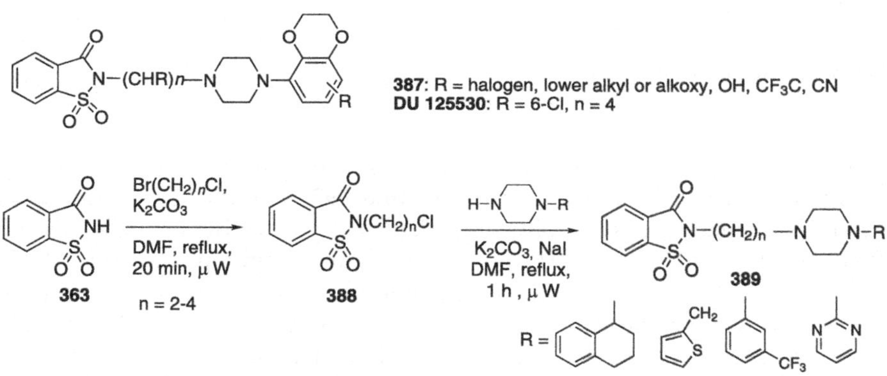

Scheme 93 *N*-(4-Substituted piperazin-1-yl-alkyl)benzisothiazol-3-one *S*,*S*-dioxides

tion [125] and were obtained according to *Method U* starting from saccharin **363** (R = H), which was deprotonated with NaH and alkylated with a functionalised piperazine. This way, 6-chloro compound **DU 125530** was obtained and was claimed as the most active. Compounds **389** were obtained according to the *Method V* by alkylation of saccharin **363** (R = H) with dihaloalkanes to afford **388**, which was then condensed with 4-alkyl-piperazines to obtain the final products **389**. Syntheses were performed by microwave heating to obtain the compounds in better yields (80–95%) than those obtained by conventional heating [126, 127].

A concise asymmetric synthesis of the 2-(aminomethyl)chroman derivative repinotan (**392**), a potent 5-HT_{1A} receptor agonist, has been described (Scheme 94) [128]. In this case, the synthetic pathway is also based on the alkylation of the saccharin nitrogen followed by reaction with a suitable amine as outlined in *Method V*. The synthesis started from allylphenol. Rearrangement of the double bond ($PdCl_2(MeCN)_2$, CH_2Cl_2, Δ) afforded the corresponding styrene, which underwent a Mitsunobu reaction with the commercially available (*S*)-2-hydroxy-3-buten-1-yl *p*-tosylate providing the requisite diene **390** (64%). An RCM reaction was performed using 0.2 equiv. of Grubb's catalyst and 2-(hydroxymethyl) chromene **391a** (78%) was obtained as *p*-toluenesulfonyl ester and then transformed into 2-(aminomethyl)chroman **391b** (83%). Its alkylation with *N*-4-bromobutylsaccharin followed by conversion to the hydrogen chloride salt provided repinotan **392** as a pure enantiomer.

Scheme 94 Synthesis of Repinotan

Compounds **393** and **394** were prepared from 5- or 6-substituted *N*-bromobutylsaccharin and the proper amine **397**, **400** and the commercial available meperidine **401**, following *Method V* (*i*-Pr_2NEt or Et_3N, DMF) (Scheme 95) [129]. Amine **397** was obtained from 2-aminophenol **395**, which was reductively alkylated with *N*-Boc-piperidone affording **396** (90%). Its cyclisation with triphosgene and deprotection of nitrogen provided piperidine **397** (90%). Concerning the preparation of amine **400**, the synthesis consisted first in a palladium mediated coupling of either phenylboronic acid or phenyltrimethyl stannane **398** with enoltriflate **399**, which gave the 4-phenylsubstituted racemic amine *cis*-**400** (80–90% overall yield) after reduction of the alkene intermediate and deprotection of nitrogen. The base catalyzed epimerisation of *cis*-**400** to the *trans* epimer at carboxy linked carbon was reported. The chiral HPLC separation of racemic *cis*-**400** was done as well as the chemical resolution of the corresponding *cis*-acid using (*S*)-α-methylbenzylamine as the resolving agent. Racemic **393a** was also separated by chiral HPLC (Chiralcel OD column).

The direct alkylation of the nitrogen atom of salt **402a** with 1-(3-iodopropyl)-3,7-dimethylxanthine afforded compound **403** (80%), which antagonised glutamate induced neurotoxicity [130] (Scheme 96).

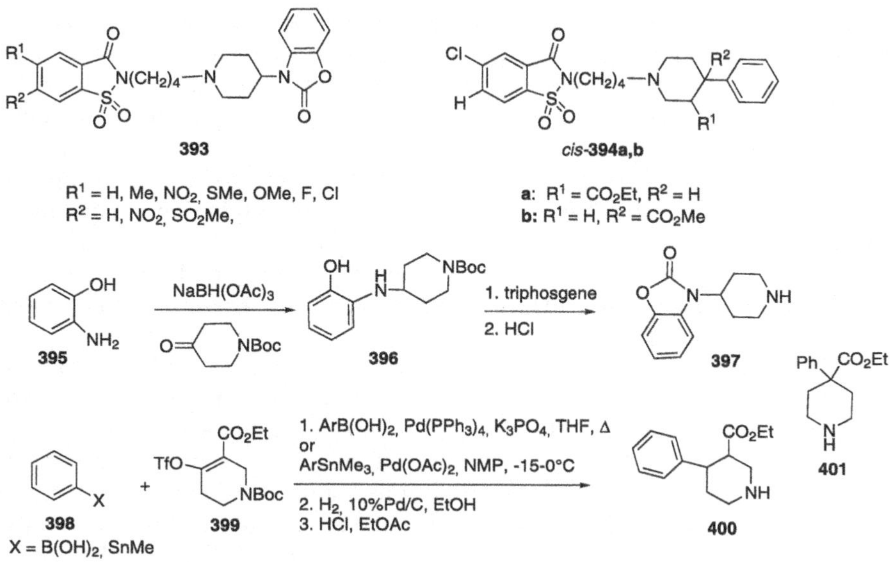

Scheme 95 Synthesis of saccharin derivatives containing a piperidine side chain

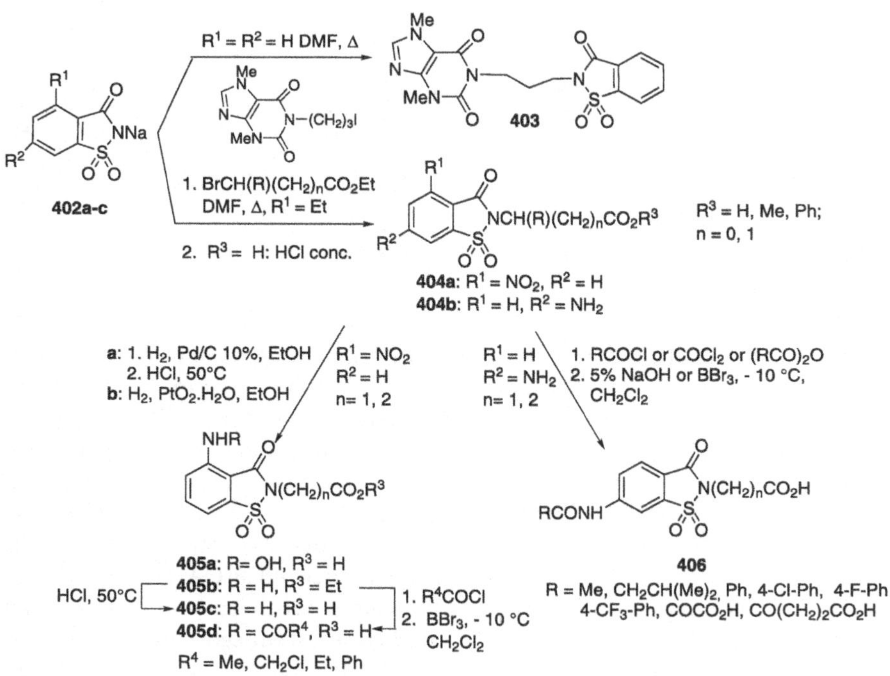

Scheme 96 Synthesis of *N*-saccharinylalkylcarboxylic acids

There is a particular interest in the study of several alkanoic acid derivatives of benzo- and naphto[1,2-*d*]- or naphto[2,3-*d*]isothiazole-3-one *S,S*-dioxides as ARIs (Aldoso reductase inhibitors) [131, 132]. As usual, the synthesis of target benzo derivatives was performed starting from the sodium salt of saccharins **402b,c** which were alkylated with an halogeno ester as outlined in Scheme 96 affording compounds **404a,b.** Different substituents on the benzene ring such as the 4-nitro group transformed both into hydroxylamine derivative **405a** or into amine **405b**, depending on the reductive conditions. The latter was hydrolised to acid **405c** or acylated to give amides **405d.** The 6-amino group in compound **404b** was also acylated and a series of amides **406** were obtained (Scheme 96).

Other alkanoic acid derivatives, i.e. the naphto[1,2-*d*]- and naphto[2,3-*d*]isothiazole-3-one *S,S*-dioxide derivatives **309** and **310**, as well as the nitro and the amino compounds **311**, have already been cited in Scheme 75.

N-Acetylsaccharinyl acid derivatives **408**, which are structurally related to COX-2 inhibitor celecoxib, were designed and synthesised [133] from *N*-saccharinyl acetate **407a**, prepared via the reaction of ethyl bromoacetate with sodium saccharin by heating the reactants in DMF (see [133]). Its transformation into the corresponding hydrazide **407b** and subsequent reaction with ethyl acetoacetate, β-diketones and maleic anhydride, afforded the heterocyclic compounds **408** [134] (Scheme 97).

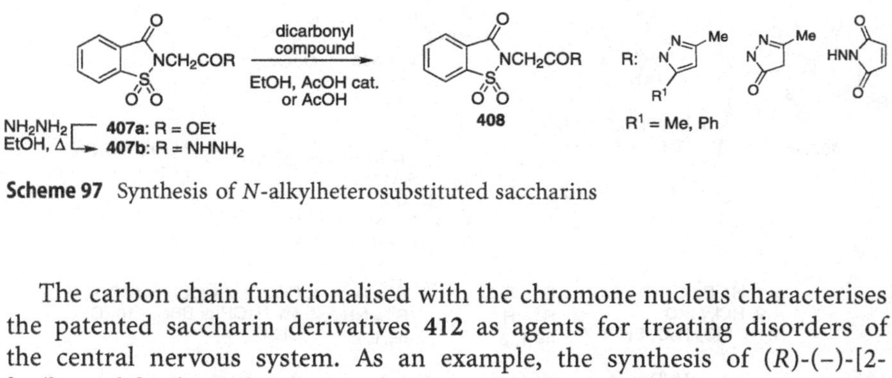

Scheme 97 Synthesis of *N*-alkylheterosubstituted saccharins

The carbon chain functionalised with the chromone nucleus characterises the patented saccharin derivatives **412** as agents for treating disorders of the central nervous system. As an example, the synthesis of (*R*)-(–)-[2-[4-(benzyl [7-(benzyloxy)-6-methoxy-3,4-dihydro-2*H*-chromen-2-yl]methyl-amino)-2-butynyl]-1,2-benzisothiazol-3(2*H*)-one *S,S*-dioxide (**412:** $R^1 = H$, $R^2 = OMe$, $R^3 = OH$) is reported. The reaction of (*R*)-(–)-[*N*-benzyl-*N*-[7-(benzyloxy)-6-methoxy-3,4-dihydro-2*H*-chromen-2-yl]methylamine (**410**) was performed in the presence of paraformaldehye/$Cu(OAc)_2$ and alkynyl saccharin **409**, prepared from the sodium salt of saccharine and the alkynyl bromide. Compound **411** (90%) was obtained and reduced with H_2 and10% Pd/C in MeOH/HCl affording **412** ($R^1 = H$, $R^2 = OMe$, $R^3 = OH$) (64%) (Scheme 98).

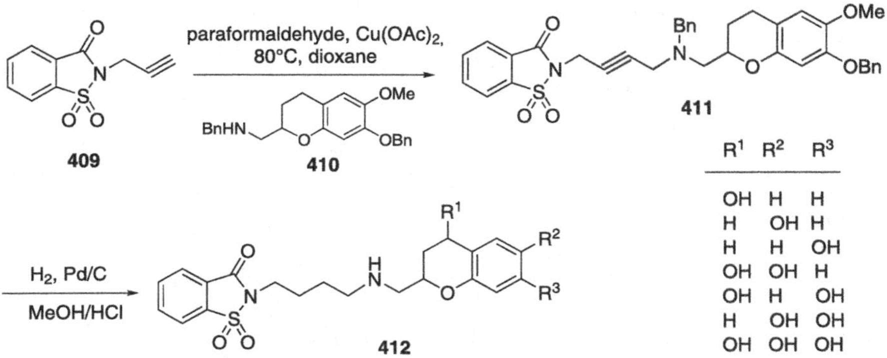

Scheme 98 Synthesis of (chromenyl)methylaminobutyryl-1,2-benzisothiazol-3(2*H*)-one *S*,*S*-dioxides

To increase the hydrophilicity of compounds **329** and of the corresponding thio compounds **330**, the hydroxy group was transformed and benzisothiazolones **413** and thiones **414**/**414′** functionalised with a carbamate chain were prepared (Fig. 7). Their activity against representative bacterial and fungal strains was tested. The reaction of **329** with a series of isothiocyanates was done in presence of DABCO in xylene, and compounds **413** were obtained in 75–90% yields [104]. The reaction of **330** with the same reagent was performed using DABCO or $Fe(acac)_3$ as catalyst in benzene at reflux and the mixture of carbamates **414**/**414′** was obtained and separated by column chromatography. Equilibration between **414**/**414′** takes place in DMSO or acetone/water [105].

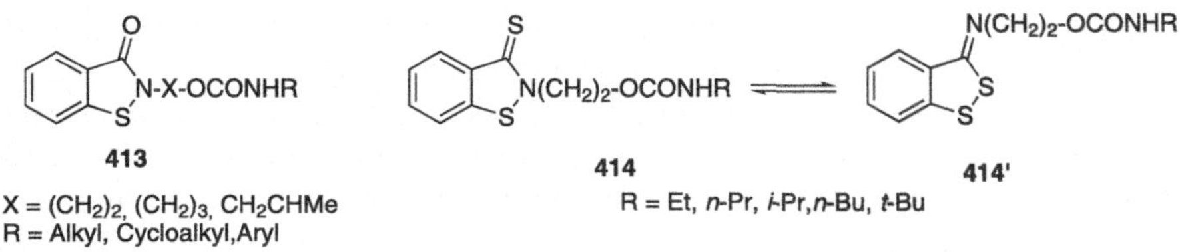

Fig. 7 *N*-2-Hydroxyalkyl-benzisothiazol-one and -thione carbamic esters

4-Arylpiperazin-1-ylalkyl chains are the common feature of a series of isothiazolo[5,4-*b*]pyridine derivatives **418** and **419** and of the corresponding *S*,*S*-dioxides **420**, tested as antimicrobials. Two reagents were used to alkylate **415a,b**, i.e. the chloroalkylpiperazine **416** and the quaternary salt **417** (*Method U*). A mixture of compounds of *N*-alkylation **418a–c** (47–53%) and *O*-alkylation **419a–c** (8–11%) was isolated from **415a**. Instead, using isothiazolopyridine *S*,*S*-dioxide **415b**, only isomers **418a–c** (42–65%) were formed. Compounds **418d** (42%) and **419d** (11%) were obtained using **417** as chain donor (Scheme 99) [135].

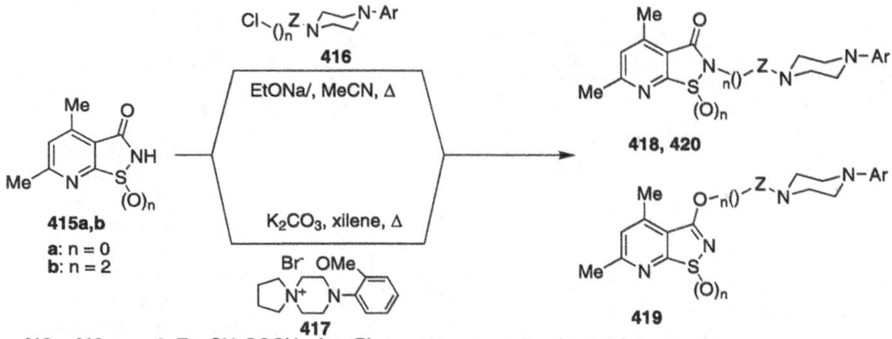

418a, 419a: n = 0, Z = CH_2COCH_2, Ar = Ph
418b, 419b: n = 0, Z = $(CH_2)_4$, Ar = 3-Cl-Ph
418c, 419c: n = 0, Z = $(CH_2)_4$, Ar = 3-CF_3-Ph
418d, 419d n = 0, Z = $(CH_2)_4$, Ar = 3-OMe-Ph
420a: n = 2, Z = CH_2COCH_2, Ar = Ph
420b: n = 2, Z = $(CH_2)_4$, Ar = 3-Cl-Ph
420c: n = 2, Z = $(CH_2)_4$, Ar = 3-CF_3-Ph

Scheme 99 Synthesis of *N*-(4-arylpiperazin-1-ylalkyl)-3-oxo-isothiazolo[5,4-*b*]pyridines

Another strategy was used to obtain a large number of saccharin analogues substituted at the nitrogen atom with polyfunctionalised chains (Scheme 100). In this case, compounds **421** [136], **422** [137], **423** [138] and **424a** containing a saturated chain [139] were first prepared according to known procedures by substitution of the nitrogen atom of **415** with a proper linker. Instead, **424b** (40%) was obtained by reaction of **415** with *trans* 1,4-dibromo-2-propene in MeCN and using K_2CO_3 as the base [135].

Compounds **425** (50–85%) were obtained from intermediate **421** (not isolated) by reaction with different cyclic amines [135, 136]. Indeed, epoxide **422** afforded compounds **426** (40–70%) containing hydroxy amine functionalised chains [139]. For the preparation of **427** (11%), **423** was first transformed into the corresponding acyl chloride using $SOCl_2$, then reacted with 1(hydroxyethyl)-4-phenylpyperazine. Instead, from **424a,b**, containing the unsaturated and saturated chin, respectively, compounds **428a** (70%) and **428b** (55%) were formed by reaction with *N*-(2-pyridinyl)pyperazine and *o*-methoxyphenyl-pyperazine.

The alkylation of the ITQ nucleus of general formula **429** is reported in a patent as well as the parent compounds **357** (see Sect. 5.2) [140]. They were prepared as inhibitors of bacterial DNA synthesis and replication. As the parent benzisothiazolones, a mixture of *N*- and *O*-alkyl derivatives **430** and **431** (major isomer) was obtained performing the reaction in DMF and $CsCO_3$ at 25 °C (Scheme 101).

The reaction of *N*-amino compounds **335** with a series of substituted benzaldehydes, 1-furanaldehyde and cinnamaldehyde, operating in the classical conditions (H_2O/EtOH/AcONa, 70 °C) afforded 2-benzilidene-amino compounds **432** (57–89%) [70–72] which were tested as antimicrobials and antifungals (Fig. 8).

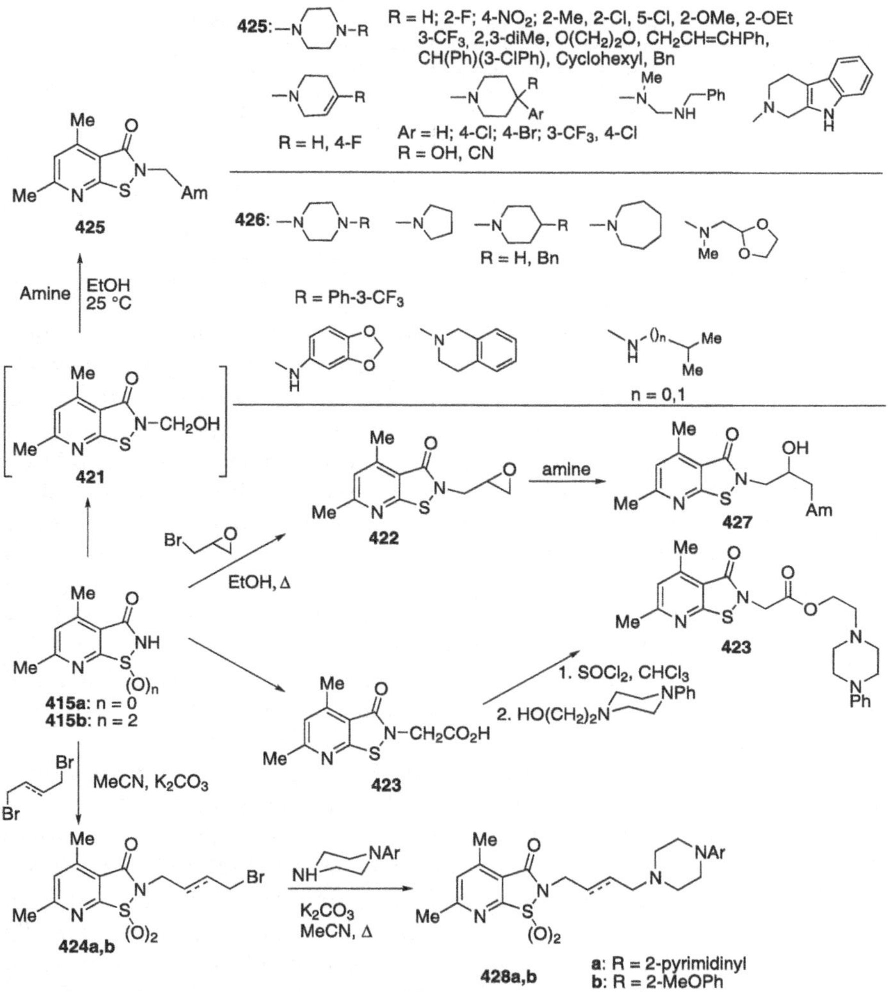

Scheme 100 General procedures for the alkylation of 3-oxo-isothiazolo[5,4-*b*]pyridines with polyfunctionalised chains

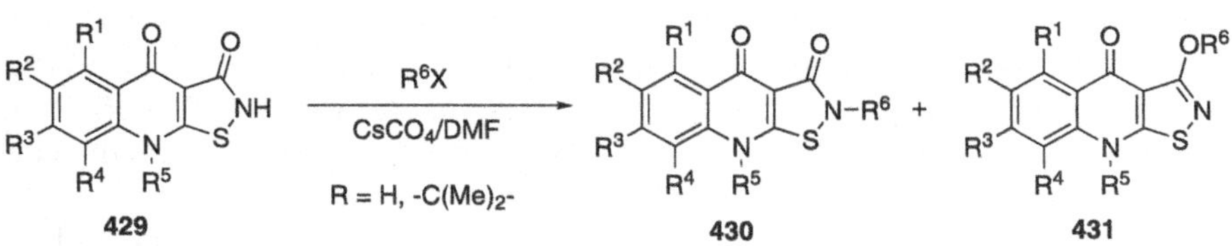

Scheme 101 Alkylation of ITQs

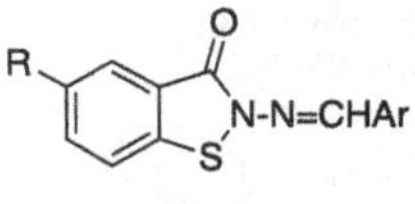

432a,b

a: R = H; Ar = 2-F-Ph, 3-F-Ph, 4-F-Ph, 3-HO-Ph, 4-HO-Ph
Ph, 2-Cl-Ph, 3-Cl-Ph, 4-Cl-Ph, 2-NO_2-Ph, 3-NO_2-Ph, 4-NO_2-Ph, 2-HO-Ph, 4-MeO-Ph, 4-HO-3-MeO-Ph, 3,4-OCH_2O-Ph, 2-furyl, CH=CHPh
b: R = Me, Ar = Ph, 2-Cl-Ph, 3-Cl-Ph, 4-Cl-Ph, 2-NO_2-Ph, 3-NO_2-Ph, 4-NO_2-Ph, 2-HO-Ph, 4-MeO-Ph, 4-HO-3-MeO-Ph, 3,4-OCH_2O-Ph, 2-furyl, CH=CHPh

Fig. 8 *N*-(Benzylidene-amino)-benzisothiazolones

6 Biologically Active Compounds

While saccharin remains one of the most important, widely used and best known isothiazole derivatives, many other isothiazole derivatives manifest a broad spectrum of useful properties and have applications in several fields. In these compounds the isothiazole ring can be both the pharmacophore or a scaffold with no direct biological activity. In order to obtain biologically active compounds, different approaches were used, which are represented by random library screenings of different heterocycle containing molecules, or by rational design. In this last case, bioisosteric replacement of isoxazole ring in known pharmacological active compounds was widely used and, although the isothiazole moiety is markedly less acidic and more lipophilic, this approach very often led to active compounds.

Below isothiazole derivatives are classified according to their activity and for each section the most active compounds are cited. When important, a short account of the mechanism of the action is discussed.

Pesticides. Isothiazoles are known as harmful organism-controlling agents and have been widely cited as agrochemicals since the 1970s [22, 141]. Recent developments in this field are represented by 5-thio-isothiazole derivatives **433**, which are claimed as termite controlling agents [142] and by 5-amino-isothiazole derivatives **434**. Among them, derivative **434a** possesses pesticidal activity against the larvae of *Spodoptera littoralis, Heliothis armigera, Plutella xylostella* [143], compound **434b** is claimed as a pesticide against *Plasmopara viticola* on vines [144] and compounds of general formula **434c** (R^1 = Cl, R^2 = substituted alkylaromatic group) are claimed as strongly active insecticides with good compatibility with crops [145].

Substituted *N*-(5-isothiazolyl)phenylacetamides **110** (R^1 = CH_2PhOPh-4-CF_3) possess an excellent broad spectrum of activity against insects by acting on mitochondrial respiration as inhibitors of MET at Complex 1 (NADH dehydrogenase), but they have been found to be toxic to fish. In order to improve safety to non-target organisms various *N*-alkylated derivatives **111a** were synthesised as proinsecticides [32], while derivatives **114** (X = NHR, R = OMe, $OCH_2CH{=}CH_2$, $NHCONH_2$, NHTs) substituted at the α-methylene position were prepared to enhance selectivity [33].

Antimicrobial. The antimicrobial activity of 1,2-benzisothiazoles and of 3-isothiazolones has been intensively studied and various derivatives are known to be effective against a wide range of bacteria and fungi. The biological activity of such compounds arises from their ability to readily pass into membranes and fungal cell walls and then react with important intracellular sulphur containing proteins, or simpler molecules, such as glutathione, causing impairment of the cell function. The S – S bond formation with a biological target is strictly related to the lability of the S – N bond [105]. The mechanism of this reaction involves a nucleophilic attack by the sulphur atom at the sulphur atom of the isothiazolone, followed by the cleavage of the S – N bond to give the ring opened amidodisulfide, which reacts further with the same nucleophile to give the mercaptoacrylamide (Scheme 102). This causes the cell's death.

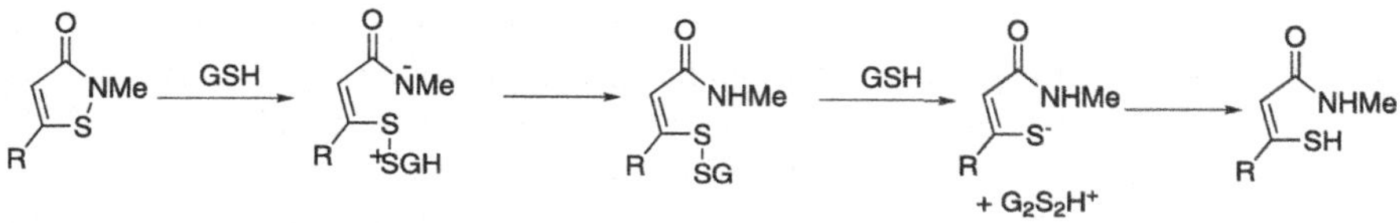

Scheme 102 Mode of action isothiazolones

The extent of activity of mononuclear isothiazoles, is strictly dependent on the nature and position of substituents on the heterocyclic ring. 5-Chloro-*N*-methylisothiazolone **21a** (R^1 = Cl, R^2 = H, R^3 = Me), the main active component of the commercial biocide Katon [2], is several orders of magnitude more active against bacteria and fungi with respect to the corresponding 4-Cl-*N*-alkylisothiazolones. Unfortunately, **21a** is a strong skin sensitiser and this unlikely characteristic is due to its ability to react with histidine and lysine, two nucleophilic amino acids that are present in epidermal proteins. This interaction causes chemical modifications of such proteins, leading to allergic contact dermatitis. Curiously, the 5 unsubstituted derivative is a very weak sensitiser, as it reacts exclusively with thiol nucleophiles [146]. 4-Benzylamino-2-methylisothiazolones unsubstituted (**37b**) or chloro substituted on C-5 (**38b**) were prepared, but their activity is lower with respect to **21a** [13].

The general interest in BIT derivatives has shifted to their industrial application as biocides even if the parent BIT compound is not recommended for pharmaceutical, cosmetic, and toiletry preparation because it is a skin sensitiser. Nevertheless, many BIT derivatives have been prepared in the last years and extensive studies on the influence of substituents in the molecule on biological activity led to the conclusion that their activity results from a concurrence of steric effects and S – N bond reactivity induced by electronic effects of the substituents [105]. As examples, hydrazones **432** were assayed against gram-positive bacteria and gram negative bacteria and on PRS but they were less active or comparable to BIT and possess an activity equal or superior to the reference drugs ampicillin and miconazole [71]. The substitution of sulphur for oxygen in the keto-benzisothiazole system (compound **414**) increases the strength of the S – N bond and lowers the antibacterial activity with respect to keto derivatives [105, 147, 148]. These compounds were found to be less active than reference compounds (cefotaxime, gentamicin). A series of compounds **413** were prepared in order to enhance lipophilicity and to facilitate membrane crossing but they were less active than BIT [149]. Some of them were active in vitro against anti-*Mycobacterium avium* [105].

In the last years other isothiazole derivatives showing antimicrobial activity and with a different mode of action have been prepared. 4-Arylpiperazin-1-ylalkyl chains are the common feature of a series of thiazolo[5,4-*b*]pyridine derivatives **418** (Ar = Ph, 3-Cl – Ph) characterised by both antimicrobial and CNS activity. [135]. 3-Amidinobenzisothiazole compounds **252** (R^1 = Me, $R^2 = CH_2 - CH = CH_2$) are active against gram positive bacteria and yeasts. The unsaturated aliphatic moiety is essential for antimicrobial activity [83]. A small library of minor groove binding ligands **57** ($R^1 = H_2N(CH_2)_3$, $R^2 = (CH_2)_4NMe_2$) consisting of a four heterocyclic ring core was synthesised. The position of basic groups in the molecule strongly influences the antimicrobial activity and DNA binding affinity [17].

ITQs have tricyclic structures comprising a quinolone nucleus with an annelated isothiazolone ring, which replaces the archetypal 3-carboxylic group of typical quinolones. ITQs inhibit type II topoisomerases, such as DNA girase and topoisomerase IV. The first examples were synthesised in the late 1980s but they exerted mammalian cytotoxicity [150]. Recently compounds of type **429a,b** (X = CH, N) have been prepared and tested on MRS. Considering the ITQ nucleus, the addition of a methoxy group at C-8 on compounds **429a** increased the activity, while the removal of fluorine at C-6 or replacement of C-8 carbon with a nitrogen (compounds **429b**) compromised activity against MRS. For amino groups linked to C-7 in compound **429**, the activity decreased in the order 6-isoquinolinyl > 4-piridinyl > 5-dihydroisoindolyl > 6-tetrahydroisoquinolinyl. The most active compound is **429a** (R^1 = H, R^2 = F, R^3 = 2-methyl-4-pyridyl, R^4 = OMe). By modulating the substituent at C-7 it is possible to reduce cytotoxicity [115, 151].

Carbapenem derivatives of general formula **434** acylate a broad spectrum of PBP with high affinity, and they are rapidly bactericidal, with potent activity against methicillin sensitive staphylococci. In addition the carbapenem nucleus is resistant to most serine β-lactamases [152]. Attached to the carbapenem nucleus is a lipophilic side chain, which is further substituted with a cationic group. The lipophilic component provides for potent binding to the target penicillin binding proteins, including the MRS. Unfortunately, first generation derivatives provoked immune responses in toxicological tests. In compound **277** ($n = 2$, R = CH_2CONH_2) the presence of a naphtosultam side chain, attached to the carbapenem through a metylene linker, lowers the immunotoxicity. The expulsion of the side chain occurs when the β-lactam has acylated the surface of red blood cells ("Releasable Hapten" hypothesis). Consequently, the red blood cells are not labelled with the potential immunogenic naphtosultam side chain [94, 95, 152]. Also carbapenem **17a** exerts potent antibacterial activity, excellent stability and a good pharmacokinetics profile [11].

Antiviral and antitumor. Purine nucleosides analogues have been investigated as antitumor and antiviral agents. Bioisosteric derivatives of nucleobases have been proposed. In particular, as the sulphur atoms is analogous to a – CH = CH – group because of its steric and electronic properties, different imidazo[4,5-*d*]isothiazoles **292** (R^1 = H, Me; R^2 = H, SMe, SBu) have been synthesised. All compounds were cytotoxic at micromolar concentrations, but showed no antiviral activity on human cytomegalovirus and herpes simplex virus type 1 [96]. The nucleoside analogue **12** was also prepared, but showed none antiviral activity nor cytotoxicity [10].

Aza-TSAO is a bioisosteric derivative of the spiro nucleoside TSAO (Fig. 9), which is a potent HIV-1 inhibitor. The bioisosteric replacement leads to less active compounds [49]. The binding mode of several aza-TSAO derivatives of type **150** which show HIV RT inhibitory activity, was in-

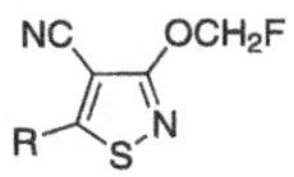

433

RS: Me, thiocyclopentyl, trimethylsilylmethylthio, OEt, OPh, SPh, pyrimidylthio, isopropylamino, 3-methylphenoxy, p-chlorophenylthio,

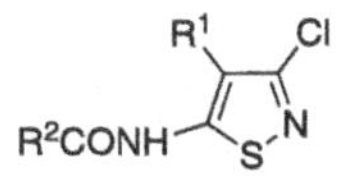

434a,c

434a: R^1 = CN, R^2 = 3-Cl-Ph
434b: R^1 = Me, R^2 = 2-propyl-benzofuran-5-yl
434c: R^1 = Cl, R^2 = substituted alkylaromatic group

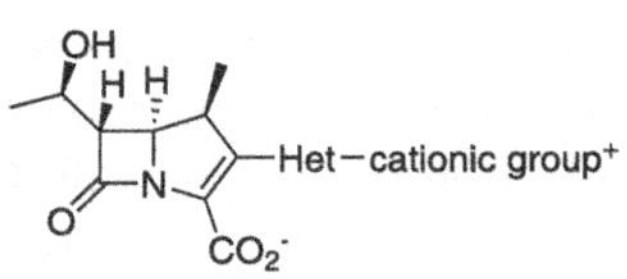

435

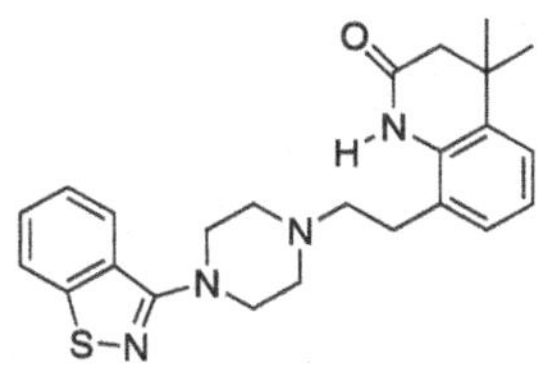

436

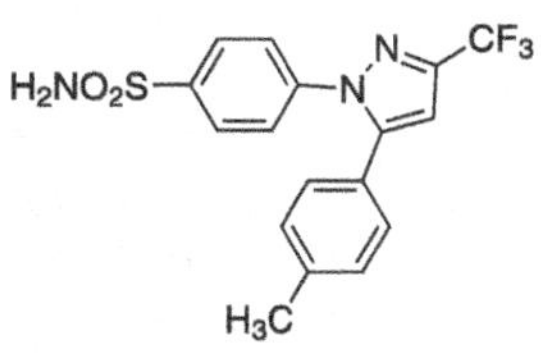

celecoxib

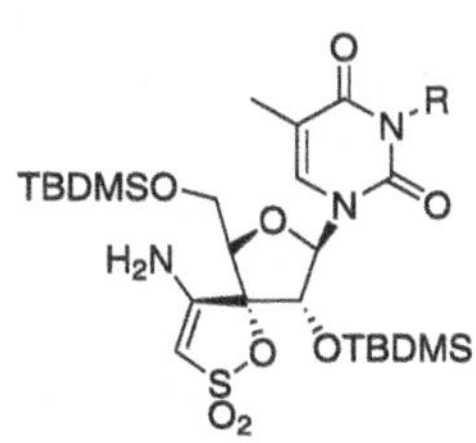

TSAO

Fig. 9 Biological active compounds

vestigated by a thorough conformational search at the MM, HF and DFT level [153]

Compound **332** is an antriretroviral agent that interferes with the HIV-1 nucleocapsid protein, a highly conserved zinc finger protein. It affects the zinc finger region causing rapid extrusion of zinc and subsequent denaturation of the viral protein [106].

Compound **53** is a potent HIV protease inhibitor, which is more active than indinavir [16].

Isothiazole-4-carbonitrile derivatives exert antiviral activity, showing different selectivity depending on the substituents on the ring [25]. Compound **86** represents a novel class of active-site inhibitors of hepatitis C virus (HCV) NS5B polymerase. They act by preventing proper positioning of natural template in the active-site, by disrupting the suitable entry path of initiating rNTP substrate and by locking the C-terminus into an inactive conformation [27, 154, 155]. The most active compound is the *N*-3,5-dichlorophenyl compound, which has an IC_{50} of 200 nM and EC_{50} of 100 nM.

3-Thio derivatives **96** (R = Ar, R^1 = H, Me) are active against picornaviruses, such as rhinoviruses and enteroviruses, but also against HIV. The biological activity depends on the presence of bulky substituents at the para position of the phenyl ring and of the presence of a low molecular thioalkyl chain at 3. These compounds act by interfering with early events of viral replication and it has been postulated that they have a capsid-binding activity and that they induce some conformational changes in the binding site [27].

Antifungal and antimycobacterial. Brassilexin and the sinalexin (see Scheme 58) are the most potent antifungal phytoalexins produced by crucifer plants. Many fungal pathogens, i.e. *Leptospheria maculans*, have evolved enzymatic systems that are able to detoxify phytoalexins. Potential inhibitors of brassilexin detoxification were used to protect plants against the fungal invader. The indole-isothiazole fused-ring system of brassilexin was replaced with quinoline-isothiazole **227**, benzothiophene-isothiazole **230**, isothiazoles **7** and benzisothiazole **232**. Compound **227** and **230** displayed the strongest growth inhibitor activity [9].

The equilibrium mixture of compounds **342** and **343** shows antimicotic and antibacterial activity in vitro [110].

Dithiocarbamate **386a** and dithiocarbonate **386b** show activity against *Mycobacterium tuberculosis* and antitumor activity [118].

Compounds of type **135** and **141** have been synthesised as inhibitors of PFT, an enzyme that catalyses the transfer of the farnesyl group from farnesyl pyrophosphate to the cysteine SH in the protozoan parasite *Trypanosoma brucei*, the causative agent of African sleeping sickness. The parent compound **140** (Ar = (2,4,6-trimethylphenyl) is the most active (ED_{50} 10 μM) [44, 45]. This class of compounds has been tested also on mammalian PFTase and some of them showed inhibitory activity. In particular, none of the dihydroderivatives affects the enzyme in a concentration-dependent manner, while compounds with a C4-C5 double bond and sulfanyl substituents (**142**) showed both inhibitory and antiproliferative activity by inhibiting G0/G1 phase of the cell cycle. The activity in such compounds is affected by two principals factors such as the planarity of the isothiazole ring and the nature of substituent at the 5-position [25].

Anti-inflammatory. The conventional NSAID's exert they activity by inhibiting the COX enzyme, which synthesises prostaglandins, the major mediators of inflammation. Two isoforms of this enzyme were identified: the constitutive COX-1 and the inducible COX-2. Celecoxib is a selective COX2 inhibitor, which was recently removed from the market for its cardiovascular toxicity. Compound **408** (R = 3-methyl-5-phenyl-pyrazol-1-yl) is a *N*-benzesulfonamide analogue of celecoxib with higher analgesic and anti-inflammatory activity [133]. Instead the benzisothiazolyl analogue **215** of celecoxib is a selective inhibitor of COX-1 [67]. Compound **208** ($R^1 = R^2 =$ t-Bu, R = Et) is a potent inhibitor of cyclooxygenase-2 and 5-lipooxigenase as well as of production of IL 1 [64, 65]. Compound **246** (R^1 = 4-Cl-Ph, R^2 = CH_2CO_2H) is an arylacetic acid derivative that exerts high anti-inflammatory activity [76].

Compound **107** (R^1 = H, $R^2 = R^2$ = 4-Cl – Ph) possesses antiviral activity, and different derivatives containing its scaffold have been tested as anti-inflammatory and immunosuppressant agents. Electron withdrawing, lipophylic and small substituents such as Cl or Me of the aromatic group in the carboxamide fragment increase immunosuppressant activity [28, 29]. Compounds **255** and **234** have been tested in vitro and in vivo for anti-inflammatory activity. Hydrazones are more active than Schiff bases [73]. Pyridoisothiazolones represent a class of compounds with high analgesic activity. For example, derivatives **426** (R = Ar) have analgesic action two to ten times more potent than acetylsalicylic acid. [136]

The aryl derivatives **347b** inhibit the IL-1β induced cartilage breakdown associated with osteoarthritis. These compounds are relatively resistant to reductive metabolism by liver microsomial preparations and act by interfering with the proteolytic activation of matrix metalloproteinases [112].

Inhibitor of H^+/K^+ ATPase. H^+/K^+ ATPase is responsible of the acidic secretion in stomach and is a pharmacological target in peptic ulcers. Compounds **351** and **352** show a high inhibitory activity in vitro, but no action in vivo. The potent in vitro activities are due to its high-thiophilic properties, which are not selective for the target enzyme. Prodrugs **350** have been prepared that are converted in the active isothiazolopyridines only within the acidic compartment of the parietal cells, avoiding interaction with thiol groups on other proteins except H^+/K^+-ATPase (see Scheme 87) [113].

Human leukocyte elastase inhibitors. HLE belongs to the chymotrypsin family of serine proteinases: the enzyme consists of a single polypeptide chain of 218 amino acid residues and four disulfide bridges. HLE has been proposed to be a primary mediator of many pulmonary disorders such as emphysysema, acute respiratory distress syndrome and chronic bronchitis. Orally bioavailable inhibitors of HLE belong to the BIT class and are characterised by their specificity and incorporation of a built-in mechanism for regeneration of enzyme activity [98, 124]. Thus, rapid acylation of the active site serine is followed by ring opening and simultaneous departure of the leaving group, yielding a reactive electrophilic species. Deacylation gives active enzyme, while subsequent reaction with an active site nucleophile yields inactive enzyme (Scheme 103).

The heterocyclic scaffold allows the attachment and optimal spatial orientation of peptidyl and nonpeptidyl recognition elements, leading to exploitation of favourable binding interactions with specific subsites of HLE. Different leaving groups have been proposed and studied for compound of general formula **369**. The most active are reported in Scheme 103. Tetrahydrobenzisothiazolone derivative **162** (R = Et, $n = 1$) has been tested as an HLE inhibitor, but it showed less activity than the BIT containing compound [53].

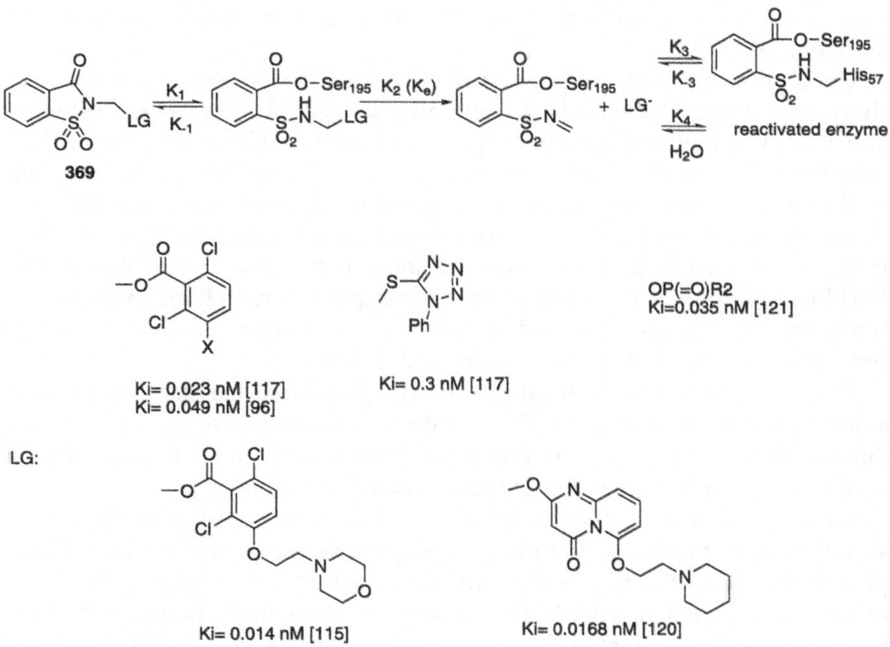

Scheme 103 Mode of action of HLE inhibitors

Recently, closely related isothiazolidin-3-one *S*,*S* dioxide **167** was prepared and tested in vitro [54]. The R^1 residue binds to primary specificity pockets of HLE and by modulating the R1 substitutent it is possible to obtain highly potent HLE inhibitors or compounds that are highly selective for other types of serine proteases (R^1 = isobutyl for HLE inhibitors, R^1 = Ph for cathepsin

G inhibitors, R^1 = small group for PR3 proteinase inhibitors). In all cases the leaving group is represented by carboxylic acid derivatives [54].

Other derivatives tested on HLE are the ortophosphoric derivatives **168** [55] and compound **46** ($R^1 = R^2 = Ph$, $R^3 = R^5 = Cl$, R^4 = O-*i*-Pr) [15].

Inhibitor of mast cell tryptase. Starting from the consideration that BIT derivatives act on HLE, a library of BIT was tested on HMCT. This is a trypsin-like serine protease which is the major product secreted from mast cells during their activation. Elevated tryptase levels have been observed in various diseases such as asthma and inflammatory skin diseases. The most active BIT compound is **371** (R = C_6H_4-4-NHCbz, $IC_{50}(\mu M) = 0.0643$). Modeling studies suggest that such compounds recognise a hydrophobic binding pocket on the S' side of tryptase that prefers a benzyloxycarbamate group approximately nine atoms from the electrophilic benzisothiazolone *S*,*S*-dioxide carbonyl [120].

Inhibitors of Aldoso reductase. Aldoso reductase catalyses the NADPH-dependent reduction of glucose to sorbitol, whose activity is higher in hyperglycaemic conditions leading to elevated intracellular concentration of sorbitol and, as a consequence, high cellular osmolarity. This fact is dramatically relevant in the development of long-term diabetic complications. NiT **309a** belong to a novel class of aldoso reductase inhibitors, possessing benzisothiazoles scaffold. The presence of a planar aromatic moiety and of an acidic function is important for activity. Also the disposition of the second aromatic ring is crucial for the biological activity. In fact, compound **310**, having a linear tricyclic moiety, is not active. The functionalisation with a second carboxylic group on the phenyl ring (**309b,c**) leads to an enhancement of the activity [101]. This class of compounds is selective for Aldoso reductase, while is not active against other enzymes, e.g. aldehyde dehydrogenase.

Kinase inhibitors. Protein kinases are key regulators of different biological pathways. Different kinases are present in the cell and represent an intriguing biological target for several diseases. MEK has a central role in regulating cell growth and survival, differentiation and angiogenesis. Overexpression and activation of these enzymes are associated with various human cancers. Compound **86** was found as lead for in vitro inhibitors of MEK and different derivatives have been synthesised in order to evaluate structure–activity relationship. In particular, a free hydroxyl group at C-3 and, a hydrogen bond donor at C-5 are essential for activity (bioisosteric derivatives with S and O are not active) and the inclusion of a spacer between 5-alkyl or aromatic group and isothiazole core lead to no activity [23]. The replacement of the cyano group with an amidino group in compound **88** (R^1 = 2,5-Cl_2, R^2 = H, $R^3 = CH_2CH(Me)OH$) and the introduction of a bulky *para*-substituent on the aromatic ring produced an improvement of oral activity [26].

ChK 1 and ChK 2 are the major effectors of the replication checkpoint: the fail-safes, which ensure that the cell cycle does not progress to the next stage until the previous step is completed. Carboxamidine isothiazole derivatives **87** (R^1= cyclohexyl, pyrazolyl, *i*Pr, alkylalcohol) are potent ChK2 inhibitors. They act by direct binding to the ATP site of ChK2 and are ATP-competitors. They possess cellular activity to regulate the ChK2 mediated cell cycle arrest and apoptosis [24].

Isothiazoles have been found to be active also as inhibitors of TrKA, kinases receptors for the neurotrophin family of ligands. Compound **117** (R^1 = 4-Cl – Ph, R^2 = H, R^3 = Me) is a potent inhibitor of this enzyme and its derivatives have been prepared. *N*-Amino-heteroaryl substituted compounds **118** (R^1 = 4-Cl – Ph, R^2 = H) were shown to be good urea surrogates, whereas thioindanyl substituted at C-3 **117** were shown to be very potent inhibitors [35].

CP-547,632 is a potent inhibitor of the VEGFR-2 and basic fibroblast growth factor kinases at nanomolar concentration. It is undergoing clinical trials (2007) [34].

Tyrosine phosphatase inhibitors. PTPs in concert with tyrosine kynases, control the phosphorilation state of many proteins involved in signal transduction pathways. Deregulation of these signalling pathways is involved in numerous diseases, such as cancer and diabetes. PTP1B was the first purified PTP and it is considered one of the best validated drug targets in type2 diabetes and obesity. The design of inhibitors of PTP1B has focused on binding to the active site and on discovery of mimetics of pTyr. Isothiazolidinone peptides have been designed as inhibitors of PTP1B and they show binding activity at nanomolar concentration. In particular, compounds containing a saturated isothiazolidinone scaffold bind the active site much more strongly than the corresponding unsaturated derivatives [156, 157].

In order to improve cellular permeability nonpeptidic pTyr mimetics have been designed and compound **179** was found as a potent, cell permeable inhibitor of PTP1B [57].

Interaction with GABA receptors. GABA is an inhibitory neurotransmitter in the CNS that operates through different receptors: ionotropic $GABA_A$ and $GABA_C$ receptors, and metabotropic $GABA_B$ receptors. $GABA_A$ receptors have been implicated in different neurological diseases and represent important therapeutic targets [14]. A number of different heterocyclic $GABA_A$ agonists, such as muscimol, THIP, isoguvacine have been synthesised (see Scheme 11). $GABA_A$ agonists are zwitterionic compounds and the structure of the heterocyclic ring is a factor of critical importance for the interaction with receptors. In fact compounds with highly delocalised negative charges on the ring have low affinity for the receptors. Other important factors that contribute to biological effects are steric effects, acidity, and tautomeric equilibria. All these factors were examined in depth using ab initio calculations [158]. Another critical factor is represented by the ability of such compounds to penetrate the blood–brain barrier and it is determined by the concentrations of the ionised and unionised form. (I/U ratio) [159]. The isothiazolyl derivatives thiomuscimol and thio-THIP exert a lower activity with respect to the isoxazolyl compounds. Furthermore the bioisosteric substitution of sulphur for the oxygen atom of THIP converts an agonist to a competitive antagonist. Other isothiazole analogues of THIP have been been prepared. They show affinity for $GABA_A$ receptors at low-micromolar range, which is higher for the satured compound 41 [14, 160, 161].

Interaction with glutamic acid receptors. S – Glu is the main excitatory neurotransmitter in CNS and activates two different types of receptors: ionotropic and metabotropic receptors. Glu is involved in many physiological processes, such as learning, memory, and vision. An altered Glu function is implicated in various neuropathologies, such as epilepsy, stroke, cognitive disorders, and neurogenerative deseases. Ionotropic receptors are subdivided into three groups: NMDA, AMPA, and KA receptors. Furthermore Ibo is a naturally occurring excitotoxin isolated from the mushroom *Amanita muscaria*, which activates KA, NMDA, and several metabotropic receptors. Considering the presence in both AMPA and Ibo structure of an isoxazolyl ring, isothiazolyl bioisoteric derivatives have been prepared in the last years, and in most cases the isothiazolyl derivatives show the same potency as the isoxazolyl ones [19–21]. Only thio-ATPA is considerably less active than ATPA on AMPA receptors, but it is a potent and selective agonist at homomerically expressed ionotropic GluR5. In particular, it has been demonstrated that the activity is solely due to the *S*-enantiomer. Thio-AMPA, as well as AMPA, penetrates the blood-brain barrier as a net uncharged diprotonated species (see Scheme 15) [19].

A **67a** series was designed as structural hybrids between agonist and antagonists ligands showing different receptor selectivity. Such compounds are bioisosteric modified analogues of the dipeptide Glu-Gly [20].

Thioibotenic acid is active both on ionotropic and cloned metabotropic Glu receptors. The different behaviour between isoxazolyl and isothiazolyl derivatives has been studied by using computational and experimental methods [162]. Thio-ibo analogues (see Scheme 20) have been synthesised in order to investigate the structure–activity relationship at both iGlu and mGlu receptors. Analogues containing smaller substituents retained affinity similar to thio-Ibo at NMDA receptors, while by increasing the bulkiness of the substituent the activity is lost. It is notable that the change in efficacy is different between individual subtypes [21]. The xantine derivative **403** significantly antagonised glutamate induced neurotoxicity [130].

Interaction with muscarinic receptors. Deficit in central cholinergic transmission causes the learning and memory impairments seen in patients with Alzhiemer's disease ("cholinergic hypothesis"). The natural alkaloid arecoline is an agonist of muscarinic AChRs, and has been shown to improve cognition when administered to Alzheimer's patients, although the pharmacological effects are shortlived. Compounds **124** and **125** (R = Me, Et, allyl, CH_2CN; $R^1 = R^2 = R^3$ = H, Me) represent conformationally restricted bicyclic analogues of arecoline, with potent inhibitor activity of the binding of mAChR radioligands [37].

Interaction with serotonin receptors. Serotonin is involved in several diseases, such as depression, schizophrenia, nad psychoses. Seven families of receptors have been discovered (5-HT1-7). Compounds such as buspirone and ipsapirone **389** (R = Ar, $n = 4$), which belong to the class of heteroarylpiperazines, are clinically effective dugs for the treatment of anxiety and depression. It is now generally accepted that the clinical effects are due to the interaction with 5-HT receptors. This class comprises five different receptors (5-HT1_{a-f}). 5-HT1a receptors are involved in anxiety and depression. Ipsaspirone is a partial agonist of 5-HT1a, while **DU1255530** (see Scheme 93) is an antagonist [126].

Repinotan (**392**) is a 5-HT1a agonist, in which the 2-(aminomethyl)-1,4-benzodioxan group bioisosterically replaces the arylpiperazine. Repinotan is being developed by Bayer as a potential treatment for ischemic stroke and traumatic brain injury [128].

It is thought that 5-HT2a antagonism together with relatively weaker dopamine antagonism are principal features that differentiate the side-effect profile of atypical antipsychotic agents of the first generation treatments. 5-HT2a antagonist with benzisothiazole scaffold has the general formula **258** (see Tables 2 and 3). Among this class compound **436** is a new atypical antipsychotics for the treatment of schizophrenia [163]. Other benzisothiazolyl substituted piperazines have been prepared, but the activity is low [87, 88].

Interaction with adrenergic receptors. Adrenergic receptors belong to the family of G-protein couplet receptors and have been classified as α (1–3) and β (1–3). Selective α-1 anatagonists represent a target for the treatment of benign prostatic hyperplasia. Ipsaspirone, a 5-HT1 partial agonist, has a modest affinity for α-1 receptors and different derivatives have been prepared in order to gain a potent antagonist activity. As an example, a chlorine atom has been introduced on the benzene ring of saccharin and the piperazine group was replaced with a piperidine derivative, leading to the high selective compound **394** ($R^1 = CO_2Et$, $R^1 = H$). The stereochemistry is important for the activity [129]. Compound **393** ($R^1 = Cl$, $R^1 = H$), is selective for α-1 receptors with no interaction with other protein G-coupled receptors [100].

References

1. Clerici F, Gelmi ML, Pellegrino S (2008) Comp Heterocycl Chem III, (in press)
2. Chapman RF, Peart BJ (1996) Comp Heterocycl Chem II 3:319
3. Kaberdin RV, Potkin VI (2002) Russ Chem Rev 71:673
4. Taubert K, Kraus S, Schulze B (2002) Sulf Rep 23:79
5. Taubert K, Siegemund A, Schulze B (2002) Sulf Rep 24:279
6. Hamad Elgazwy A-SS (2003) Tetrahedron 59:7445

7. Ager DJ, Pantaleone DP, Henderson SA, Katritzky AR, Prakash I, Walters DE (1998) Ang Chem Int Ed 37:1802
8. Adams A, Slask R (1956) Chem Ind 42:1232
9. Soledade M, Pedras C, Suchy M (2006) Bioorg Med Chem 14:714
10. Luyten I, De Winter H, Busson R, Lescrinier T, Creuven I, Durant F, Balzarini J, De Clercq E, Herdewijn P (1996) Helv Chim Acta 79:1462
11. Kang YK, Lee KS, Yoo KH, Shin KJ, Kim DC, Lee C-S, Kong JY, Kim DJ (2003) Bioorg Med Chem Lett 13:463
12. Khalaj A, Adibpour N, Shahverdi AR, Daneshtalab M (2004) E J Med Chem 39:699
13. Nádel A, Pálinkás J (2000) J Het Chem 37:1463
14. Ebert B, Mortensen M, Thompson SA, Kehler J, Wafford KA, Krogsgaard-Larsen P (2001) Bioorg Med Chem Lett 11:1573
15. Gütschow M, Pietsch M, Taubert K, Freysoldt THE, Schulze B (2003) Z Naturforsch B: Chem Sci 58:111
16. Lu Z, Raghavan S, Bohn J, Charest M, Stahlhut MW, Rutkowski CA, Simcoe AL, Olsen DB, Schleif WA, Carella A, Gabryelski L, Jin L, Lin JH, Emini E, Chapman K, Tata JR (2003) Bioorg Med Chem Lett 13:1821
17. Bürli RW, Ge Y, White S, Baird EE, Touami SM, Taylor M, Kaizerman JA, Moser HE (2002) Bioorg Med Chem Lett 12:2591
18. Kaizerman JA, Gross MI, Ge Y, White S, Hu W, Duan J-X, Baird EE, Johnson KW, Tanaka RD, Moser HE, Bürli RW (2003) J Med Chem 46:3914
19. Matzen L, Engesgaard A, Ebert B, Didriksen M, Frølund B, Krogsgaard-Larsen P, Jaroszewski JW (1997) J Med Chem 40:520
20. Matzen L, Ebert B, Stensbøl TB, Frølund B, Jaroszewski JW, Krogsgaard-Larsen P (1997) Bioorg Med Chem 5:1569
21. Jørgensen CG, Clausen RP, Hansen KB, Bräuner-Osborne H, Nielsen B, Metzler BB, Kehler J, Krogsgaard-Larsen P, Madsen U (2007) Org Biomol Chem 5:463
22. Perronnet J, Taliani L, Girault P, Poittevin A (1976) US Patent 3 983 129
23. Varaprasad CVNS, Barawkar D, El Abdellaoui H, Chakravarty S, Allan M, Chen H, Zhang W, Wu JZ, Tam R, Hamatake R, Lang S, Hong Z (2006) Bioorg Med Chem Lett 16:3975
24. Larson G, Yan S, Chen H, Rong F, Hong Z, Wu JZ (2007) Bioorg Med Chem Lett 17:172
25. Yan S, Appleby T, Gunic E, Shim JE, Tasu T, Kim H, Rong F, Chen H, Hamatake R, Wu JZ, Hong Z, Yao N (2007) Bioorg Med Chem Lett 17:28
26. El Abdellaoui H, Varaprasad CVNS, Barawkar D, Chakravarty S, Maderna A, Tam R, Chen H, Allan M, Wu JZ, Appleby T, Yan S, Zhang W, Lang S, Yao N, Hamatake R, Hong Z (2006) Bioorg Med Chem Lett 16:5561
27. Cutri CCC, Garozzo A, Siracusa MA, Castro A, Tempera G, Sarva MC, Guerrera F (2002) Antiviral Res 55:357
28. Lipnicka U, Regiec A, Sulkowski E, Zimecki M (2006) Arch Pharm Chem Life Sci 339:401
29. Schieweck F, Tormo I Blasco J, Blettner C, Mueller B, Gewehr M, Grammenos W, Grote T, Gypser A, Rheinhimer J, Schaefer P, Schwoegler A, Wagner O, Rack M, Baumann E, Strathmann S, Schoefl U, Scherer M, Stierl R, Treacy MF, Culbertson DL, Bucci T (2005) WO Patent 2 005 040 162
30. Watanabe Y, Yamazaki D, Otsu Y, Shibuya K (2003) WO Patent 2 003 051 123
31. Samaritoni JG, Arndt L, Bruce TJ, Dripps JE, Gifford J, Hatton CJ, Hendrix WH, Schoonover JR, Johnson GW, Hegde VB, Thornburgh S (1997) J Agric Food Chem 45:1920

32. Samaritoni JG, Babcock JM, Schlenz ML, Johnson GW (1999) J Agric Food Chem 47:3381
33. Larson ER, Noe MC, Gant TG (1999) WO Patent 9 962 890
34. Beebe JS, Jani JP, Knauth E, Goodwin P, Higdon C, Rossi AM, Emerson E, Finkelstein M, Floyd E, Harriman S, Atherton J, Hillerman S, Soderstrom C, Kou K, Gant T, Noe MC, Foster B, Rastinejad F, Marx MA, Schaeffer T, Whalen PM, Roberts WG (2003) Cancer Res 63:7301
35. Lippa B, Morris J, Corbett M, Kwan TA, Noe MC, Snow SL, Gant TG, Mangiaracina M, Coffey HA, Foster B, Knauth EA, Wessel MD (2006) Bioorg Med Chem Lett 16:3444
36. Falch E, Perregaard J, Frølund B, Søkilde B, Buur A, Hansen LM, Frydenvang K, Brehm L, Bolvig T, Larsson OM, Sanchez C, White HS, Schousboe A, Krogsgaard-Larsen P (1999) J Med Chem 42:5402
37. Pedersen H, Bräuner-Osborne H, Ball RG, Frydenvang K, Meier E, Bøgesø KP, Krogsgaard-Larsen P (1999) Bioorg Med Chem 7:795
38. Marco JL, Ingate ST, Chinchòn PM (1999) Tetrahedron 55:7625
39. Dodge JA, Evers B, Jungheim LN, Muehl BS, Ruehter G, Thrasher KJ (2002) WO Patent 2 002 032 888
40. Evers B, Ruehter G, Tebbe MJ, Martin de la Nava EM (2003) WO Patent 2 003 087 069
41. Evers B, Ruehter G, Martin de la Nava EM, Tebbe MJ (2003) WO Patent 2 003 087 070
42. Clerici F, Marazzi G, Taglietti M (1992) Tetrahedron 48:3227
43. Clerici F, Gelmi ML, Yokoyama K, Pocar D, Van Voorhis WC, Buckner FS, Gelb MH (2002) Bioorg Med Chem Lett 12:2217
44. Clerici F, Contini A, Corsini A, Ferri N, Grzesiak S, Pellegrino S, Sala A, Yokoyama K (2006) Eur J Med Chem 41:675
45. Clerici F, Erba E, Gelmi ML, Valle M (1997) Tetrahedron 53:15859
46. Beccalli EM, Clerici F, Gelmi ML (1999) Tetrahedron 55:2001
47. Clerici F, Contini A, Gelmi ML, Pocar D (2003) Tetrahedron 59:9399
48. Van Nhien AN, Tomassi C, Len C, Marco-Contelles JL, Balzarini J, Pannecouque C, De Clercq E, Postel D (2005) J Med Chem 48:4276
49. Alvarez R, Jimeno M-L, Gago F, Balzarini J, Perez-Perez M-J, Camarasa M-J (1998) Antiviral Chem Chemother 9:333
50. Marco JL, Ingate ST, Jaime C, Bea I (2000) Tetrahedron 56:2523
51. Evers B, Ruehter G, Berg M, Dodge JA, Hankotius D, Hary U, Jungheim LN, Mest H-J, de la Nava E-MM, Mohr M, Muehl BS, Petersen S, Sommer B, Riedel-Herold G, Tebbe MJ, Thrasher KJ, Voelkers S (2005) Bioorg Med Chem 13:6748
52. Subramanyam C, Bell MR, Ghose AK, Kumar V, Dunlap RP, Franke C, Mura AJ (1995) Bioorg Med Chem Lett 5:325
53. Kuang R, Epp JB, Ruan S, Chong LS, Venkataraman R, Tu J, He S, Truong TM, Groutas WC (2000) Bioorg Med Chem 8:1005
54. Kuang R, Venkataraman R, Ruan S, Groutas WC (1998) Bioorg Med Chem Lett 8:539
55. Yue EW, Wayland B, Douty B, Crawley ML, McLaughlin E, Takvorian A, Wasserman Z, Bower MJ, Wei M, Li Y, Ala PJ, Gonneville L, Wynn R, Burn TC, Liu PCC, Combs AP (2006) Bioorg Med Chem 14:5833
56. Combs AP, Zhu W, Crawley ML, Glass B, Polam P, Sparks RB, Modi D, Takvorian A, McLaughlin E, Yue EW, Wasserman Z, Bower M, Wei M, Rupar M, Ala P, Reid BM, Ellis D, Gonneville L, Emm T, Taylor N, Yeleswaram S, Li Y, Wynn R, Burn TC, Hollis G, Liu PCC, Metcalf B (2006) J Med Chem 49:3774
57. Greig IR, Tozer MJ, Wright PT (2001) Org Lett 3:369

58. Cherney RJ, Mo R, Meyer DT, Hardman KD, Liu R-Q, Covington MB, Qian M, Wasserman ZR, Christ DD, Trzaskos JM, Newton RC, Decicco CP (2004) J Med Chem 47:2981
59. Cherney RJ, King BW, Gilmore JL, Liu R-Q, Covington MB, Duan JJ-W, Decicco CP (2006) Bioorg Med Chem Lett 16:1028
60. Spaltenstein A, Almond MR, Bock WJ, Cleary DG, Furfine ES, Hazen RJ, Kazmierski WM, Salituro FG, Tung RD, Wright LL (2000) Bioorg Med Chem Lett 10:1159
61. Alcaraz L, Baxter A, Bent J, Bowers K, Braddock M, Cladingboel D, Donald D, Fagura M, Furber M, Laurent C, Lawson M, Mortimore M, McCormick M, Roberts N, Robertson M (2003) Bioorg Med Chem Lett 13:4043
62. Bakshi RK, Hong Q, Tang R, Kalyani RN, MacNeil T, Weinberg DH, Van der Ploeg LHT, Patchett AA, Nargund RP (2006) Bioorg Med Chem Lett 16:1130
63. Inagaki M, Tsuri T, Jyoyama H, Ono T, Yamada K, Kobayashi M, Hori Y, Arimura A, Yasui K, Ohno K, Kakudo S, Koizumi K, Suzuki R, Kato M, Kawai S, Matsumoto S (2000) J Med Chem 43:2040
64. Inagaki M, Haga N, Kobayashi M, Ohta N, Kamata S, Tsuri T (2002) J Org Chem 67:125
65. Valente C, Guedes RC, Moreira R, Iley J, Gut J, Rosenthal PJ (2006) Bioorg Med Chem Lett 16:4115
66. Singh SK, Reddy PG, Rao KS, Lohray BB, Misra P, Rajjak SA, Rao YK, Venkateswarlu A (2004) Bioorg Med Chem Lett 14:499
67. Dehmlow H, Aebi JD, Jolidon S, Ji Y-H, Von der Mark EM, Himber J, Morand OH (2003) J Med Chem 46:3354
68. Brown BS, Jinkerson TK, Keddy RG, Lee C-H (2006) WO Patent 2006065646
69. Vicini P, Zani F, Cozzini P, Doytchinova I (2002) Eur J Med Chem 37:553
70. Zani F, Vicini P, Incerti M (2004) Eur J Med Chem 39:135
71. Vicini P, Fisicaro E, Lugari MT (2000) Arch Pharm 333:135
72. Geronikaki A, Vicini P, Incerti M, Hadjipavlou-Litina D (2004) Arzneim Forsch 54:530
73. Vicini P, Incerti M, Amoretti L, Ballabeni V, Tognolini M, Barocelli E (2002) Farmaco 57:363
74. Morini G, Pozzoli C, Adami M, Poli E, Coruzzi G (1999) Farmaco 54:740
75. Sharma PK, Sawhney SN (1997) Bioorg Med Chem Lett 7:2427
76. Yevich JP, New JS, Smith DW, Lobeck GW, Catt JD, Minielli LJ, Eison MS, Taylor DP, Riblet LA, Temple DL (1986) J Med Chem 29:359
77. Su W, Chen Z, Lou F, Xu Y, Wang B (2006) CN Patent 1850811
78. Kraus H, Krebs A, Sollner R (1998) EU Patent 846692
79. Vicini P (1986) Farmaco 41:808
80. Böshagen H, Geiger W (1977) J Liebigs Ann Chem, p 20
81. Panico A, Vicini P, Incerti M, Cardile V, Gentile B, Ronsisvalle G (2002) Farmaco 57:671
82. Vicini P, Zani F (1997) Farmaco 52:21
83. Navas FIII, Tang FLM, Schaller LT, Norman MH (1998) Bioorg Med Chem 6:811
84. Ishizumi K, Kojima A, Antoku F, Saji I, Yoshigi MC (1995) Chem Pharm Bull 43:2139
85. Orjales A, Mosquera R, Toledo A, Pumar C, Labeaga L, Innerarity A (2002) Eur J Med Chem 37:721
86. Perrone R, Berardi F, Colabufo NA, Leopoldo M, Tortorella V (2000) Bioorg Med Chem 8:873
87. Orjales A, Mosquera R, Toledo A, Pumar MC, Garcia N, Cortizo L, Labeaga L, Innerarity A (2003) J Med Chem 46:5512

88. Zbigniew M, Rusin KE, Sawicki AW, Kurowski K, Matusiewicz KJ, Stawinski T, Sulikowski D, Ktudkiewicz DD (2006) US Patent 2 006 160 870
89. Morini G, Pozzoli C, Menozzi A, Comini M, Poli E (2005) Farmaco 60:810
90. Xie W, Herbert B, Schumacher RA, Ma J, Nguyen TM, Gauss CM, Tehim A (2005) WO Patent 2 005 111 038
91. López-Rodríguez ML, Porras E, Benhamú B, Ramos JA, Morcillo MJ, Lavandera JL (2000) Bioorg Med Chem Lett 10:1097
92. Wilkening RR, Ratcliffe RW, Wildonger KJ, Cama LD, Dykstra KD, DiNinno FP, Blizzard TA, Hammond ML, Heck JV, Dorso KL, St Rose E, Kohler J, Hammond GG (1999) Bioorg Med Chem Lett 9:673
93. Ratcliffe RW, Wilkening RR, Wildonger KJ, Waddell ST, Santorelli GM, Parker DL Jr, Morgan JD, Blizzard TA, Hammond ML, Heck JV, Huber J, Kohler J, Dorso KL, St Rose E, Sundelof JG, May WJ, Hammond GG (1999) Bioorg Med Chem Lett 9:679
94. Humphrey GR, Miller RA, Pye PJ, Rossen K, Reamer RA, Maliakal A, Ceglia SS, Grabowski EJJ, Volante RP, Reider PJ (1999) J Am Chem Soc 121:11261
95. Swayze EE, Drach JC, Wotring LL, Townsend LB (1997) J Med Chem 40:771
96. Zask A, Gu Y, Albright JD, Du X, Hogan M, Levin JI, Chen JM, Killar LM, Sung A, Di Joseph JF, Sharr MA, Roth CE, Skala S, Jin G, Cowling R, Mohler KM, Barone D, Black R, March C, Skotnicki JS (2003) Bioorg Med Chem Lett 13:1487
97. Desai RC, Dunlap RP, Farrell RP, Ferguson E, Franke CA, Gordon R, Hlasta DJ, Talomie TG (1995) Bioorg Med Chem Lett 5:105
98. Xu L, Shu H, Liu Y, Zhang S, Trudell ML (2006) Tetrahedron 62:7902
99. Nerenberg JB, Erb JM, Thompson WJ, Lee H-Y, Guare JP, Munson PM, Bergman JM, Huff JR, Broten TP, Chang RSL, Chen TB, O'Malley S, Schorn TW, Scott AL (1998) Bioorg Med Chem Lett 8:2467
100. Da Settimo F, Primofiore G, La Motta C, Sartini S, Taliani S, Simorini F, Marini AM, Lavecchia A, Novellino E, Boldrini E (2005) J Med Chem 48:6897
101. Nagasawa HT, Kawle SP, Elberling JA, DeMaster EG, Fukuto JM (1995) J Med Chem 38:1865
102. Groutas WC, Chong LS, Venkataraman R, Epp JB, Kuang R, Houser-Archield N, Hoidal JR (1995) Bioorg Med Chem 3:187
103. Carmellino ML, Pagani G, Terreni M, Pregnolato M, Borgna P, Pastoni F, Zani F (1997) Farmaco 52:359
104. Borgna P, Carmellino ML, Natangelo M, Pagani G, Pastoni F, Pregnolato M, Terreni M (1996) Eur J Med Chem 31:919
105. Fiore PJ, Puls TP, Walker JC (1998) Org Proc Res Dev 2:151
106. Turpin JA, Song Y, Inman JK, Huang M, Wallqvist A, Maynard A, Covell DG, Rice WG, Appella E (1999) J Med Chem 42:67
107. Borgna P, Pregnolato M, Invernizzi AG, Mellerio G (1993) J Het Chem 30:1079
108. Pagani G, Pregnolato M, Ubiali D, Terreni M, Piersimoni C, Scaglione F, Fraschini F, Rodríguez Gascón A, Pedraz Munoz JL (2000) J Med Chem 43:199
109. Pregnolato M, Borgnan P, Terreni M (1995) J Het Chem 32:847
110. Wright SW, Corbett RL (1993) Org Prep Proc Int 25:247
111. Wright SW, Petraitis JJ, Batt DG, Corbett RL, Di Meo SV, Freimark B, Giannaras JV, Orwat MJ, Pinto DJ, Pratta MA, Sherk SR, Stampfli HF, Williams JM, Magolda RL, Arner EC (1995) Bioorg Med Chem 3:227
112. Terauchi H, Tanitame A, Tada K, Nishikawa Y (1996) Heterocycles 43:1719
113. Wiles JA, Song Y, Wang Q, Lucien E, Hashimoto A, Cheng J, Marlor CW, Ou Y, Podos SD, Thanassi JA, Thoma LC, Deshpande M, Pucci MJ, Bradbury BJ (2006) Bioorg Med Chem Lett 16:1277

114. Wiles JA, Wang Q, Lucien E, Hashimoto A, Song Y, Cheng J, Marlor CW, Ou Y, Podos SD, Thanassi JA, Thoma LC, Deshpande M, Pucci MJ, Bradbury BJ (2006) Bioorg Med Chem Lett 16:1272
115. Hlasta DJ, Ackerman JH, Court JJ, Farrell RP, Johnson JA, Kofron JL, Robinson DT, Talomie TG, Dunlap RP, Franke CA (1995) J Med Chem 38:4687
116. Hlasta DJ, Subramanyam C, Bell MR, Carabateas PM, Court JJ, Desai RC, Drozd ML, Eickhoff WM, Ferguson EW, Gordon RJ, Dunlap RP, Franke CA, Mura AJ, Rowlands A, Johnson JA, Kumar V, Maycock AL, Mueller KR, Pagani ED, Robinson DT, Saindane MT, Silver PJ, Subramanian S (1995) J Med Chem 38:739
117. Guzel O, Salman A (2006) Biorg Med Chem 14:7804
118. Hlasta DJ, Bell MR, Court JJ, Cundy KC, Desai RC, Ferguson EW, Gordon RJ, Kumar V, Maycock AL, Subramanyam C (1995) Biorg Med Chem Lett 5:331
119. Combrink KD, Gülgeze HB, Meanwell NA, Pearce BC, Zulan P, Bisacchi GS, Roberts DGM, Stanley P, Seiler SM (1998) J Med Chem 41:4854
120. Yu K-L, Civiello R, Roberts DGM, Seiler SM, Meanwell NA (1999) Bioorg Med Chem Lett 9:663
121. Varga M, Kapui Z, Batori S, Nagy LT, Vasvari-Debreczy L, Mikus E, Urbán-Szabó K, Arányi P (2003) Eur J Med Chem 38:421
122. Desai RC, Court JC, Ferguson E, Gordon R, Hlasta DJ (1995) J Med Chem 38: 1571
123. Groutas WC, Epp JB, Venkataraman R, Kuang R, Truong TM, McClenahan JJ, Prakash O (1996) Biorg Med Chem 4:1393
124. Hartog J, Van Steen BJ, Mos J, Schipper J (1995) EU Patent 0 633 260
125. Van Steen BJ, Van Wijngaarden I, Ronken E, Soudijn W (1998) Bioorg Med Chem Lett 8:2457
126. Caliendo G, Fiorino F, Perissutti E, Severino B, Scolaro D, Gessi S, Cattabriga E, Borea PA, Santagada V (2002) Eur J Pharm Sci 16:15
127. Gross JL (2003) Tetrahedron Lett 44:8563
128. Patane MA, Di Pardo RM, Price RAP, Chang RSL, Ransom RW, O'Malley SS, Di Salvo J, Bock MG (1998) Bioorg Med Chem Lett 8:2495
129. Zlatkov A, Peikov P, Rodriguez-Alvarez J, Danchev N, Nikolova I, Mitkov J (2000) Eur J Med Chem 35:941
130. Primofiore G, Da Settimo F, La Motta C, Simorini F, Minutolo A, Boldrini E (1997) Farmaco 52:583
131. Da Settimo A, Primofiore G, La Motta C, Da Settimo F, Simorini F, Boldrini E, Bianchini P (1996) Farmaco 51:261
132. Aal EHA, El-Sabbagh OI, Youssif S, El-Nabtity (2002) Monats Chem 133:255
133. Scherling D, Karl W, Seidel D, Weinz C, Schohe-loop R, Mauler F (2003) WO Patent 2 003 029 250
134. Malinka W, Sieklucka-Dziuba M, Rajtar G, Zgodzinski W, Kleinrok Z (2000) Pharmazie 6:416
135. Malinka W, Swiatek P, Flipek B, Sapa J, Jezierska A, Koll A (2005) Farmaco 60:961
136. Malinka W, Sieklucka-Dziuba M, Rajtar G, Marowska D, Kleinrok Z (1995) Farmaco 50:769
137. Malinka W, Rutkowska M (1997) Farmaco 52:595
138. Malinka W, Ryng S, Sieklucka-Dziuba M, Rajtar G, Głowniak A, Kleinrok Z (1998) Farmaco 53:504
139. Bradbury BJ, Pais G, Wang Q, Deshpande M, Pucci MJ, Song Y, Lucien E, Wiles JA, Hashimoto A (2006) WO Patent 2 006 089 054
140. Singerman GM (1976) US Patent 3 997 548

141. Ikeda K, Abe N, Katoh C, Kanaoka A (1994) EU Patent 578 246
142. Davis RH, Krummel G (1994) EU Patent 623 282
143. Pilkington BL, Armstrong S, Barnes NJ, Barnett SP, Clarke ED, Crowley PJ, Fraser TEM, Hughes D, Methews CJ, Salmon R, Smith SC, Viner R, Whittingham WG, Williams J, Whittle AJ, Mound WR, Urch CJ (2001) WO Patent 2 001 090 105
144. Watanabe Y, Yamazaki D, Otsu Y, Shibuya K (2003) (2003) WO 2 003 051 123
145. Morley JO, Oliver AJ, Charlton MH (1998) J Mol Struct 429:103
146. Alvarez-Sanchez R, Basketter D, Pease C, Lepoittevin J-P (2004) Bioorg Med Chem Lett 14:365
147. Zani F, Pregnolato M, Zampollo F (1999) Farmaco 54:643
148. Pagani G, Borgna P, Piersimoni C, Nista D, Terreni M, Pregnolato M (1996) Arch Pharm 329:421
149. Chu DTW, Fernandes PB, Claiborne AK, Shen L, Pernet AG (1988) Drugs Exp Clin Res 14:379
150. Wang Q, Lucien E, Hashimoto A, Pais GCG, Nelson DM, Song Y, Thanassi JA, Marlor CW, Thoma CL, Cheng J, Podos SD, Ou Y, Deshpande M, Pucci MJ, Buechter DD, Bradbury BJ, Wiles JA (2007) J Med Chem 50:199
151. Yasuda N, Yang C, Wells KM, Jensen MS, Hughes DL (1999) Tetrahedron Lett 40:427
152. Soriano E, Marco-Contelles J, Tomassi C, Van Nhien AN, Postel D (2006) J Chem Inf Model 46:1666
153. Cutri CCC, Garozzo A, Siracusa MA, Sarva MC, Tempera G, Geremia E, Pinizzotto MR, Guerrera F (1998) Biorg Med Chem 6:2271
154. Cutri CCC, Garozzo A, Siracusa MA, Sarva MC, Castro A, Geremia E, Pinizzotto MR, Guerrera F (1999) Biorg Med Chem 7:225
155. Ferri N, Clerici F, Yokoyama K, Pocar D, Corsini A (2005) Biochem Pharmacol 70:1735
156. Combs AP, Yue EW, Bower M, Ala PJ, Wayland B, Douty B, Takvorian A, Polam P, Wasserman Z, Zhu W, Crawley ML, Pruitt J, Sparks R, Glass B, Modi D, McLaughlin E, Bostrom L, Li M, Galya L, Blom K, Hillman M, Gonneville L, Reid BG, Wei M, Becker-Pasha M, Klabe R, Huber R, Li Y, Hollis G, Burn TC, Wynn R, Liu P, Metcalf B (2005) J Med Chem 48:6544
157. Yue EW, Wayland B, Douty B, Crawley ML, McLaughlin E, Takvorian A, Wasserman Z, Bower MJ, Wei M, Li Y (2006) Bioorg Medl Chem 14:5833
158. Brehm L, Ebert B, Kristiansen U, Wafford KA, Kemp JA, Krogsgaard-Larsen P (1997) Eur J Med Chem 32:357
159. Frolund B, Kristiansen U, Brehm L, Hansen AB, Krogsgaard-Larsen P, Falch E (1995) J Med Chem 38:3287
160. Ebert B, Mortensen M, Thompson SA, Kehler J, Wafford KA, Krogsgaard-Larsen P (2001) Bioorg Med Chem Lett 11:1573
161. Krogsgaard-Larsen P, Mikkelsen H, Jacobsen P, Falch E, Curtis DR, Peet MJ, Leah JD (1983) J Med Chem 26:895
162. Hermit MB, Greenwood JR, Nielsen B, Bunch L, Jorgensen CG, Vestergaard HT, Stensbol TB, Sanchez C, Krogsgaard-Larsen P, Madsen U, Brauner-Osborne H (2004) Eur J Pharm 486:241
163. Singer JM, Barr BM, Coughenour LL, Gregory TF, Walters MA (2005) Bioorg Med Chem Lett 15:4560

Top Heterocycl Chem (2007) 9: 265–276
DOI 10.1007/7081_2007_089

Published online: 7 August 2007

Structure and Biological Activity of Furocoumarins

Roberto Gambari[1,3] (✉) · Ilaria Lampronti[1] · Nicoletta Bianchi[1] · Cristina Zuccato[1] · Giampietro Viola[2] · Daniela Vedaldi[2] · Francesco Dall'Acqua[2]

[1]ER-GenTech, Department of Biochemistry and Molecular Biology, Section of Molecular Biology, University of Ferrara, via Fossato di Mortara, 74, 44100 Ferrara, Italy
gam@unife.it

[2]Department of Pharmaceutical Sciences, University of Padova, via Marzolo, 5, 35131 Padova, Italy

[3]*Present address:*
GenTech-for-Thal, Laboratory for the Development of Pharmacological and Pharmacogenomic Therapy of Thalassaemia, Biotechnology Centre, via Fossato di Mortara, 64, 44100 Ferrara, Italy

Abstract In this review we summarize the structure and biological effects of linear and angular psoralens. These compounds exhibit very interesting biological effects on cell cycle, apoptosis and differentiation. These molecules should be considered as promising drugs in the therapy of several diseases, including psoriasis, mycosis fungoides, cancer. In addition, pre-clinical data demonstrate a possible employment of these molecules for the treatment of β-thalassemia.

Keywords Apoptosis · Cell cycle · Erythroid differentiation · Erythroid precursors · K562 cells · Psoralens

Abbreviations

5-MOP 5-methoxypsoralen
8-MOP 8-methoxypsoralen

PUVA psoralens plus ultraviolet A
MF mycosis fungoides
PI propidium iodide
FACS fluorescence activated cell sorter
PS phophatidylserine

1 Introduction

Psoralens, also known as furocoumarins, are naturally occurring or synthetic tricyclic aromatic compounds and are derived from the condensation of a coumarin nucleus with a furan ring [1–10]. Several new furocoumarins have been isolated from natural sources [10, 11]. In addition, the synthetic methods for production of these molecules have been described and reviewed [10]. The synthetic methods can be organized on the basis of the key step used for the formation of the two different oxygenated rings. In this respect, there are three possibilities: (i) formation of the furan ring onto the coumarin, (ii) formation of the pyrone ring onto the benzofuran and (iii) the simultaneous formation of both oxygenated rings onto a benzene unit. Psoralens have been extensively studied and demonstrated to retain interesting biological effects on eukaryotic cells, allowing biomedical applications and the development of clinical trials [12, 13, 20]. The biological importance of furocoumarins mainly focuses on their relevant applications in photochemotherapy, as pointed out in several reviews [21–23].

2 Linear Psoralens

Linear psoralens widely used in therapy are 5-methoxypsoralen (5-MOP or bergapten) and 8-methoxypsoralen (8-MOP) (see Fig. 1) [24–29]. In addition, several analogues have been described [25, 26]. It is generally accepted that these molecules cause cell damage by covalent binding to DNA following UVA irradiation; they in fact exhibit a planar tricyclic structure with two photoreactive sites (3,4-pyrone and 4′,5′-furan double bonds). The initial intercalation and interaction with double stranded DNA is not characterized by covalent bonds, but, upon absorption of a photon of UVA, a pyrimidine residue (preferentially a thymine) of the DNA covalently binds to the first photoreactive site with a 5,6-double bond. The resulting monoadduct can form a diadduct by absorbing a second photon, if a new pyrimidine on the opposite strand of DNA is available for an interstrand cross-link. In conclusion, the planar structure of psoralens helps them to intercalate between nucleic acid base pairs. UVA irradiation activates the intercalated complex, result-

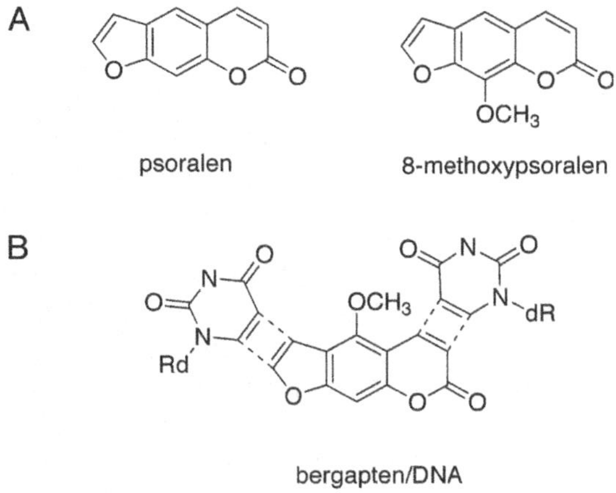

Fig. 1 **A** Structure of linear psoralens. **B** Complex between bergapten and DNA

ing in the formation of photoadducts with pyrimidines in cellular DNA. The psoralen monoadducts formed in the DNA can further react photochemically with a pyrimidine base on the complementary strand of the DNA, thus leading to interstrand cross links (ICL), that are believed to be the primary cause of photoinduced cell killing [1].

The crosslinking process depends on the structure of the psoralen derivative. Linear furocoumarins form crosslinks efficiently with a yield up to 50% of the overall adducts formed.

Substitution of the psoralen molecule by alkyl groups, like methyl(s), which increase lipophilicity of the compound and, within certain limits also the intercalation capacity, is likely to modify this picture. In particular, the photoreaction of 4-methyl psoralen derivatives with DNA yields a lower percentage of pyrone-side monoadducts. This may be rationalized in term of hindering effect due partly to the methyl group of thymine, although the cross-linking capacity is not affected. The photoreactivity of one of the double bonds of the furan and pyrone moieties may be reduced or completely eliminated if the substituting groups are bulkier or exhibit electron-withdrawing properties. Also this could be achieved by introducing a fourth aromatic ring, fused at the 4′,5′ or 3,4 position. As a result, linear monofunctional compounds, including carbethoxy-, pyrido-benzopsoralen and allopsoralen, were developed [11, 25].

2.1
Biological Activity of Linear Psoralens

Nowadays, many human skin diseases, such as psoriasis, T-cell lymphoma (cutaneous T-cell lymphoma, CTCL; mycosis fungoides MF), and vitiligo, are commonly treated with a combination of psoralens and UVA radiation commonly referred as PUVA (psoralens plus UVA) therapy.

Psoriasis

As far as treatment of psoriasis, PUVA has been compared in several review articles with the current practice of phototherapy with ultraviolet (UV) radiation without sensitizers [18–20]. Both treatment modalities are well established in therapy of psoriasis. Phototherapeutic regimens use repeated controlled UV exposures to alter cutaneous biology, aiming to induce remission of skin disease. Although UVB has been used for a longer time than PUVA, the latter has been evaluated and validated in a more detailed and coordinated fashion [18]. It is widely accepted that lesion clearance is obtained with oral psoralens (such as 5- or 8-methoxypsoralen), plus UVA exposure in patients with vitiligo or psoriasis, although differences were found in respect to the total UV exposure needed to obtain clinical outcomes [20]. Interestingly, differences among the employed psoralens were described. For instance, the incidence and severity of adverse events was generally lower in PUVA 5-MOP than in PUVA 8-MOP recipients. Nausea and/or vomiting, pruritus and erythema were the most commonly reported adverse events in the short term; they occurred about 2 to 11 times more frequently in 8-MOP than 5-MOP recipients. Adverse hepatic events after oral administration of the drug were uncommon. By contrast, long term tolerability data for PUVA are still scarce; however, carcinogenicity was not reported during a 14-year observation period of 413 patients with psoriasis [20].

T-cell Lymphoma

Linear psoralens have been extensively studied in the treatment of T-cell lymphoma (mycosis fungoides, MF) [22, 30–35]. Treatment of MF is indicated to reduce symptoms, improve clinical appearance, prevent secondary complications, and prevent progression of disease, all of which may have an impact on survival. It has been reported that psoralen and ultraviolet A radiation is effective in early-stage MF, inducing complete remissions in most patients. Psoralens and ultraviolet A radiation may also be combined with low doses of interferon (IFN)-alpha to treat stage I/II disease. Extracorporeal photopheresis may also be used successfully, but it is not generally available [31].

3 Angular Psoralens

Among psoralens-related compounds, the angular angelicin-like isomers (the structure of angelicin is shown in Fig. 2A) are both natural or synthetic compounds, present for instance in the medicinal plant *Angelica arcangelica*, that could exhibit interesting pharmacological activity when compared with linear psoralens, including low toxicity and low DNA-binding activity. Unlike

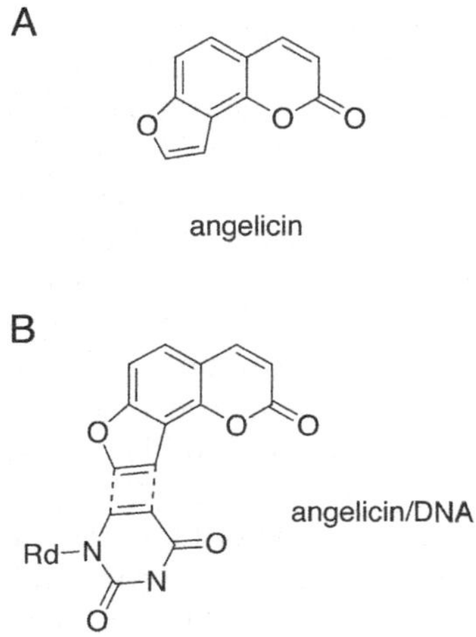

Fig. 2 Structure of angelicin (**A**) and its interaction with DNA (**B**)

linear psoralens, angelicin and its angular analogues, are monofunctional isopsoralen isomers and cannot create interstrand cross-links because of their angular geometric structure [36, 37] (see Fig. 2). In conclusion, these angular psoralen derivatives allow only monofunctional photobinding, thus reducing undesirable side effects, such as genotoxicity and risk of skin cancer.

3.1 Biological Activity of Angular Psoralens

These molecules might be of great interest to induce biological functions without (or with limited) toxicity. Recently, some of us have reported the accumulation of γ-globin mRNA in human erythroid cell treated with angelicin in the absence of UVA irradiation [38, 39]. Angelicin was able to stimulate increase of γ-globin mRNA and fetal hemoglobin (HbF) production. This feature is of potential clinical relevance. Psoralens could be proposed for the therapy of β-thalassemia and sickle-cell disease [39–46]. Indeed, even a minor increase in the production of HbF by patients affected by these diseases has been described to be beneficial. Patients treated with HbF inducers have been reported to be converted to transfusion-independent individuals [39].

4 Psoralens and the Cell Cycle

For flow cytometric analysis of DNA content, cells under investigation, in the exponential phase of cell growth, can be treated at different concentrations with psoralens. After an incubation period that can be varied at the investigator's will, the cells are centrifuged, fixed in ice-cold ethanol, then treated with lysis buffer containing RNAseA, and finally stained with propidium iodide (PI). Samples can be analyzed using standard fluorescence-activated cell sorters (FACS, such as the Becton Coulter Epics XL-MCL flow cytometer). This and similar approaches give evidence for psoralen-dependent alteration of the cell cycle. In fact, flow cytometric analysis of DNA content indicate that treatment of eukaryotic cells with increasing concentrations of both linear and angular psoralens, is associated with deep changes of cell cycle profile. For example, many papers showed that irradiation of lymphocytes, lymphoblastoid cell lines and human fibroblast in the presence of psoralens induce a G2/M arrest of the cell cycle [28, 47–50]. In addition, Jorges et al. showed that in a human keratinocyte cell line (HaCat) 8-MOP induces a cell cycle arrest in S-phase. This block may be abrogated by treating the cells with caffeine, a well know inhibitor of the ATM (ataxia-telangiectasia-mutated) and ATR (ATM and Rad3 related) protein kinases, two key enzymes involved in the down stream cellular response to DNA damage. In a recent paper, Viola et al. [50] demonstrated a differential response of linear and angular derivatives of psoralens in a human promyelocytic cell line (HL-60). They studied two linear (5-MOP, 8-MOP) and two angular derivatives (angelicin and trimethylangelicin) and the results obtained shown that psoralen derivatives efficiently induce apoptosis, as the two angular compounds are the most potent and caspase-3 is essential for this process. In the case of linear derivatives this event is preceded by a cell cycle block in G2/M phase require depending on different mechanisms involved in the phototoxicity of psoralen derivatives.

5
Psoralens and Apoptosis

Apoptosis can be analysed by several approaches, including immunohistochemistry analysis of DNA fragmentation, Annexin release assay, detection of Sub-G1 population in FACS analysis, increase of caspase activity. For instance, after treatment for 5 days with 5-MOP or angelicin, the cells can be rinsed two times with a PBS solution, fixed in paraformaldehyde at room temperature and apoptotic cells can be detected by several reagent kits, such as the DeadEnd Colorimetric Apoptosis Detection System (Promega). Measurement of apoptosis is calculated as the % of apoptotic nuclei (dark brown nuclei) versus total nuclei, as shown in the representative experiment of Fig. 3b. A dark brown signal indicates positive staining, while shades of blue-green to greenish tan indicate a non-reactive cell. The data obtained indicate that both linear and angular psoralens (5-MOP and angelicin in the representative experiment reported) induce high level of apoptosis, even without UVA

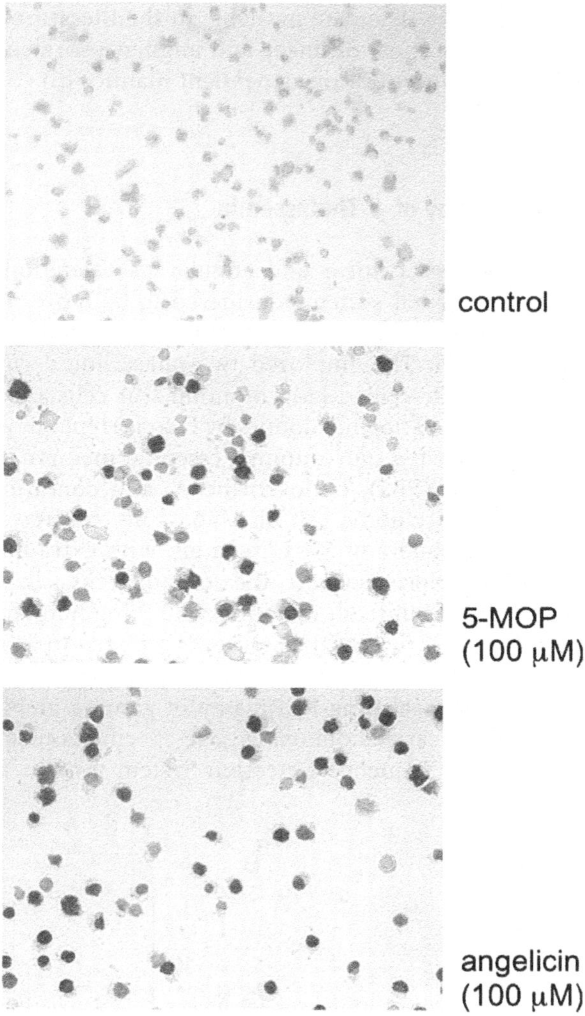

Fig. 3 Induction of apoptosis of human leukemic K562 cells treated for 5 days with 5-MOP and angelicin, as indicated

treatment of the cells. To characterize psoralen-induced apoptosis, biparametric cytofluorimetric analysis can be also performed using propidium iodide (PI) and Annexin-V-FITC, which stain DNA and phosphatidylserine (PS) residues, respectively. Annexin-V is, indeed, a Ca^{2+} dependent phospholipid binding protein with high affinity for PS. This assay represents a measurement of apoptosis, because externalization of PS occurs in the early stages of the apoptotic process. Later, analysis of the cell cycle might identify apoptotic cells as a sub-G_1 peak, evidence of the DNA fragmentation occurring at later stages of apoptosis. The data available in the literature and our own concurrently indicate that most of linear and angular psoralens activate the apoptotic pathway in a concentration-dependent manner [51–57].

6 Psoralens and the Therapy of β-Thalassemia

For assessing the activity of linear and angular psoralens on human erythroid cells, the experimental system considered to be more reliable is constituted by erythroid precursor cells isolated from normal donor or patients affected by β-thalassemia. The employed two-phase liquid culture procedure has been previously described [46]. Mononuclear cells are isolated from peripheral blood samples of normal donors by Ficoll–Hypaque density gradient centrifugation and seeded in α-minimal essential medium supplemented with fetal bovine serum (FBS), cyclosporine A, and conditioned medium from the 5637 bladder carcinoma cell line [46]. The cultures are incubated at 37 °C, under an atmosphere of 5% CO_2 in air, with extra humidity. After 7 days incubation in this phase I culture, the non-adherent cells are harvested, washed, and then cultured in fresh medium containing human recombinant erythropoietin (EPO). This part of the culture is referred to as phase II. Psoralens are added on day 6 of phase II. Cell samples are analyzed on day 9 of phase II. Quantitative real-time PCR assay of gamma-globin and alpha-globin transcripts can be carried out using gene-specific double fluorescently labeled probes in a 7700 Sequence Detection System version 1.7.1 (Applied

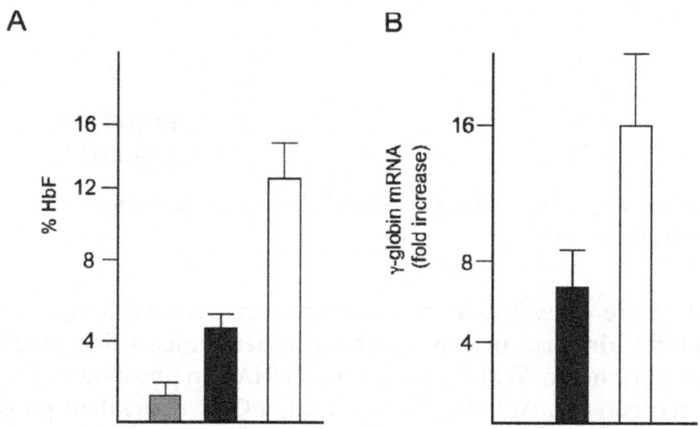

Fig. 4 Effects of angelicin (*white boxes*) on HbF production (**A**) and γ-globin mRNA expression (**B**). HbF was analysed by HPLC, γ-globin mRNA accumulation by real-time quantitative RT-PCR. The effects of angelicin were compared to those of hydroxyurea (*black boxes*; *grey box* = control untreated cells)

Biosystems, Warrington Cheshire, UK). The results obtained indicate that psoralens should be included in the class of HbF inducers, as shown in Fig. 4, which demonstrates high level of HbF induction of HbF by angelicin, quantified by HPLC (High Performance Liquid Chromatography).

7
Conclusion and Future Perspectives

The conclusion of this review is that linear and angular psoralens [1–11, 58, 59] exhibit very interesting biological effects on eukaryotic cells and should be considered as promising drugs in the therapy of several diseases, including psoriasis, mycosis fungoides, cancer.

Of great interest is the possibility that this class of molecules could be a source of lead compounds. In fact, it has been recently shown that some products of photolysis of psoralens may exhibit interesting biological effects in the dark. In this context, Potapenko et al. suggested that the photobiological activity of psoralens derives from their photooxidized products formed during pre-irradiation (POP) [60]. This author in fact analyzed the efficacy of crude pre-irradiated solutions of psoralens in a variety of biological models [61–64]. First, they proved that the solutions are active only when irradiation is carried out in the presence of oxygen and concluded that the activity should be ascribed to photooxidized psoralen species (POP). It was found that POP induces hemolysis of erythrocytes [65], modifies the "respiratory burst" of phagocytes and increases the permeability of their membrane [66]. Chemically oxidized or photooxidized psoralens inhibit chemotactic activity of polymorphonucleated cells and induce mutagenic and lethal effects in microorganism. The products of photooxidation of psoralens administered to mice have been shown to induce modulation of the T-cell mediated immune response and inhibit growth of grafted EL-4 lymphoma [67]. Recently, Caffieri et al. [68] isolated from a preirradiated solution of psoralens three cytotoxic molecules. These compounds induce apoptosis in a lymphoblastoid cell line that was preceded by mitochondrial dysfunction caused by the opening of the mitochondrial transition pore. Whether these molecules are involved in the mechanism of PUVA action remains to be clarified.

Acknowledgements R.G. is granted by AIRC, Fondazione Cariparo (Cassa di Risparmio di Padova e Rovigo), Cofin-2002, by STAMINA Project (University of Ferrara) and by UE ITHANET Project (eInfrastructure for Thalassemia Research Network). This research was also supported by Regione Emilia-Romagna (Spinner Project) and by Associazione Veneta per la Lotta alla Talassemia (AVLT), Rovigo.

References

1. Dall'Acqua F, Viola G, Vedaldi D (2004) Cellular and molecular target of psoralen. In: Hoorspool WM, Lenci F (eds) CRC Handbook of Organic Photochemistry and Photobiology. CRC Press, Boca Raton USA, pp 1–17
2. Chilin A, Marzano C, Guiotto A, Manzini P, Baccichetti F, Carlassare F, Bordin F (1999) J Med Chem 42:2936
3. Bordin F, Marzano C, Baccichetti F, Carlassare F, Vedaldi D, Falcomer S, Lora S, Rodighiero P (1998) Photochem Photobiol 68:157
4. Conconi MT, Montesi F, Parnigotto PP (1998) Pharmacol Toxicol 82:193
5. Marzano C, Caffieri S, Fossa P, Bordin F (1997) J Photochem Photobiol 38:189
6. Bordin F, Dall'Acqua F, Guiotto A (1991) Pharmacol Ther 52:331
7. Guiotto A, Rodighiero P, Manzini P (1984) J Med Chem 27:959
8. Komura J, Ikehata H, Hosoi Y, Riggs AD, Ono T (2001) Biochemistry 40:4096
9. Mosti L, Lo Presti E, Menozzi G, Marzano C, Baccichetti F, Falcone G, Filippelli W, Piucci B (1998) Farmaco 53:602
10. Curini M, Cravotto G, Epifano F, Giannone G (2006) Curr Med Chem 13:199
11. Santana L, Uriarte E, Roleira F, Milhazes N, Borges F (2004) Curr Med Chem 11: 3239
12. Naldi L, Griffiths CE (2005) Br J Dermatol 152:597
13. Okamoto T, Kobayashi T, Yoshida S (2005) Curr Med Chem Anticancer Agents 5:47
14. Scheinfeld N, Deleo V (2003) Dermatol Online J 9:7
15. Khurshid K, Haroon TS, Hussain I, Pal SS, Jahangir M, Zaman T (2000) Int J Dermatol 39:865–867
16. de Berker DA, Sakuntabhai A, Diffey BL, Matthews JN, Farr PM (1997) J Am Acad Dermatol 36:577
17. Morison WL (1995) Photodermatol Photoimmunol Photomed 11:6
18. Honigsmann H (2001) Clin Exp Dermatol 26:343
19. Abel EA (1999) Cutis 64:339
20. McNeely W, Goa KL (1998) Drugs 56:667
21. Scheinfeld N, Deleo V (2003) Dermatol Online J 9:7
22. McGinnis KS, Shapiro M, Vittorio CC, Rook AH, Junkins-Hopkins JM (2003) Arch Dermatol 139:771
23. Zarebska Z, Waszkowska E, Caffieri S, Dall'Acqua F (2000) Farmaco 55:515
24. Stolk LM, Siddiqui AH (1988) Gen Pharmacol 19:649
25. Dalla Via L, Gia O, Marciani Magno S, Santana L, Teijeira M, Uriarte E (1999) J Med Chem 42:4405
26. Gia O, Anselmo A, Conconi MT, Antonello C, Uriarte E, Caffieri S (1996) J Med Chem 39:4489
27. Anselmino C, Averbeck D, Cadet J (1995) Photochem Photobiol 62:997
28. Bartosova J, Kuzelova K, Pluskalova M, Marinov I, Halada P, Gasova Z (2006) J Photochem Photobiol B 85:39
29. Engin B, Oguz O (2005) Int J Dermatol 44:337
30. Zane C, Venturini M, Sala R, Calzavara-Pinton P (2006) Photodermatol Photoimmunol Photomed 22:254
31. Huber MA, Staib G, Pehamberger H, Scharffetter-Kochanek K (2006) Am J Clin Dermatol 7:155
32. Coors EA, Von den Driesch P (2005) Br J Dermatol 152:1379
33. Lundin J, Osterborg A (2004) Therapy for mycosis fungoides. Curr Treat Opt Oncol 5:203

34. McGinnis KS, Shapiro M, Vittorio CC, Rook AH, Junkins-Hopkins JM (2003) Arch Dermatol 139:771
35. Dalla Via L, Marciani Magno S (2001) Curr Med Chem 8:1405
36. Bisagni E (1992) J Photochem Photobiol B 14:23
37. Bordin F, Dall'Acqua F, Guiotto A (1991) Pharmacol Ther 52:331
38. Lampronti I, Bianchi N, Borgatti M, Fibach E, Prus E, Gambari R (2003) Eur J Haematol 71:189
39. Gambari R, Fibach E (2007) Curr Med Chem 14:199
40. Fibach E, Premakala P, Rodgers GP, Samid D (1993) Blood 82:2203
41. Perrine SP, Ginder GD, Faller DV, Dover GH, Ikuta T, Witkowska HE, Cai SP, Vichinsky EP, Olivieri NF (1993) N Engl J Med 328:81
42. Rodgers GP, Dover GJ, Uyesaka N, Noguchi CT, Schechter AN, Nienhuis AW (1993) N Engl J Med 328:73
43. Rodgers GP, Rachmilewitz EA (1995) Br J Haematol 91:263
44. Steinberg MH, Lu LZ, Barton FB, Terrin ML, Charache S, Dover GJ (1997) Blood 89:1078
45. Olivieri NF, Rees DC, Ginder GD, Thein SL, Waye JS, Chang L, Brittenham GM, Weatherall DJ (1998) Ann NY Acad Sci 850:100
46. Fibach E, Bianchi N, Borgatti M, Prus E, Gambari R (2003) Blood 102:1276
47. Viola G, Fortunato E, Cecconet L, Disaro S, Basso G (2007) Toxicol In Vitro 21:211
48. Ma W, Hommel C, Brenneisen P, Peters T, Smit N, Sedivy J, Scharffetter-Kochanek K, Wlaschek M (2003) Exp Dermatol 12:629
49. Wlaschek M, Ma W, Jansen-Durr P, Scharffetter-Kochanek K (2003) Exp Gerontol 38:1265
50. Viola G, Facciolo M, Vedaldi D, Disarò S, Basso G, Dall'Acqua F (2004) Photochem Photobiol Sci 3:237
51. Canton M, Caffieri S, Dall'Acqua F, Di Lisa F (2002) FEBS Lett 522:168
52. Santamaria AB, Davis DW, Nghiem DX, McConkey DJ, Ullrich SE, Kapoor M, Lozano G, Ananthaswamy HN (2002) Cell Death Differ 9:549
53. Efferth T, Fabry U, Osieka R (2001) Anticancer Res 21(4A):2777
54. Kacinski BM, Flick M (2001) Ann NY Acad Sci 941:194
55. Coven TR, Walters IB, Cardinale I, Krueger JG (1999) Photodermatol Photoimmunol Photomed 15:22
56. Enomoto DN, Schellekens PT, Yong SL, ten Berge IJ, Mekkes JR, Bos JD (1997) Photochem Photobiol 65:177
57. Yoo EK, Rook AH, Elenitsas R, Gasparro FP, Vowels BR (1996) J Invest Dermatol 107:235
58. Kanne D, Straub K, Rapoport H, Hearst JE (1982) Biochemistry 21:861
59. Dalla Via L, Mammi S, Uriarte E, Santana L, Lampronti I, Gambari R, Gia O (2007) J Med Chem 49:4317
60. Potapenko AY (1991) J Photochem Photobiol B Biol 9:1
61. Mizuno N, Esaki K, Sakakibara J, Murakami N, Nagai S (1991) Photochem Photobiol 54:697
62. Esaki K, Mizuno N (1992) Photochem Photobiol 55:783
63. Kyagova AA, Korkina LG, Snigireva TV, Lisenko EP, Tomashaeva SK, Potapenko AY (1991) Photochem Photobiol 53:633
64. Potapenko AY, Kyagova AA, Bezdetnaya LN, Lysenko EP, Chernyakhovskaya Y, Bekhalo VA, Nagurskaya EV, Nesteerenko VA, Korotky NG, Akhtyamov SN, Lanshcikova TM (1994) Photochem Photobiol 60:171

65. Lysenko PE, Melnikova VO, Andina SE, Wunderlich Pliquett SF, Potapenko AY (2000) J Photochem Photobiol B Biol 56:187
66. Caffieri S (2002) Photochem Photobiol Sci 1:149
67. Kagoya AA, Zhuravel NN, Malakhov MV, Lysenko EP, Adam W, Saha-Möller CR, Potapenko AY (1997) Photochem Photobiol 65:694
68. Caffieri S, Di Lisa F, Bolesani F, Facco M, Semenzato G, Dall'Acqua F, Canton M (2007) Blood 109(11):4988

Top Heterocycl Chem (2007) 9: 277
DOI 10.1007/7081_2007_091

Published online: 18 September 2007

Erratum to Xanthones in *Hypericum*: Synthesis and Biological Activities

Ozlem Demirkiran

Department of Chemistry, Faculty of Science and Arts, Trakya University, Gullapoglu Campus, 22030 Edirne, Turkey
ozlemdemirkiran@yahoo.com

Due to an oversight of the author, Figs. 9 and 11 of the above-mentioned article were published online with mistakes and have to be corrected as follows:

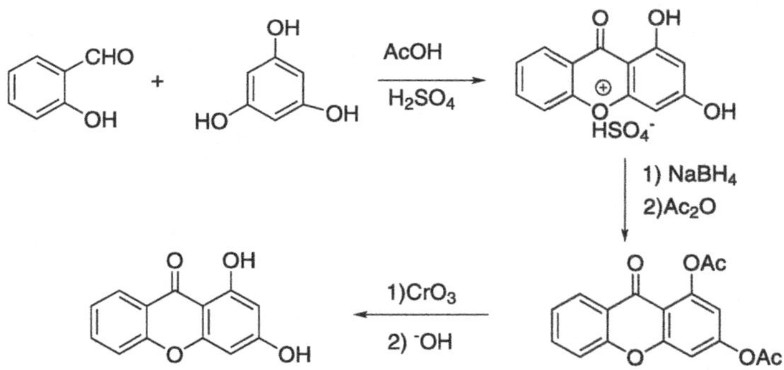

Fig. 9 Synthesis of 1,3-dihydroxyxanthone with the Tanase method [53]

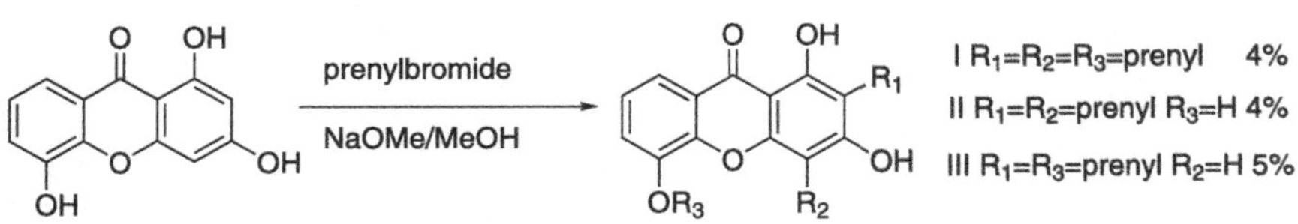

Fig. 11 C prenylation with prenyl bromide in the presence of strong base [58]

Author Index Volumes 1–9

The volume numbers are printed in italics

Subject Index

The manufacturer's authorised representative in the EU is Springer Nature Customer Service Centre GmbH, Europaplatz 3, 69115 Heidelberg, Germany. If you have any concerns regarding our products, please contact ProductSafety@springernature.com

Printed and bound by CPI Group (UK) Ltd, Croydon, CR0 4YY
30/09/2026
02225561-0005